T0395892

Nanoscience and Nanotechnology in Foods and Beverages

Nanoscience and Nanotechnology in Foods and Beverages

V. Chelladurai
Digvir S. Jayas

CRC Press is an imprint of the
Taylor & Francis Group, an **informa** business

CRC Press
Taylor & Francis Group
6000 Broken Sound Parkway NW, Suite 300
Boca Raton, FL 33487-2742

Printed on acid-free paper

International Standard Book Number-13: 978-1-4987-6063-8 (Hardback)

Visit the Taylor & Francis Web site at
http://www.taylorandfrancis.com

and the CRC Press Web site at
http://www.crcpress.com

Dedicated to our spouses

Mrs. Anusuya Marimuthu
Mrs. Manju Jayas

Contents

Preface

Materials behave uniquely when their particle size is reduced to nano level (10^{-9} m). These unique properties have led to extensive applications of nanoscience and nanotechnology in many fields such as electronics, electrical engineering, communication, construction, manufacturing, pharmaceuticals, cosmetics, and water treatment sectors, and the increasing rate of applications has been accelerating in recent years. Similar to these industries, nanoscience and nanotechnology are being used in the food industry for developing innovative products and processes. For example, nanoparticles can be used as food ingredients to enhance food characteristics, can be used in food packaging to preserve food products, can be used as nanoemulsions for decontamination of equipment and facilities, can be used in nanosensors to detect incipient spoilage of food and alert consumers, and can be used as nanobarcodes for identification. Research agencies and food industries are also investing significant funds to improve the awareness among consumers about the benefits and drawbacks of using nanotechnology in food processing. National and international regulatory agencies are working to develop a framework with a balanced approach to promote the applications of nanotechnology in the food and beverages industries based on science that does not hinder its applications based on public fear and emotions. Since many people from different sectors are working on nanotechnology-based products and process development and implementation in the food and beverages industry, the amount of scientific output is becoming large and spreading widely. The main objective of this book is to cover the basic understanding of nanoscience and nanotechnology and their applications to different food industry sectors with details and benefits to the consumers, as well as the customers' concerns and regulations about application of nanotechnology in the food industry. This book is a collection and synthesis of the recent technical documents related to the nanomaterials and nanotechnology-based processes applied in the food and beverages industry. Thus, this book contains the material that will be critical to expose students in the classrooms of food science and food engineering programs at both the undergraduate and postgraduate level. The material will be of use for food industry personnel to assess the application potential of nanoscience and nanotechnology within their own industry. The book will also serve the needs of the researchers as a reference book and of consumers to get an overview of the field.

The first chapter of this book introduces the fundamentals of nanoscience and nanotechnology, types of nanomaterials, and their fabrication techniques to the readers. Chapters 2 and 3 discuss the impact of nanomaterial incorporation on the food functionality and processing operations. Different types of nanosensors and their applications in food and beverages industries for qualitative and quantitative analysis of chemical and nutritional constituents and microorganisms in the food products are discussed in Chapter 4. Applications of nanotechnology in various sectors of food and beverages industries (i.e., beverages, baking, dairy, meat) are elaborated in Chapters 5 through 8. Industrial wastewater from food and beverages manufacturing plants may have large amounts of chemical residues as well as biological matters, which may harm the environment, and the application of nanotechnology-based systems for treatment of the food and beverages industry wastewater is explained in Chapter 9. Finally, Chapter 10 deals with the consumer perspectives and benefits and regulatory aspects related to application of nanoscience and nanotechnology-based products and processes in food and beverages.

Acknowledgments

The field of nanoscience and nanotechnology in foods and beverages is in its infancy but is expanding at an enormous rate. New applications of nanoscience and nanotechnology are being developed and implemented regularly to enhance food production, processing, and preservation. Significant advances have been made in the use of many ingredients in foods and beverages at nanoscale to enhance appearance, taste, texture, and bioavailability. Advances have also been made towards detection and control of food spoilage, detection of adulterants in foods, purification of water, and treatment of wastewater from food and beverages industries. To bring this expanded field into a single book, we relied heavily on the published literature (listed at the end of every chapter) and gratefully acknowledge the contributions of each researcher and their teams around the globe whose work we have cited. We took great care in paraphrasing their works and hope we have succeeded in doing so. We ask for their forgiveness if we have not done it correctly. We also were not able to cite all of the published articles, but these are also critical to the growth of the field. We wish each researcher and their teams a great success in their research and in advancing the field. It is their work that will benefit global consumers who we hope will embrace the applications of nanoscience and nanotechnologies in their foods and beverages. We also hope that researchers will always keep consumers' interests in mind when planning, conducting, and disseminating their research. We thank the consumers for their willingness to accept the applications of nanoscience and nanotechnologies in their foods and beverages and encourage them to support the research community by bringing their concerns to researchers. The trusted partnership between researchers and consumers is critical for the growth of applications of nanoscience and nanotechnologies in the food and beverages industries.

Writing a book requires dedicated time to complete the project and when focusing on writing, the family gets less attention. We sincerely thank our spouses (Mrs. Anusuya Marimuthu and Mrs. Manju Jayas) and our children (Kanishkaa Chelladurai and Drs. Rajat Jayas, Ravi Jayas, and Mr. Rahul Jayas) for their understanding and support. Dr. Jayas also thanks his grandchildren (Priya Jayas, Isabella Jayas, and Rohan Jayas) for not spending as much time he would like to spend with them. During busy times of writing, we may have, unconsciously, not completed some of our assigned duties at work in a timely fashion, and for this we thank our employers (the University of Manitoba and Bannari Amman Institute of Technology), our immediate supervisors, and close working colleagues (particularly Drs. Gary Glavin, Jay Doering, David Barnard, and Janice Ristock; Mr. John Kearsey; Ms. Lynn Zapshala-Kelln; Mr. Jeff Leclerc; and Ms. Kerry McQuarrie-Smith) for their understanding and encouragement. We acknowledge the help we received from Dr. Ravikanth Lankapalli and Mr. Kishna Phani Kaja during preparation of this book.

We acknowledge gratefully the assistance we received from Darya Crockett in copyediting the book, Stephen Zollo, John Shannon and Todd Perry for spearheading the project through the production process, and CRC Press/Taylor & Francis Group for giving us an opportunity to write this book.

Last but not the least, we thank the readers of the book in whole or in part and hope they will find it useful and will let us know how we could have done it better.

V. Chelladurai

Digvir S. Jayas

Authors

V. Chelladurai was educated at the Tamil Nadu Agricultural University and the University of Manitoba and is currently working as an Associate Professor of Agricultural and Food Processing Engineering in the Department of Agriculture Engineering, Bannari Amman Institute of Technology, Sathyamangalam, Tamil Nadu, India. Before joining the Bannari Amman Institute of Technology, Dr. Chelladurai worked as a post-doctoral fellow and research engineer in the department of biosystems engineering, University of Manitoba, for six years. He conducted research on applications of imaging and spectroscopy for agricultural and food products; hermetic storage of cereal grains and oilseeds; and drying and non-chemical methods for stored grain management. He authored or co-authored 21 technical articles in peer-reviewed journals, 11 technical articles in conferences, and 1 book chapter. He also served as a technical expert in Canadian International Grain Institute's training programs for various stakeholders of the agri-food industry. He is the recipient of the "2016 Superior Paper Award" from the American Society of Agricultural and Biological Engineers (ASABE) and also recipient of the University of Manitoba Graduate Fellowship, W. E. Muir Fellowship, Edward R. Toporeck Graduate Fellowship, and Gordon P. Osler Graduate Scholarship for his academic and research excellence during his time at the University of Manitoba. Dr. Chelladurai served as vice president (regional) and Manitoba regional director for the Canadian Society of BioEngineering (CSBE) from 2010 to 2016.

Distinguished Professor **Dr. Digvir S. Jayas** was educated at the G. B. Pant University of Agriculture and Technology in Pantnagar, India; the University of Manitoba; and the University of Saskatchewan. Before assuming the position of vice president (research and international), he held the position of vice president (research) for two years and associate vice president (research) for eight years. Prior to his appointment as associate vice president (research), he was associate dean (research) in the faculty of Agricultural and Food Sciences, department head of Biosystems Engineering, and interim director of the Richardson Centre for Functional Foods and Nutraceuticals. He is a registered professional engineer and a registered professional agrologist.

Dr. Jayas is a former tier I (senior) Canada research chair in stored-grain ecosystems. He conducts research related to drying, handling, and storing grains and oilseeds and digital image processing for automation grading and processing operations in the agri-food industry. He has authored or co-authored over 900 technical articles in scientific journals, conference proceedings, and books dealing with issues of storing, drying, handling, and quality monitoring of grains and processed foods.

Dr. Jayas has received awards from several organizations in recognition of his research and professional contributions. He is the recipient of the 2017 Sukup Global Food Security Award from the American Society of Agricultural and Biological Engineers (ASABE) and the 2008 Natural Sciences and Engineering Research Council (NSERC) Brockhouse Canada Prize. In 2009, he was inducted as a fellow of the Royal Society of Canada. He has received professional awards from the Agriculture Institute of Canada, Applied Zoologists Research Association (India), American Society of Agricultural and Biological Engineers, Association of Professional Engineers and Geoscientists of Manitoba (Engineers and Geoscientists Manitoba), Canadian Institute of Food Science and Technology, Canadian Academy of Engineering, Canadian Society for BioEngineering, Engineers Canada, Engineering Institute of Canada, Indian Society of Agricultural Engineers, Manitoba Institute of Agrologists, National Academy of Agricultural Sciences (India), National Academy of Sciences (India), and Sigma Xi.

Dr. Jayas serves on the boards or committees of many organizations, including ArcticNet, Cancer Care Manitoba Projects Grants and Awards Committee, Churchill Marine Observatory (CMO), Composite Innovation Centre, Engineers Canada, Centre for Innovative Sensing of Structures (SIMTReC), Genome Prairie, GlycoNet, Manitoba Centre for Health Policy, National Coordinating Centre for Infectious Diseases (NCCID), North Forge Technology Exchange, NSERC Council, Oceans Research in Canada Alliance Council, Research Manitoba, Research Institute of Oncology and Hematology, and TRIUMF. He has served as the president of Agriculture Institute of Canada, Association of Professional Engineers and Geoscientists of Manitoba (Engineers and Geoscientists Manitoba), Canadian Institute of Food Science and Technology, Canadian Society for BioEngineering, Engineers Canada, and Manitoba Institute of Agrologists. He currently chairs the board of directors of RESOLVE, a prairie research network on family violence; TRIUMF; Smartpark Advisory Committee; and NSERC Council.

1

Nanoscience and Nanotechnology

Introduction

Nanoscience and nanotechnology are the emerging fields that deal with the materials in "nano" (10^{-9}) scale. Developments in nanoscience and nanotechnology are increasing day by day, and the application of nanotechnology-based products and processes are also expanding. Nanoscience and nanotechnology are widely applied in electronics, electrical, communication, construction, manufacturing, pharmaceutical, cosmetic, agriculture, food and beverages, and wastewater treatment sectors. Nanoscience may be broadly defined, based on the length scale, as understanding and manipulation of matter at the nanometer scale. In other words, it is defined by the European Union as "the study of phenomena and manipulation of materials at atomic, molecular, and macromolecular scales, where properties differ significantly from those at a large scale." Although both physical and chemical properties of a materials change with the size, there is no accurate way of defining nanoparticles based on their size. Nanoparticles definition is more related to the behavior of the particles at the nano size than the size itself. Nanoscience is one of the most important developments in the recent decades (Whitesides 2005).

Nanotechnology involves using the particles with varied properties in their nano size for the development of products with practical applications. Hence, the nanotechnology is generally based on the nanoscience insights. During the recent years, the discussion and research about nanotechnology and nanoscience moved so far that most of the scientific and engineering conferences and meetings dedicate some special area for the discussion of advances and applications related to their fields. To give a more insight, the nanotechnology is not a single technology or cannot be restricted to a particular manufacturing process or design of devices. Rather it is a broad area that involves research on nanomaterials fabrication, understanding their properties, and utilizing these peculiar properties of nanomaterials for various applications (Filipponi and Sutherland 2010; Sugunan 2012).

In some cases, the nanotechnology is defined based on the nanoscale as the construction and use of the functional structure of the material designed from the molecular or atomic scale in such a manner that at least one characteristic dimension of this material measures in nanometers. Because of the size, these particles exhibit significant improvement in the physical, chemical, biological properties, and processes. Hence, if the nanoparticle is defined as something in between the bulk materials and the atom, its properties are considered to be different from both the bulk materials and its atomic form (Murthy et al. 2013). The behavior of the material at the nanoscale is very different, and it should be understood that this behavior cannot be easily predicted. In the recent years, government and the private research organizations are spending billions of dollars to understand this behavior, which is beyond the scope of this book. Fundamentals of nanotechnology and nanoscience are discussed in this chapter, and the other chapters of this book are mainly focused on the applications of nanoscience and nanotechnology in food and beverages industry, which is gaining more attention among researchers and industry and booming enormously in recent times.

Nanoscale

The origin of the word "nano" is from Greek meaning dwarf. It is hard to imagine how dwarf a nano is in the real world. The nanometer scale is always defined conventionally as 1–100 nm. One nanometer is one billionth of a meter, which is 10^{-9} of a meter. The small size range is generally set to be as small as 1 nm to avoid single atoms or very small groups of atoms (e.g., three and a half gold atoms measure 1 nm) being designated as nano-objects (EU 2015). Hence, it can be said that nanotechnology and nanoscience deal with at least clusters of atoms of 1 nm size. The general upper limit for defining the nanoparticle is generally 100 nm, but it should be made clear that this limit is a "fluid limit," as certain particles with larger size (200 nm) are also recognized as nanomaterials. In the minds of students, researchers, and public, there is always a question why only nanomaterials are in between "1–100 nm" and "why not 1–150 nm or maybe 1–1000 nm?" It is because nanoscience and nanotechnology focus on the effect of the dimension range (1–100 nm) on the certain materials—for example, the insurgence of the quantum phenomena—rather than at what exact dimension this effect arises (EU 2015). The nanoscience is not just the science of small particles but the science where the materials with the smaller dimension show new physical properties, which are collectively called the quantum mechanisms. These quantum mechanisms of the material are size dependent and are dramatically different from the properties of microparticles of the same material. Hence, the nanoscience based on their size properties can be defined as the study of materials that exhibit extraordinary properties, functionality, and mechanisms when they are used in extremely small sizes (Alagarasi 2011; EU 2015; Pokropivny et al. 2007).

To give a more realistic view of nano unit, below are a few examples:

1. A single sheet of paper is about 100,000 nm thick.
2. A human hair is about 80,000 nm wide.
3. A red blood cell is approximately 7000 nm wide.
4. In one inch there are 25,400,000 nm.

Figure 1.1 shows the nanometer scale and various particles measured within the scale. The comparisons to provide the larger picture of the particles in the nanometer scale are fascinating and are provided in Figure 1.2. The nanomaterials have incredibly higher surface area than their conventional form (Cox 1999). The increase of surface area results from creating smaller particles where the surface-to-volume ratio of the particles increases, or by creating porous materials where the void surface of the materials is higher than its bulk support material. Materials such as high-dispensed supported metal crystals and gas phase clusters come under the former category and nano-porous materials like zeolites, other high-surfaced-area inorganic oxides, amorphous silica, and porous carbons come under the latter category (Cox 1999). In general, the reaction of the chemical compounds takes place at the surface of the material; the greater the surface for the same volume, the greater the reactivity. For instance, let us consider the sugar cube getting dissolved in the water. The outer surface of the

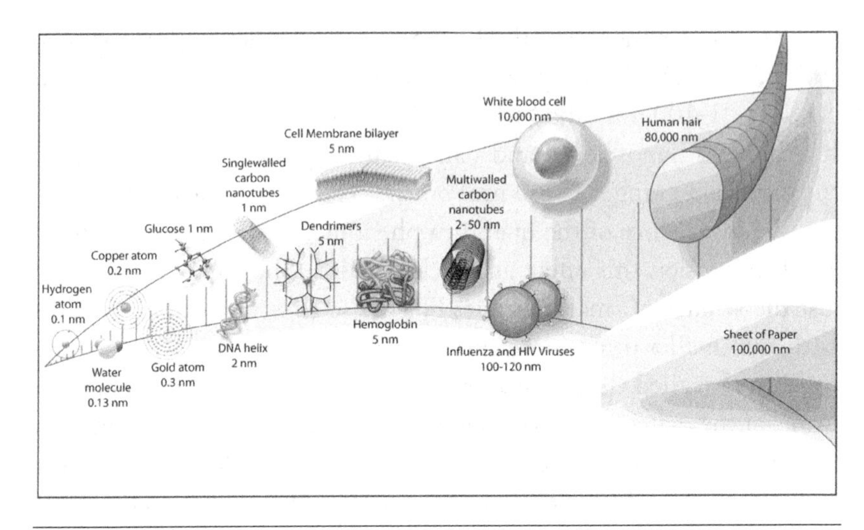

Figure 1.1 Size comparison of various materials. (Reproduced from Yokel, R.A., and MacPhail, R.C., *J. Occup. Med. Toxicol.*, 6, 7, 2011.)

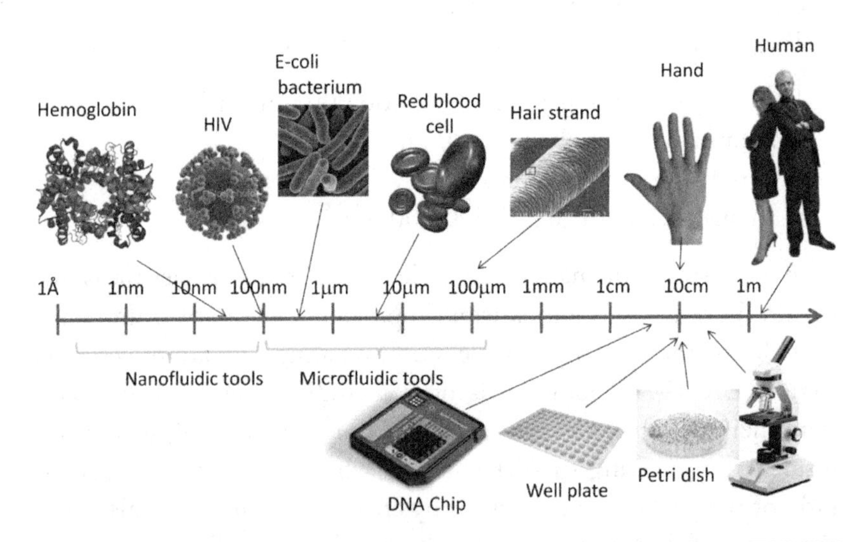

Figure 1.2 **(See color insert.)** Scale of different materials. (Reproduced from Nguyen et al. *Adv. Drug Del. Rev.*, 65, 1403–1419, 2013.)

sugar cube that is in contact with the water reacts rapidly. Once the same sugar cube is cut into many little pieces, each particle has its own surface area available for reaction with the water. Thus, the smaller cubes have greater reaction sites than the particles with the low surface area and thus increasing its chemical reactivity. One another example that deals with the size reduction and its influence on the reactivity of the material is gold. Gold when present in material form is one of the inert metals, but when the size of the gold is reduced to the nanoscale, the gold nanoparticles become extremely reactive and often this form of gold is used as a catalyst to speed up certain reactions.

History of Nanoscience and Nanotechnology

The fact that the small particles of different substances have variation in their properties from those that of the larger particles of the same sample was well-known for a long time, but the reason for this was not clear during that time. Thus, for a long time, the world was engaged with nanotechnology without knowing the real science behind it. For instance, fabric materials like cotton, flax, wool, and silk were used by humans for various uses for many years. The main concept behind this is that these fibers are developed with nano-sized (1–20 nm) pores. These pores can instantly absorb liquids, making them a very good raw material for manufacturing various fabrics (Tolochko 2009).

Nanotechnology for food products was explored by the Egyptians for the manufacturing of bread, cheese, and wine. During this Egyptian era, the fermented foods were the leading food products, and the fermentation process occurring at the nanoscale was of prime importance to provide the final product with desired texture and flavor. Ancient hair dye dating back to two thousand years, that is, from the Greco-Roman era was the earliest example of the use of nanotechnology (Tomassoni et al. 2013). The particles used to develop this dye were nanoscale lead dioxide and calcium hydroxide. This compound mixture when reacted with the keratin protein of the hair produced lead sulfide nanoparticles, also commonly called as the "galena." The size of galena particles was about 5 nm; repeated application of this substance to the hair penetrated deep into the hair and darkened the hair.

The Lycurgus Cup from the Roman era glass vessel was an extraordinary example of the ancient use of nanotechnology (Freestone et al. 2007). Though the earliest history of the cup was unknown, the cup was first known to the public when it was published in print in 1845 (Harden and Toynbee 1959; Harden et al. 1987), later moved with different owners and finally acquired by the British Museum. The cup was observed to be made by the glass by Dr. G. F. Claringbull,

Keeper of the Department of Mineralogy in the British Museum (Natural History), and these observations were later confirmed by the X-ray diffraction (British Museum Research Lab 1959). The fascinating aspect of the glass cup was the spectacular color changes that can be exhibited by the glass coating. Further analysis of the glass identified that the glass was originally a soda-lime-silica type, similar to most of the present-day window and bottle glasses. The glass was made by adding traces of gold, silver, manganese, copper, and sodium metals in nanoparticles sizes of 50–100 nm (Freestone et al. 2007). The presence of these metals at the nanoparticles level in glass gave the Lycurgus Cup astonishing peculiar colors. This cup stands as an example of one of the ancient uses of nanotechnology and the basis for the present-day holograms.

Few other examples of the use of nanotechnology in the ancient world include, the Maya Blue dye, used as a corrosion resistant pigment, was developed by introducing blue stable dye into the non-porous clay matrix (Eglash 2011). The Damascus steel used in the manufacture of the Middle East sword during AD 300 to AD 1700 was known for its combination of sharpness and strength, which was possible by the presence of carbon nanowires and nanotubes like structures along with the steel blade (Eglash 2011). Although these artifacts were the result of hundreds of years of trial and error, the fact was that the prehistoric people who used these artifacts might not have been exactly sure that they were working with nanoparticles. Present-day scientific analysis of these artifacts enlightened that the use of nanotechnology in their manufacture is the key for the peculiar properties of these materials.

Nanotechnology is a very vast field, and it is in its present state due to the contributions of many scientists and engineers. This subject has its existence for a long time in the field of chemistry, material sciences, and various areas of physics. The existence in the recent times was a revolution by high-level technical research and designing of various materials in nanoscale (Allhoff et al. 2009). The concept of nanotechnology was first popularized by the noble laureate, Richard Feynman. In his talk "There's plenty of room at the bottom" at the California Institute of Technology in 1959, Feynman spoke about the atomic level particles and their precision and also explained how these particles accurately follow laws of physics and would have an enormous number of real-life applications (Feynman 1992). He discussed reducing the information on the page of *Encyclopaedia Britannica* on to a head of a pin, which will approximately be 1/25,000 of its original size. This accomplishment needs development of high-magnification electron microscopes. The challenge stood unachieved for twenty-six years until it was achieved by Stanford graduate student Tom Newmen in 1985 by reducing the first page of Charles Dickens book *A Tale of Two Cities* from a page to an area 5.9 × 5.9 micrometers (Feynman 1992). Although the idea of nanotechnology was introduced by Feynman, it was not until 1974 that the term "nanotechnology" was first used. A Tokyo Science University Professor Norio Taniguchi, in his 1974 conference paper entitled "On the basic concept of nanotechnology" (Taniguchi 1974) used the term nanotechnology and discussed the designing of various engineering materials at the nanoscale. According to Taniguchi, these kinds of designs will end up providing extra high accuracy and precision at the order of 1 nm or 10^{-9} m in length. He also strongly believed that the ion sputtering would be the most promising process for this technology (Allhoff et al. 2009). But, during recent times, many processes have been applied to this technology. Later in 1987, K. Eric Drexler, graduate from MIT, published a book, *Engines of Creation: The Coming Era of Nanotechnology*, which was directed towards the non-technical audience. This book primarily concentrated on what we can build by careful arranging the atoms. It is possible to build assembly machines that are very strong but much smaller than a living cell and lighter than anything available today. The arrangement of atoms in almost any reasonable arrangement follows laws of nature. In this book and in the later years, various scholars discussed and debated about the importance of possibilities, promises, and hurdles with manufacturing machinery at the molecular level.

Since the 1980s, the International Business Machines (IBM) corporation stood as a pioneer in conducting groundbreaking research in relation to nanotechnology. In 1981, the IBM Zurich research laboratory scientists Gerd K. Binnig and Heinrich Rohrer developed scanning tunneling microscope (STM), which gave access for the first time to the nanoscale world. This invention made it possible to understand the behavior of individual atoms and molecules when they are presented on an electrically conducting substrate. This invention of the STM gave these two scientists the Nobel Prize in Physics in 1986.

The IBM scientists studied the phenomena of luminescence and fluorescence in the nanoscale by observing the photon emission from the nanometer size area using STM. The first single-walled nanotube was independently developed by the IBM Zurich lab and the NEC Corporation of Japan in 1993. The advancement in nanotechnology had moved further with greater pace in the 2000s. The invention of the single-molecule computer circuit revolutionized the computer industry and laid the path for the development of various palmtops and laptops. In 2006, IBM developed the complete carbon nanotube integrated circuit. This circuit was built around a single carbon nanotube molecule, and this technology was the basis for the current-day standard silicon semiconductors. Although the visualization of nanoscale process was primarily involved using the STM for observing in the nanoscale, IBM developed the magnetic resonance imaging (MRI) technique that can help to visualize the nanoscale objects. This was a major breakthrough and laid the path for the development of microscopes that can visualize atomic structures in three dimensions. Recently (in 2008), the IBM Zurich and ETH Zurich collaborated for the demonstration of nanoscale printing. This technique was very precise, as it used 60 nm particles for printing, and this gave a lead for the advancement of nanotechnology into the nanoscale biosensors and the fabrication of nanowires that will be the future of tomorrow's computer chips (IBM 2008).

Different Types of Nanomaterials and Their Fabrication Processes

Type of Nanomaterials

Nanomaterials can be classified into various categories (i) based on dimensionality and (ii) based on the type of material used.

Based on Dimensionality The classification of nanomaterials based on dimensionality basically depends on the number of dimensions of an object or material, which are not in the nanoscale (<100 nm) range. The nanomaterials are broadly classified into four groups (Vengatesan and Mittal 2015):

1. Zero-Dimensional (0D) nanomaterials
2. One-Dimensional (1D) nanomaterials

3. Two-Dimensional (2D) nanomaterials
4. Three-Dimensional (3D) nanomaterials

Zero-Dimensional Nanomaterials The zero-dimensional (0D) nanomaterials are the nanomaterials that have all the dimensions in the nanoscale (i.e., none of the dimensions is >100 nm). The common 0D nanomaterials are nanoparticles and nanoclusters. These nanoparticles and nanoclusters can be an amorphous or crystalline material. If it is a crystalline material, it can be single crystalline or polycrystalline (Tiwari et al. 2012). The nanoparticles can also be made up of a single chemical element or multi-chemical elements, and these nanoparticles are available in different shapes and forms. Commonly, the 0D nanomaterials like nanoparticles are metal, ceramic, or polymer materials. The 0D nanomaterials can be fabricated through different chemical and physical methods. The 0D nanoparticles like quantum dots, core-shell quantum dots, nanolenses, and the hollow spheres are produced by researchers using various techniques and detailed in an elaborate manner by Tiwari et al. (2012).

One-Dimensional Nanomaterials The one-dimensional (1D) nanomaterials have at least one of their dimensions more than the nanoscale. Common examples for 1D nanomaterials are nanorods, nanotubes, nanowires, nanoribbons, and nanobelts. In nanorods and nanotubes, the diameter and thickness are in nanoscale range, but the length is greater than 100 nm. Most of the 1D nanomaterials have the needle-like shape. Similar to 0D nanomaterials, 1D nanomaterials are also made up of amorphous or crystalline material and can be standalone materials (single element materials) or mixed/embedded with other elements (Huang et al. 2002; Kuchibhatla et al. 2007; Tiwari et al. 2012).

Thin films (sizes 1–100 nm) or monolayer is now commonplace in the field of solar cells, and in different technological applications, such as chemical and biological sensors, information storage systems, magneto-optic and optical devices, and fiber-optic systems (Hussain and Palit 2018). The 1D nanomaterials have been widely applied as nanocomposite materials, and in nanoelectronics, and fabrication of nanodevices.

Two-Dimensional Nanomaterials The two-dimensional (2D) nanomaterials have at least two of their dimensions outside the nanoscale range. Because of this capability, 2D nanomaterials generally have a plate or sheet-like shape. The nanofilms, nanocoating, nanolayers, and graphene are the commonly available/produced 2D nanomaterials. The amorphous or crystalline materials, which are all made up of various chemical compositions, can be used in a single or multilayer structure in 2D nanomaterials. The 2D nanomaterials like nanocoating or nanolayers can be deposited on a substrate to improve property or functionality of the substrate. The other common 2D nanomaterials are nanowalls, nanoplates, nanoprisms, and nanodisks (Jun et al. 2005; Kim et al. 2009; Tiwari et al. 2012).

Three-Dimensional Nanomaterials Three-dimensional (3D) nanomaterials are the bulk materials, and none of their dimensions fall inside the nanoscale range (<100 nm). The 3D nanomaterials have all their three arbitrary dimensions above 100 nm. The 3D nanomaterials have nanoparticles present or the presence of nanocrystalline. In the 3D nanomaterials containing nanocrystalline, structures have the nanocrystals organized in multiple orientations and arrangements. The common examples for 3D nanomaterials are bundles of nanowires and nanotubes, bulk materials (generally powders) with the dispersions of nanoparticles, and multi-nanolayer materials.

The 3D nanomaterials are widely applied as magnetic and electrode material in batteries, and also in the field of catalysis. Due to the higher surface area per unit mass, and enhanced energy absorbance capacity, the research interest as well as the application in the real-time processes are increasing every day of 3D nanomaterials. The common 3D nanomaterials are nanostructured materials (NSMs), such as nanoballs, nanocones, nanopillars, nanocoils, and nanoflowers. Figures 1.3 and 1.4 show various types of nanomaterials with examples (Dong et al. 2010; Mann and Skrabalak 2011; Shen et al. 2008; Tiwari et al. 2010).

Major Differences between Nanomaterials and Bulk Materials The main question among all of us is, how the properties of nanomaterial differ from the bulk material, and what makes these useful. The two major factors that make the nanomaterials so special are the increased

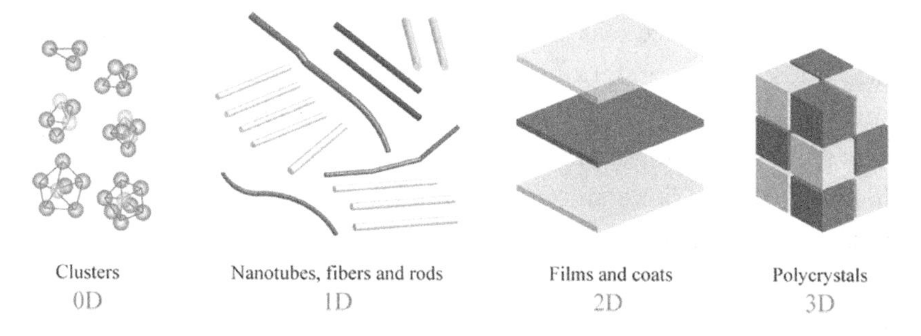

Figure 1.3 (See color insert.) Types of nanomaterials based on dimensions. (Reproduced from Gusev, A.I., and Rempel, A.A., *Nanocrystalline Materials*, Cambridge International Science Publishing, Cambridge, UK, pp. 351, 2004.)

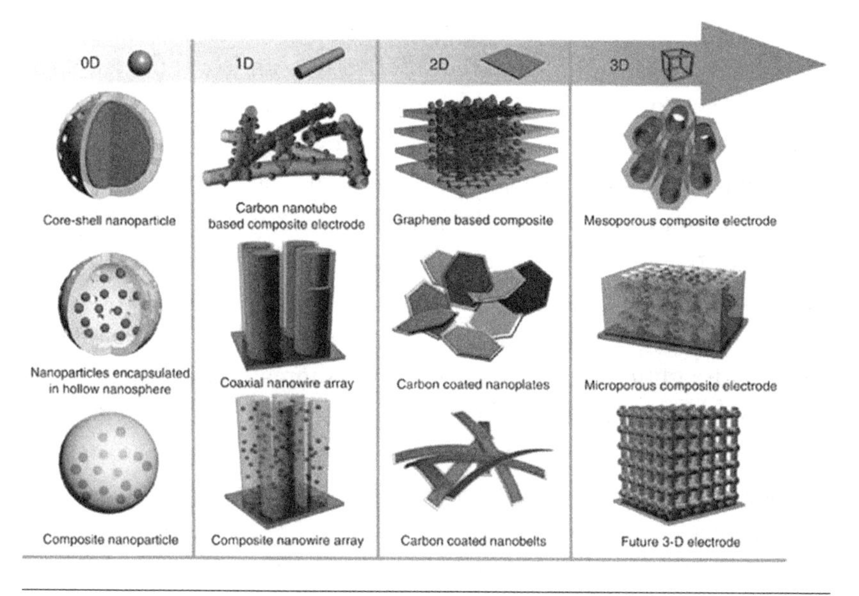

Figure 1.4 (See color insert.) Examples of different types of nanomaterials based on the dimension. (Reproduced from Lukatskaya, M.R. et al., *Nat. Commun.*, 7, 1–13, 2016.)

surface area and the quantum effects of nanomaterials. Because of these, the way nanomaterial reacts with other materials, and the relative strength of the material get altered or enhanced. When the particle size decreases, the amount or number of atoms on the surface area of the material increases. For example, a material with an average particle size of 30 nm contains 5% of its atoms on the surface, while if the particle size is decreased to 10 nm, the number of atoms on the surface are 20%. In the same manner, the nanomaterial with an average particle size of 3 nm has nearly 50% of its atoms on its surface. Most of the catalytic and other chemical reactions happen at the surfaces of the materials. Since the nanoparticles have higher surface area per one-unit mass, the reaction rate with nanomaterials are much higher than the same amount of bulk material with larger size particles (Tiwari et al. 2012).

Classification Based on the Composition of the Material Nanomaterials are classified mainly into four categories based on their composition: carbon-based nanomaterials, metal-based nanomaterials, dendrimers, and nanocomposites (Figure 1.5).

Carbon-Based Nanomaterials The carbon-based nanomaterials are fabricated from carbon or carbon derivatives and contain nanoparticles of carbon as a major component. The examples of carbon-based materials are fullerenes, carbon nanotubes, and graphene. Carbon-based nanomaterials have a major role in the nanotechnology development history and the developments in the various applications of nanotechnology-based products as well as processes. Carbon has been used as a nanomaterial for a long time. The reports about the application of carbon nanomaterials as fullerenes, and other compounds from mid-eighties, show the history of nanomaterials application in semiconductor and other fields. Mainly, carbon has been used in the nanoscale level for the fabrication of carbon nanotubes, nanoscale diamond, carbynes, and carbon onions. The developments in the nanoscience to explore the chemistry and other properties of carbon nanomaterial helped to extend the applications of nanomaterials in various fields such as medicine (diagnosis, drug delivery), agriculture

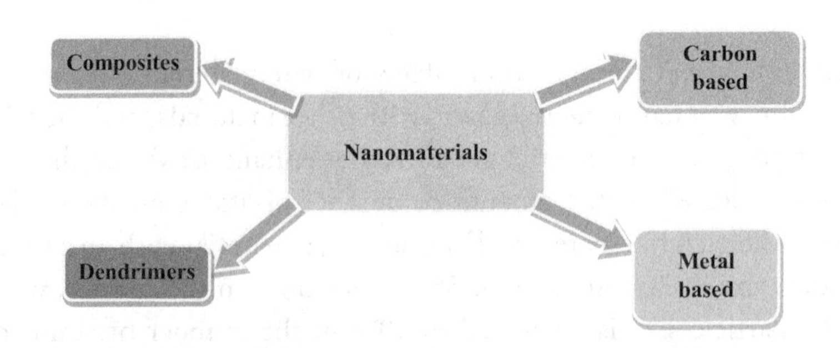

Figure 1.5 Types of nanomaterials based on the composition.

and food, water treatment, environmental sciences, and sensor development in last three decades (Domingo et al. 2007; Dumitrache et al. 2005; Jäger et al. 2007; Schmidt 1998).

The carbon nanotubes (CNT) are the newer generation of carbon-based nanomaterials, which made revolutionary changes in the application of nanoparticles and nanomaterials. These were developed by Sumio Iijima, an electron microscopist from NEC Corporation, Tokyo, Japan, in 1991 by winding the graphite sheets with the honeycomb structures into long, thin tubes. CNT has the highest mechanical strength and is applied in many sectors such as fabrication of machine parts (gears and bearings), sports equipment, aircraft parts, and food processing equipment due to its high thermal and fracture resistance (Moraru et al. 2003). The CNTs have been successfully tested for biological applications like fabrication of biosensors, bioreactors, and protein crystallization (Bandyopadhyaya et al. 2002; Dagani 2002).

Metal-Based Nanomaterials Metal-based nanomaterials are mainly fabricated from metals or metal oxides like silver, gold, and titanium dioxide. These metal-based nanoparticles are extensively used as coating agents and sensing agents, as well as the antimicrobial agents. The thin films manufactured with the nanoparticles of metal-oxides helped to enhance the detectability sensitivity and the selectivity of the gas sensors used for detection of gases like CO and CO_2. The porous films with the titanium dioxide nanoparticles enhanced the surface area and higher transmission rate when these were applied in solar cells. In the electronics industry, nanomaterials of metals (metal nanopowders) have been extensively used for metal-metal bonding like welding due to their higher ductility and the cold-welding capability. Metallic nanopowders have also been one of the major components in the porous coating and gas-tight materials manufacturing. In the sensors industry, metal nanoparticles have been preferred for their higher surface area and the affinity towards certain compounds. Ferrous oxide nanoparticles have been extensively used for magnetic nanosensors manufacturing (Ngo and Van De Voorde 2014; Schrand et al. 2010).

Dendrimers The dendrimers are the macromolecules in which the monomers are assembled in a tree structure with branch units around the central core. The word "dendron" comes from Greek, which means tree. Dendrimers are versatile structures and have extensive functionality due to the branched 3D structure. Dendrimers are mostly used in drug delivery systems and encapsulation of nanoparticles. Dendrimers can be fabricated by bottom-up and top-down technologies, and the bottom-up fabrication technique is most commonly used for dendrimers synthesis since bottom-up methods grow dendrimers from single-core molecules. Using bottom-up technology, the multifunctional core molecules react with the monomer molecule and form the first-generation dendrimer. These dendrimers contain one reactive group and two dormant groups. When the dormant group of the first-generation dendrimer is activated, the monomer molecule reacts again with the dormant group, and second-generation dendrimers are formed. Using the top-down fabrication method, two types of dendrimers are produced, that is, surface-block dendrimers, and segment-block dendrimers. If the surface of the dendrimers has two different chemical components, then it is called surface-block dendrimer. If it has more than one chemical volume parts, then it is called as segment-block dendrimer. The major application of the dendrimers is as nanocarriers for drug and nutrient (functional compound) delivery systems (Ngo and Van De Voorde 2014).

Nanocomposites Nanocomposites or composite nanomaterials are the bulk materials in which nanoparticles are incorporated in order to improve the physical, chemical, mechanical, and gas transmission properties of the bulk materials. A good example of nanocomposite is the nanoscale clay particles incorporated polymer composites, which have better mechanical strength and thermal properties, as well as better gas and moisture transmission properties when compared with the regular polymers. Nanocomposite materials have several phases (several compounds), in which at least one phase (one compound) has at least one of its dimension in nanoscale (<100 nm). Incorporation of nanoparticles with bulk materials not only improves the mechanical properties but it also increases the ductility of the nanocomposites. Another good example of the nanocomposite is the nanocarbon particles incorporated in rubber for the production of tires. These nanocomposite tires have improved black color, mechanical strength, enhanced hardness, and resistance to abrasion and tear. Nanocomposites can be fabricated through sol-gel technique, gas vapor deposition, spinodal decomposition, and self-assemble techniques. Due to the enhanced mechanical strength, resistance against heat, lighter in weight, and improved gas and moisture transmission properties, the application of nanocomposites is increasing day by day in many fields (Handford et al. 2014; Huang et al. 2003; Kelarakis et al. 2007; Ramanathan et al. 2008; Ray and Okamoto 2003).

Nanomaterial Fabrication Techniques

As mentioned in previous sections, nanomaterial is a material with at least one dimension within nanometer range usually between 1 and 100 nanometers. Naturally occurring nanomaterials can be found everywhere, which is out of scope for this chapter. Fabrication of nanomaterials can be broadly classified into two groups based on the processing approach: top-down and bottom-up methods. In recent times, a method with the combination of top-down and bottom-up is also becoming popular in the fabrication of nanomaterials. In the top-down method, a nanomaterial with a desired size and shape is obtained from a bulk large substrate through progressive removal or progressive size reduction. It is pretty much similar to the concept of carving a statue from a big block of stone. The nanoprinting method is the best example of top-down technology. The bottom-up fabrication method is directly opposite to top-down technology. In the bottom-up fabrication method, the fabrication process is started from a small atomic or molecular level of the material, and the desired size and shape is obtained by the progressive assembly. This method is pretty much similar to building a structure using bricks. In the hybrid method, part of nanomaterial fabrication is done by bottom-up, and another part is carried out using top-down technology or vice-versa (Filipponi and Sutherland 2010; Sugunan 2012). Figure 1.6 graphically illustrates the top-down and bottom-up nanomaterial fabrication techniques. In all three methods, controlling the process parameters (like energy level of electron beam applied) and the environmental parameters (like contaminants, dust present in the surrounding area) are so critical in the success of nanomaterial fabrication process. Therefore, the nanomaterial fabrication is commonly carried out under well-controlled and clean laboratory facilities with advanced fabrication tools (Filipponi and Sutherland 2010).

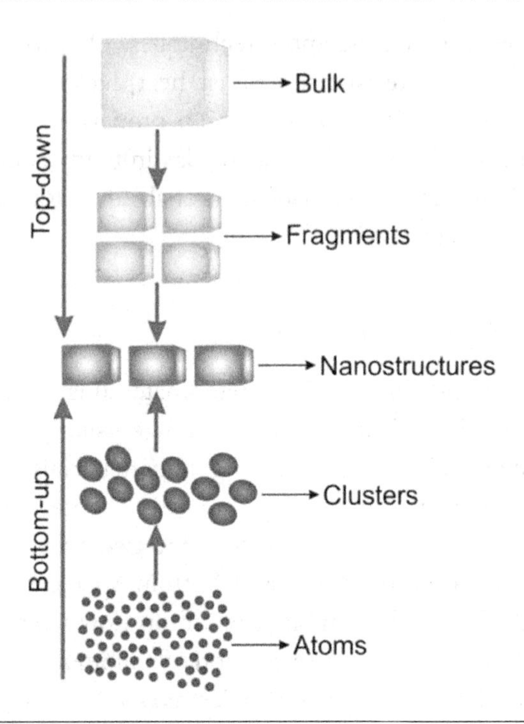

Figure 1.6 Illustration of top-down and bottom-up nanomaterial fabrication techniques. (Reproduced from Galstyan, V. et al., *Chemosensors*, 6, 1–21, 2018.)

Top-Down Fabrication Methods

In the top-down manufacturing of nanomaterials, the size of bulk or large materials is reduced to even smaller particles. For reducing the size of bulk materials into the nanoscale, commonly ball mills are used, which is the main top-down technique. Major parts of nanomaterials are fabricated using the top-down (from large bulk precursor to small nanoparticles) based on the technology used by the semiconductor manufacturing industry for fabricating integrated circuit chips for computers. An electron beam or a light source is used to remove the parts or reduce the size of precursor material to achieve the nanolevel materials by this fabricating technology, and the common terminology used for this fabrication technique is the "lithography" technique. Conventional lithography involves a series of steps, such as polymer coating of the substrate, exposing the resist to remove extraneous material in precursor, and developing the resist image with the chemical. During exposure of the resist in the lithographic process, the resist is irradiated in a large portion through a physical mask or point-by-point, using a computer program known as software mask. The software mask is also known as scanning lithography, which is a slow and serial process, but high resolution and precision can be obtained through the software masking process. The physical mask methodology is another lithography technique, which is a fast and parallel process. The scanning lithography uses highly energetic radiation for removal of precursor material, which requires sophisticated and expensive equipment. Photolithography and E-beam lithography are the practical world examples using the physical mask and scanning lithographic techniques, respectively (Filipponi and Sutherland 2010). Figures 1.6 through 1.9 illustrate various top-down nanomaterial fabrication methods, nanoimprint lithography, and dip pen and scanning lithography techniques, respectively.

Photolithography Photolithography is the best example of a physical mask lithographic process. In this process, light (in the range of X-rays and UV rays) is passed through a physical mask to expose and remove a portion of resist depending on the objective of the final product. The physical mask is made using light passing material with a pattern of opaque or light-absorbing materials; both materials depend on the type of light being used (Filipponi and Sutherland 2010).

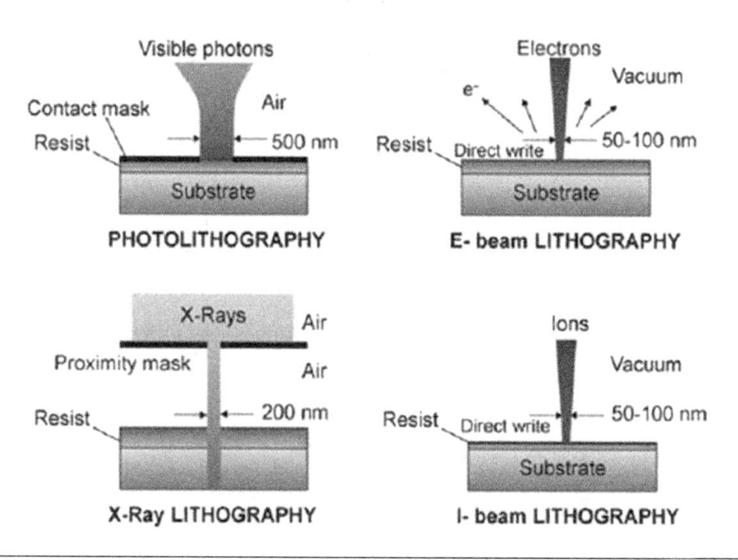

Figure 1.7 Different top-down fabrication techniques. (Reproduced from Daraio, C. and Jin, S., Synthesis and patterning methods for nanostructures useful for biological applications, In *Nanotechnology for Biology and Medicine*, Springer, New York. pp. 27–44, 2012.)

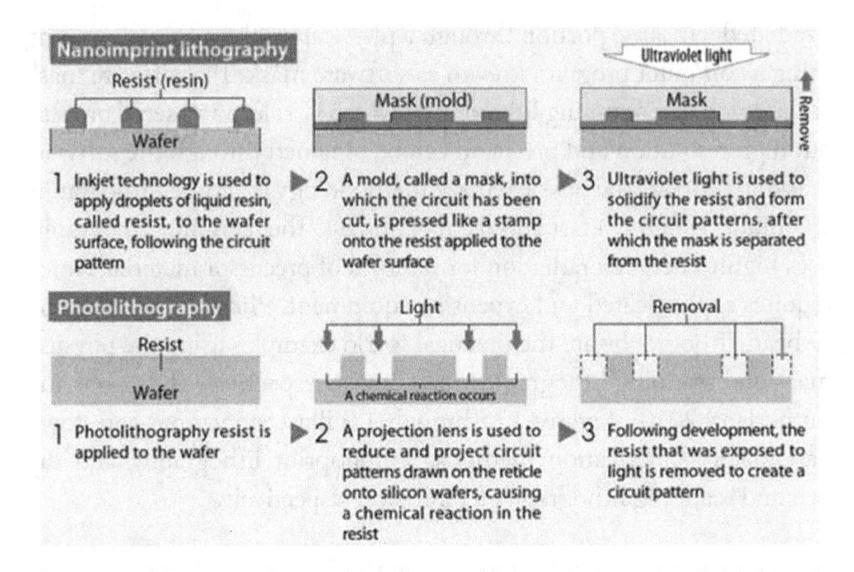

Figure 1.8 **(See color insert.)** Schematic of nanoimprint lithography technique. (Reproduced from Clarke, R., Report: Toshiba adopts imprint litho for NAND production, ee News Analog, Available at http://www.eenewsanalog.com/news/report-toshiba-adopts-imprint-litho-nand-production, 2016.)

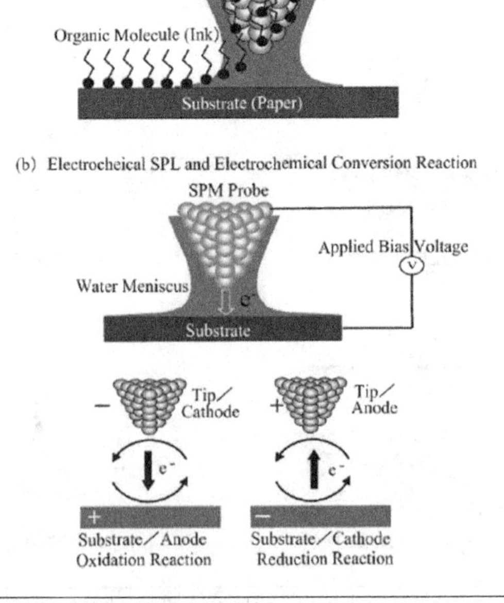

Figure 1.9 Schematic illustration of Dip pen lithography and scanning lithography. (Reproduced from Lee, S. et al., Scanning probe lithography on organic monolayers, In *Recent Advances in Nanofabrication Techniques and Applications*, InTech, London, UK, 2011.)

E-Beam Lithography E-beam lithography is a scanning lithographic process that uses highly focused electron beams on the pre-selected resist film to scathe the unnecessary material in a pattern required to obtain a desired final product. E-beam lithography facilitates to produce better resolution than photolithography. However, it is limited due to electron scattering on the resist film. This can be minimized by using heavy mass particles compared to electrons, such as H^+ and He^{++}. The process of using the ions instead of E-beam for fabricating nanomaterials is known as ion-beam lithography. Both the E-beam and the ion-beam lithography are serial processes, thus making these slow, but these can be employed where better resolution is needed over the temporal (Filipponi and Sutherland 2010).

Bottom-Up Fabrication Method

In the bottom-up fabrication technique, a nanomaterial with a desired shape and size is obtained by the gradual buildup of material from a small atom or ion. Bottom-up fabrication of nanomaterials involves the controlled crystallization of ions/atoms to form the desired structure. Bottom-up fabrication methods can be divided into two primary methods: gas-phase and liquid-phase based on the medium being used during the fabrication process (Figure 1.10).

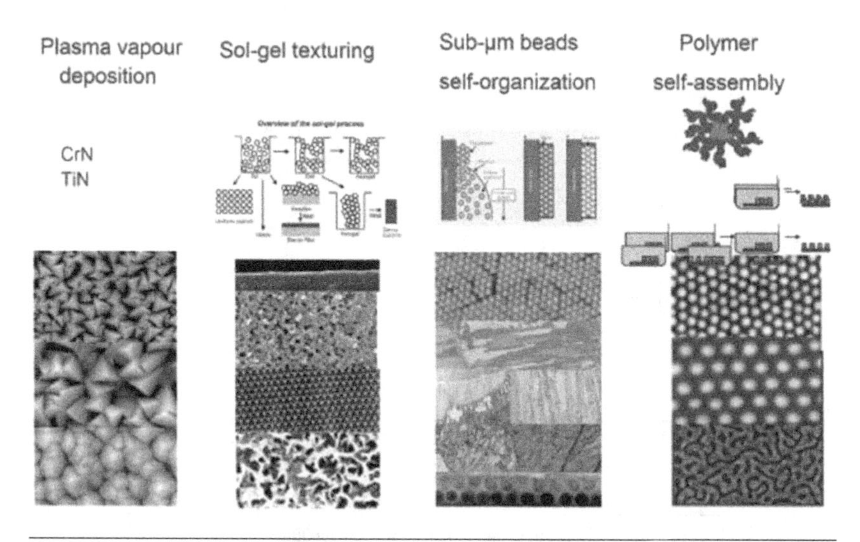

Figure 1.10 (See color insert.) Bottom-up fabrication techniques. (Reproduced from CSME, Nanostructure engineering, Accessed on November 21, Available at https://www.csem.ch/Publications, 2017.)

Gas-Phase Methods

Plasma Arcing Plasma arcing is the widely used gas-phase method for fabricating nanomaterials, particularly nanotube. This method uses ionized gas also known as plasma in a closed system, which facilitates movement of electrons between the electrodes by the potential difference. During the procedure, the anode vaporizes releasing electrons, which then deposit on the cathode. The material deposited on the cathode can be few atoms thick and can be considered as nanomaterial when it is at least 1 nm (Figure 1.11). The selection of electrodes depends on the nanomaterial to be fabricated; for example, carbon nanotubes are produced using carbon electrodes (Filipponi and Sutherland 2010; Kaur et al. 2013).

Chemical Vapor Deposition The chemical vapor deposition (CVD) technique uses the volatile properties of the material to fabricate its nanomaterials. In this method, a selected material is heated to vaporize to its gaseous state and cooled down to be deposited on a material. The chemical composition of deposited material depends on the process chosen to let the vaporized material to deposit directly or indirectly by controlled chemical reaction before deposition (Filipponi and Sutherland 2010). The chemical vapor deposition process is generally employed during the fabrication of nanopowders (Figure 1.12), and it involves five basic steps (Filipponi and Sutherland 2010):

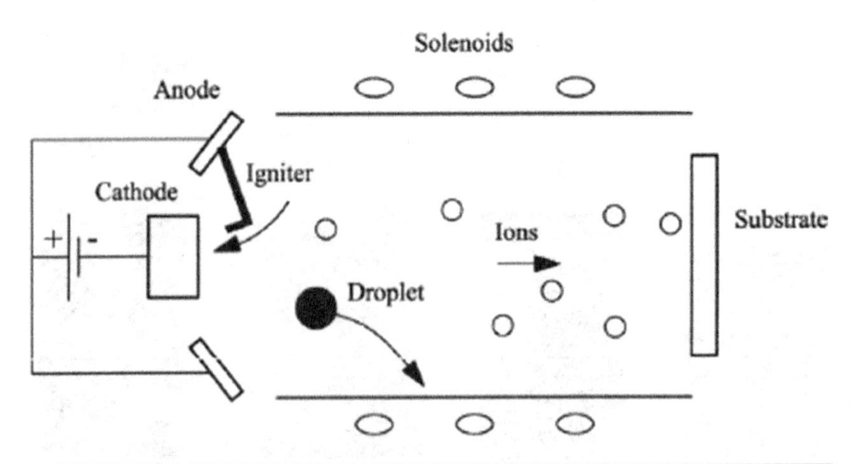

Figure 1.11 Graphical illustration of plasma arcing techniques. (Reproduced from TTU, Techniques for synthesis of nanomaterials (I), https://www.ttu.ee/public/m/Mehaanikateaduskond/Instituudid/Materjalitehnika_instituut/MTX9100/Lecture11_Synthesis.pdf, Accessed on November 23, 2017.)

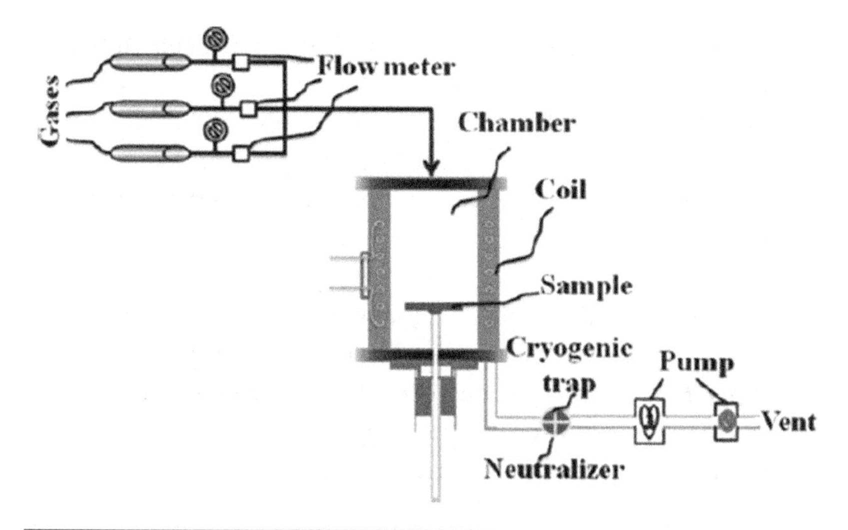

Figure 1.12 Schematic diagram of chemical vapor deposition technique. (Reproduced from Tiwari, J.N. et al., *Prog. Mater. Sci.*, 57, 724–803, 2012.)

1. The inert gas and vapor of the target material (reactant gas) are mixed using flow controllers and sent to the main chamber.
2. The mixed gases (inert and reactant) move towards the sample surface.
3. The reactant gas gets absorbed by the sample surface.
4. The chemical reaction between sample substrate and the reactant gas forms the nanomaterials on the substrate.
5. Then the resultant gases are desorbed and emptied from the chamber.

Sol-Gel Technique The sol-gel fabrication technique involves the transformation of a liquid-phase colloid (commonly called as Sol) into a gel form with continuous networks. Further processing of this gel using thermal treatment converts the gel into solid nanoparticles. The sol-gel process involves four processes (Rahman and Padavettan 2012):

1. Hydrolysis reaction: The precursor materials (generally metal oxides) dispersed with liquid (water, alcohol), and the -OR group of the precursor is replaced with the -OH group through the hydrolysis reaction. The hydrolysis can happen with or without the catalyst (either base or acid), and the presence of catalyst increases the hydrolysis reaction speed and effectiveness.

2. Condensation: After the hydrolysis reaction, the colloid form occurs due to the condensation.
3. Polymerization/gelation: After condensation, the particles in the colloid start agglomeration by creating a link between the polymer chains. This network forms throughout the colloid resulting in thickening of the liquid medium, which forms the gel.
4. Drying/thermal treatment: Finally, the solid nanoscale particles are obtained from the gel through precipitation and supercritical drying techniques. The thermal technique like calcination also helps in improving the mechanical strength of the nanomaterials and further polycondensation.

Figure 1.13 graphically shows the steps involved in sol-gel fabrication technique. The common application of sol-gel technique is for making silica gels (with nanoscale silica particles). The sol-gel technique is also used to produce aluminosilicate gels, which can be used to form nanotubular structures and to obtain nanoscale metal and metal oxide particles for various applications like nanocoating and nanomaterial fabrication. The sol-gel technique is commonly used to produce metal oxide or metal carbide nanopowders. Sol-gel methods also can

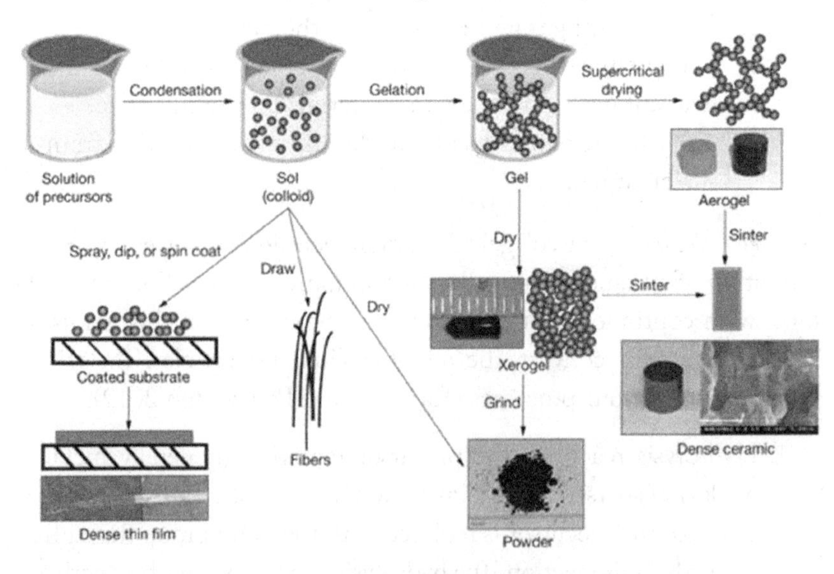

Figure 1.13 Schematic diagram of sol-gel fabrication technique. (Reproduced from Lawrence Livermore National Laboratory, *Novel Materials from Solgel Chemistry*, Lawrence Livermore National Laboratory, University of California, Livermore, CA, 2005.)

be used to produce pure metal nanopowders, but the process is little complicated than the metal oxide nanopowder production (Rahman and Padavettan 2012; Rani et al. 2008; Xu et al. 2001; Yang et al. 2005).

Combination of Top-Down and Bottom-Up Technologies

The nanofabrication industry is using either top-down or bottom-up fabrication technologies for producing various nanomaterials, but both of them have their own merits and demerits. The top-down techniques are the best suited for fabricating nanostructures having macroscopic connections and long-range order structures. The bottom-up fabrication methods are well suited for making short-range structures with nanoscale level. A combination of top-down and bottom-up fabrication techniques will open up a pathway for producing structures and materials in nanoscale with enhanced properties and abilities that extend the application of nanoscience and nanotechnology-based sensing tools and circuits (Sahoo et al. 2015). Onses et al. (2014) fabricated 3D copolymer films for semiconductors and magnetic storage devices with a hybrid nanofabrication technique, which combines ultra-high-resolution jet printing with the self-assembling block copolymers. Fabrication of nanostructures from the soft materials like polymers, proteins, and DNA by the common top-down fabrication techniques like lithographic and printing technologies is impossible due to the soft nature of those materials. But combining the jet printing with the self-assembling techniques helps to overcome the issues related to the fabrication of structures at the nanoscale level using soft materials like polymers. Through this jet printing with self-assembling copolymer hybrid techniques, Onses et al. (2014) were able to fabricate block-copolymer films with well-defined shapes and sizes (the dimensions in the range of 13–20 nm). Sahoo et al. (2015) fabricated "nanobridges" by combining top-down and bottom-up approaches. In the first step, silicon nanowires (SiNWs) were formed and grown vertically using silver particles as mask material at "electro-less chemical bath deposition," and in the second step, CNTs were formed by "atmospheric chemical vapor decomposition" techniques using the silver nanoparticles on the SiNW walls as a

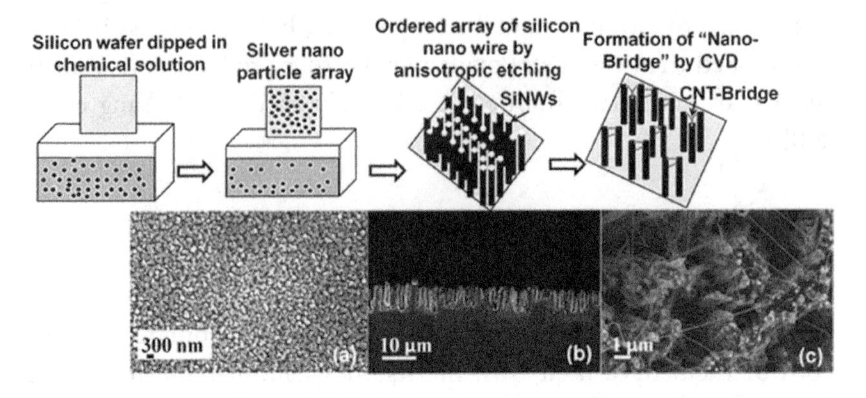

Figure 1.14 Graphical representation of steps involved in fabricating nanobridges from SiNW and CNTs. (Reproduced from Sahoo, R.K. et al., *J. Mater. Sci. Mater. Electron.*, 26, 435–440, 2015.)

catalyst in the process. The lateral growth of bridging network of this CNT ultimately produced the "nanobridges," and these nanobridges can be used for anti-reflection coating materials since the developed nanobridge samples had reduced reflectance of 65% when compared with the control samples (Figure 1.14). These nanobridge materials also can be used in nanosensors, since the surface-to-volume ratio is high in this type of nanostructures, which will result in higher efficiency when used in nanosensors (Sahoo et al. 2015).

Applications of Nanotechnology in Food and Beverages

The nanotechnology-based products and processes are used extensively in the fields of electronics (semiconductor and sensors), medicine, pharmaceuticals, defense, cosmetics, and textiles, which are out of the scope of this book. In recent times, the applications of nanotechnology and nanoscience-based processes and products in agriculture and food sector are also gaining attention from the processors as well as consumers. The nanoscience and nanotechnology-based products and processes are applied in the food and beverages industry for the enhancement of raw material properties, improving bioavailability of nutrients, development of nutrient delivery systems, tools for assessing food quality and food safety, treatment of diseases, wastewater treatment, improving nutritive value of the products (incorporation of nanoencapsulated nutrients), and smart and intelligent packaging materials (Chen et al. 2006; Maynard et al. 2006).

When the nanosize/nanoscale particles of ingredients and food additives are used in food processing, the nutrient, textural, and sensory properties of the food products can be enhanced due to the higher surface-to-mass ratio of the nanoparticles. Addition of functional components like probiotics and prebiotics in nanoscale level to the food and beverages not only improves the nutritional value of the product, but it also improves the bioavailability of those nutritional/functional components (Chaudhry et al. 2008; NIH 2016; Weiss et al. 2006; Yusop et al. 2012). Nanotechnology-based materials and methods have been widely tested for detection or inhibition of pathogenic microorganisms in the food and beverages. Since these spoilage and pathogenic microorganisms cause large amount of qualitative as well as quantitative losses to the food industry, rapid detection and control tools are becoming essential tool for the industry to save millions of dollars every year (Bhattacharya et al. 2007). The metal and metal oxide nanoparticles, like gold nanoparticles, silver nanoparticles, and titanium dioxide nanoparticles, exhibit excellent inhibition properties against major foodborne pathogens. When these nanoparticles are added to the food and beverages products as an ingredient or additive or a preservative, the microbial load of the product can be controlled, extending the shelf life of the product. The incorporation of these nanoparticles does not significantly alter or change the physical or textural or sensory properties of the product (Alfadul and Elneshwy 2010; Augustin and Sanguansri 2009; Gobi et al. 2005). The food and beverages sector is also applying advanced techniques like spectroscopy and imaging methods for food composition and microorganism analysis, but the complex nature of the operational procedure and the expensive cost of the equipment are the main disadvantages with these advanced techniques (Vadivambal and Jayas 2016). The recently developed nanosensors can overcome these major drawbacks because these are simple to use, have low cost, and can also be directly applied to the food production line, which may enhance the food safety and reduce the production cost and sample analysis time (Augustin and Sanguansri 2009; Avella et al. 2011; Baeumner 2004; Bernardes et al. 2014; Gomes et al. 2015; Ramachandraiah et al. 2015). These nanotechnology-based nanosensors and sensing techniques are not only used for detection of foodborne pathogens, but these can be also used for quality analysis like detection and measurement of chemical components of raw and processed food materials, detection of food adulterants, identification and quantification of chemical residues like pesticides in food and beverages, and monitoring food products quality from the manufacturing line to the consumer's kitchen table.

Food packaging is another major sector where the nanotechnology has been applied enormously in food and beverages sector. Regular conventional food and meat-packaging films are passive systems, which just play a role of protective barrier between the food packed and the surrounding. But the nanotechnology helped the packaging industry to develop active, smart, and intelligent packaging materials, which have improved or retained the quality of food for longer duration, at the same time keeping the microbial load of the food as low as possible from the processing plant to the consumer's hand. The nanotechnology-based packing films also serve as active packaging systems, which protect food as well as actively interact with food by removing/detecting undesirable factors, such as oxygen, moisture, and unwanted vapors released by food (like ammonia) during storage, and also releasing desirable compounds (such as antioxidants and antimicrobial agents) in order to keep quality and freshness of the food for a long time (Berekaa 2015; Duncan 2011; Sozer and Kokini 2009). Nanotechnology also helped to develop smart packaging technology for food and beverages, in which the condition of the product and the surrounding environment was monitored continuously by the nanosensors affixed in the package (Yam et al. 2005). With these smart packaging nanosensors, temperature, moisture, and gas composition inside the package or the storage environment can be monitored in order to identify the condition of the product throughout the supply chain. These nanosensors can also be used to detect the presence of foodborne pathogens and toxic substances in the product (Silvestre et al. 2011). These applications of nanotechnology in various sectors of the food and beverages industry are discussed in detail in this book in Chapters 4 through 9.

Summary

Nanoscience and nanotechnology deal with the substances in nanoscale (1–100 nm), and the application of nanotechnology is increasing in a rapid manner due to its decreasing cost, adaptability in various sectors, and increased accuracy. In the food and beverages industry also, the application of nanotechnology-based products and processes are increasing enormously ranging from simple food ingredients or additive to high-accuracy nanosensors for foodborne pathogen detection. Nanotechnology also helps to incorporate macro and micronutrients into food products without affecting physical, texture, and sensory properties of the food products and enhances the bioavailability of the nutrients by target delivery and reducing breakdown of nutrients during processing through encapsulation techniques. Nanotechnology-based packaging methods keep the food safe and fresh from the processing plant to consumer's dining table. Even though the application of nanotechnology in food and beverages provides lots of advantages for processing, packaging, and storage of food products, there are lots of questions among the consumers about the health risks involved with these applications. Leaching of nanoparticles from the packaging material to food, over intake of components due to the increased surface area to the mass property of nanomaterials, are some examples of queries consumers have in their minds about the usage of nanotechnology-based products and processes in food and beverages. Researchers throughout the world are working in order to eliminate these questions from the consumers' minds. This book mainly discusses the effect of nanotechnology-based products and processes have on the functionality of the food products and processes in Chapters 2 and 3; nanotechnology-based sensors for food and beverages sector in Chapter 4; application of nanotechnology on various products (dairy, bakery, meat, beverages) in Chapters 5 through 8; application of nanotechnology in treating wastewater from food and beverages industry in Chapter 9; and the regulations, risks, and consumer acceptance about application of nanotechnology for food products in Chapter 10.

References

Alagarasi A. (2011) *Introduction to Nanomaterials*. National Center for Environmental Research. pp. 1–76.

Alfadul S., Elneshwy A. (2010) Use of nanotechnology in food processing, packaging and safety–review. *African Journal of Food, Agriculture, Nutrition and Development* 10:2719–2739.

Allhoff F., Lin P., Moore D. (2009) *What Is Nanotechnology and Why Does It Matter? From Science to Ethics.* John Wiley & Sons, Chichester, UK.

Augustin M.A., Sanguansri P. (2009) Nanostructured materials in the food industry. *Advances in Food and Nutrition Research* 58:183–213. doi:10.1016/S1043-4526(09)58005-9.

Avella M., Errico M.E., Gentile G., Volpe M.G. (2011) Nanocomposite sensors for food packaging. In: Reithmaier J., Paunovic P., Kulisch W., Popov C., Petkov P. (Eds.) *Nanotechnological Basis for Advanced Sensors.* NATO Science for Peace and Security Series B: Physics and Biophysics. Springer, Dordrecht, the Netherlands. pp. 501–510. doi:10.1007/978-94-007-0903-4_53.

Baeumner A. (2004) Nanosensors identify pathogens in food. *Food Technology* 58:51–55.

Bandyopadhyaya R., Nativ-Roth E., Regev O., Yerushalmi-Rozen R. (2002) Stabilization of individual carbon nanotubes in aqueous solutions. *Nano Letters* 2(1):25–28.

Berekaa M.M. (2015) Nanotechnology in food industry: Advances in food processing, packaging and food safety. *International Journal of Current Microbiology and Applied Science* 4:345–357.

Bernardes P.C., de Andrade N.J., Soares N.D.F. (2014) Nanotechnology in the food industry. *Bioscience Journal* 30:1919–1932.

Bhattacharya S., Jang J., Yang L., Akin D., Bashir R. (2007) BioMEMS and nanotechnology-based approaches for rapid detection of biological entities. *Journal of Rapid Methods & Automation in Microbiology* 15(1):1–32.

British Museum Research Laboratory. (1959) File 1144, letter dated 5.2.59 to Dr Mackey, Department of Physics, Birkbeck College, London, UK.

Chaudhry Q., Scotter M., Blackburn J., Ross B., Boxall A., Castle L., Aitken R., Watkins R. (2008) Applications and implications of nanotechnologies for the food sector. *Food Additives and Contaminants* 25(3):241–258.

Chen H., Weiss J., Shahidi F. (2006) Nanotechnology in nutraceuticals and functional foods. *Food Technology* 60(3):30–36.

Clarke R. (2016) Report: Toshiba adopts imprint litho for NAND production. ee News Analog. Available from http://www.eenewsanalog.com/news/report-toshiba-adopts-imprint-litho-nand-production. Accessed on July 25, 2017.

Cox D.M. (1999) High surface area materials. In *Nanostructure Science and Technology.* Springer, Dordrecht, the Netherlands. pp. 49–66.

CSME. (2017) Nanostructure engineering. Available at https://www.csem.ch/Publications. Accessed on November 21, 2017.

Daraio C., Jin S. (2012) Synthesis and patterning methods for nanostructures useful for biological applications. In *Nanotechnology for Biology and Medicine.* Springer, New York. pp. 27–44.

Domingo C., Resta V., Sanchez-Cortes S., García-Ramos J.V., Gonzalo J. (2007) Pulsed laser deposited Au nanoparticles as substrates for surface-enhanced vibrational spectroscopy. *The Journal of Physical Chemistry C* 111(23):8149–8152.

Dong X., Ji X., Jing J., Li M., Li J., Yang W. (2010) Synthesis of triangular silver nanoprisms by stepwise reduction of sodium borohydride and trisodium citrate. *The Journal of Physical Chemistry C* 114(5):2070–2074.

Dumitrache F., Morjan I., Alexandrescu R., Ciupina V., Prodan G., Voicu I., Fleaca C., Albu L., Savoiu M., Sandu I., Popovici E. (2005) Iron–iron oxide core–shell nanoparticles synthesized by laser pyrolysis followed by superficial oxidation. *Applied Surface Science*, 247(1–4):25–31.

Duncan T.V. (2011) Applications of nanotechnology in food packaging and food safety: Barrier materials, antimicrobials and sensors. *Journal of Colloid and Interface Science* 363:1–24. doi:10.1016/j.jcis.2011.07.017.

Eglash R. (2011) Nanotechnology and traditional knowledge systems. In Maclurcan D., Radywyl N. (Eds.) *Nanotechnology and Global Sustainability*. CRC Press, Boca Raton, FL. pp. 45–68.

EU (2015) *Scientific Basis for the Definition of the Term "Nanomaterial."* Scientific Committee on Emerging and Newly Identified Health Risks. Europian Union, Brussels, Belgium.

Feynman R.P. (1992) There's plenty of room at the bottom [data storage]. *Journal of Microelectromechanical Systems* 1(1):60–66.

Filipponi L., Sutherland D. (2010) *Introduction to Nanoscience and Nanotechnologies, Interdisciplinary Nanoscience Center.* Aarhus University, Aarhus, Denmark. pp. 2–29.

Freestone I., Meeks N., Sax M., Higgitt C. (2007) The Lycurgus cup—A roman nanotechnology. *Gold Bulletin* 40(4):270–277.

Galstyan V., Bhandari M.P., Sberveglieri V., Sberveglieri G., Comini E. (2018) Metal oxide nanostructures in food applications: Quality control and packaging. *Chemosensors* 6(16):1–21.

Gobi K.V., Tanaka H., Shoyama Y., Miura N. (2005) Highly sensitive regenerable immunosensor for label-free detection of 2,4-dichlorophenoxyacetic acid at ppb levels by using surface plasmon resonance imaging. *Sensors and Actuators B: Chemical* 111–112:562–571. doi:10.1016/j.snb.2005.03.118.

Gomes R.C., Pastore V.A.A., Martins O.A., Biondi G.F. (2015) Nanotechnology applications in the food industry. A review. *Brazilian Journal of Hygiene and Animal Sanity* 9:1–8. doi:10.5935/1981-2965.20150001.

Gusev A.I., Rempel A.A. (2004) *Nanocrystalline Materials*. Cambridge International Science Publishing, Cambridge, UK. p. 351.

Handford C.E., Dean M., Spence M., Henchion M., Elliott C.T., Campbell K. (2014) *Nanotechnology in the Agri-Food Industry on the Island of Ireland: Applications, Opportunities and Challenges*. SafeFood, Cork, Ireland.

Harden D.B., Hellenkemper H., Painter K., Whitehouse D. (1987) *Glass of the Caesars*. Olivetti, Milan, Italy. pp. 245–249.

Harden D.B., Toynbee J.M.C. (1959) The Rothschild Lycurgus cup. *Archaeologia* 97:179–212.

Huang L., Wang H., Wang Z., Mitra A., Zhao D., Yan Y. (2002) Cuprite nanowires by electrodeposition from lyotropic reverse hexagonal liquid crystalline phase. *Chemistry of Materials* 14(2):876–880.

Huang Z.M., Zhang Y.Z., Kotaki M., Ramakrishna S. (2003) A review on polymer nanofibers by electrospinning and their applications in nanocomposites. *Composites Science and Technology* 63(15):2223–2253.

Hussain C.M., Palit, S. (2018). Nanomaterials, ecomaterials, and wide vision of material science. In *Handbook of Ecomaterials*. Springer International Publishing. pp. 1–29.

IBM (2008) A history of nanotechnology milestones at IBM research. *Media Conference*, Swiss Federal Institute of Technology, Zurich, June 25.

Ismail M., Gul S., Khan M.A., Khan M.I. (2016) Plant-mediated green synthesis of anti-microbial silver nanoparticles—A review on recent trends. *Reviews in Nanoscience and Nanotechnology* 5(2):119–135.

Jäger C., Huisken F., Mutschke H., Henning T., Poppitz W., Voicu I. (2007) Identification and spectral properties of PAHs in carbonaceous material produced by laser pyrolysis. *Carbon* 45(15):2981–2994.

Jun Y.W., Seo J.W., Oh S.J., Cheon J. (2005) Recent advances in the shape control of inorganic nano-building blocks. *Coordination Chemistry Reviews* 249(17–18):1766–1775.

Kaur D., Chaudhary S., Pandya D.K., Gupta R., Kotnala R.K. (2013) Magnetization reversal studies in structurally tailored cobalt nanowires. *Journal of Magnetism and Magnetic Materials* 344:72–78.

Kelarakis A., Giannelis E.P., Yoon K. (2007) Structure–properties relationships in clay nanocomposites based on PVDF/(ethylene–vinyl acetate) copolymer blends. *Polymer*, 48(26):7567–7572.

Kim K.S., Zhao Y., Jang H., Lee S.Y., Kim J.M., Kim K.S., Ahn J.H., Kim P., Choi J.Y., Hong B.H. (2009) Large-scale pattern growth of graphene films for stretchable transparent electrodes. *Nature* 457(7230):706.

Kuchibhatla S.V., Karakoti A.S., Bera D., Seal S. (2007) One-dimensional nanostructured materials. *Progress in Materials Science* 52(5):699–913.

Lawrence Livermore National Laboratory. (2005) *Novel Materials from Sol-Gel Chemistry*. Lawrence Livermore National Laboratory, University of California, Livermore, CA.

Lee, S., Ishizaki, T., Teshima, K., Saito, N., Takai, O. (2011) Scanning probe lithography on organic monolayers. In *Recent Advances in Nanofabrication Techniques and Applications*. InTech, London, UK.

Lukatskaya M.R., Dunn B., Gogotsi Y. (2016) Multidimensional materials and device architectures for future hybrid energy storage. *Nature Communications* 7(12647):1–13.

Mann A.K., Skrabalak S.E. (2011) Synthesis of single-crystalline nanoplates by spray pyrolysis: A metathesis route to Bi_2WO_6. *Chemistry of Materials* 23(4):1017–1022.

Maynard A.D., Aitken R.J., Butz T., Colvin V., Donaldson K., Oberdörster G., Philbert M.A., Ryan J., Seaton A., Stone V., Tinkle S.S., Tran L., Walker N.J., Warheit D.B. (2006) Safe handling of nanotechnology. *Nature* 444(7117):267–269.

Moraru C.I., Panchapakesan C.P., Huang Q., Takhistov P., Liu S., Kokini J.L. (2003) Nanotechnology: A new frontier in food science. *Food Technology* 57(12):24–29.

Murthy A.K., Stover R.J., Hardin W.G., Schramm R., Nie G.D., Gourisankar S., Truskett T.M., Sokolov K.V., Johnston K.P. (2013) Charged gold nanoparticles with essentially zero serum protein adsorption in undiluted fetal bovine serum. *Journal of the American Chemical Society* 135(21):7799–7802. doi:10.1021/ja400701c.

Nayak B.B., Behera D., Mishra B.K. (2010) Synthesis of silicon carbide dendrite by the arc plasma process and observation of nanorod bundles in the dendrite arm. *Journal of the American Ceramic Society* 93(10):3080–3083.

Ngô C., Van de Voorde M.H. (2014) Nanomaterials: Doing more with less. In: *Nanotechnology in a Nutshell.* Atlantis Press, Paris, France. pp. 55–70.

NIH. (2016) *Selenium: Dietary Supplement Fact Sheet.* National Institutes of Health, Bethesda, MD.

Onses M.S., Ramírez-Hernández A., Hur S.M., Sutanto E., Williamson L., Alleyne A.G., Nealey P.F., De Pablo J.J., Rogers J.A. (2014) Block copolymer assembly on nanoscale patterns of polymer brushes formed by electrohydrodynamic jet printing. *ACS Nano* 8(7):6606–6613.

Pokropivny V., Lohmus R., Hussainova I., Pokropivny A., Vlassov S. (2007) *Introduction to Nanomaterials and Nanotechnology.* Tartu University Press, Ukraine. pp. 45–100.

Rahman I.A., Padavettan V. (2012) Synthesis of silica nanoparticles by sol-gel: Size-dependent properties, surface modification, and applications in silica-polymer nanocomposites—A review. *Journal of Nanomaterials* 8:Article ID 132424:1–8. doi:10.1155/2012/132424.

Ramachandraiah K., Han S.G., Chin K.B. (2015) Nanotechnology in meat processing and packaging: Potential applications—A review. *Asian-Australasian Journal of Animal Sciences* 28:290–302. doi:10.5713/ajas.14.0607.

Ramanathan T., Abdala A.A., Stankovich S., Dikin D.A., Herrera-Alonso M., Piner R.D., Adamson D.H., Schniepp H.C., Chen X.R.R.S., Ruoff R.S., Nguyen, S.T. (2008) Functionalized graphene sheets for polymer nanocomposites. *Nature Nanotechnology* 3(6):327–331.

Rani S., Suri P., Shishodia P.K., Mehra R.M. (2008) Synthesis of nanocrystalline ZnO powder via sol–gel route for dye-sensitized solar cells. *Solar Energy Materials and Solar Cells* 92(12):1639–1645

Ray S.S., Okamoto M. (2003) Polymer/layered silicate nanocomposites: A review from preparation to processing. *Progress in Polymer Science* 28(11):1539–1641.

Sahoo R.K., Damodar D., Jacob C. (2015) A combination of "top-down" and "bottom-up" approaches in the fabrication of "nano bridges." *Journal of Materials Science: Materials in Electronics* 26(1):435–440.

Schmidt H.K., Geiter E., Mennig M., Krug H., Becker C., Winkler R.P. (1998) The sol-gel process for nano-technologies: new nanocomposites with interesting optical and mechanical properties. *Journal of Sol-gel Science and Technology* 13(1–3):397–404.

Schrand A.M., Rahman M.F., Hussain S.M., Schlager J.J., Smith D.A., Syed A.F. (2010) Metal-based nanoparticles and their toxicity assessment. *Wiley Interdisciplinary Reviews: Nanomedicine and Nanobiotechnology* 2(5):544–568.

Shen Q., Jiang L., Zhang H., Min Q., Hou W., Zhu J.J. (2008) Three-dimensional dendritic Pt nanostructures: Sonoelectrochemical synthesis and electrochemical applications. *The Journal of Physical Chemistry C* 112(42):16385–16392.

Silvestre C., Duraccio D., Cimmino S. (2011) Food packaging based on polymer nanomaterials. *Progress in Polymer Science* 36:1766–1782.

Sozer N., Kokini J.L. (2009) Nanotechnology and its applications in the food sector. *Trends in Biotechnology* 27:82–89.

Sugunan A. (2012) Fabrication and photoelectrochemical applications of II–VI semiconductor nanomaterials, Doctoral dissertation, KTH Royal Institute of Technology, Stockholm, Sweden.

Taniguchi N. (1974) On the basic concept of nanotechnology. In *Proceedings of the International Conference on Production Engineering-Part II*. Japan Society of Precision Engineering, Tokyo. pp. 18–23.

Tiwari J.N., Tiwari R.N., Chang Y.M., Lin K.L. (2010) A promising approach to the synthesis of 3D nanoporous graphitic carbon as a unique electrocatalyst support for methanol oxidation. *ChemSusChem* 3(4):460–466.

Tiwari J.N., Tiwari R.N., Kim K.S. (2012) Zero-dimensional, one-dimensional, two-dimensional and three-dimensional nanostructured materials for advanced electrochemical energy devices. *Progress in Materials Science* 57(4):724–803.

Tolochko N. (2009) History of nanotechnology. Nanoscience and nanotechnology. Encyclopaedia of life Support Systems (EOLSS), Developed under the auspices of the UNESCO-EOLSS Publication, Oxford, 3–4.

Tomassoni M.E., Matsuoka Y., Ramesh R. (2013) *How Nanotechnology Will Change Hair Styles*. Wavecloud Publishing, Aurora, CO.

TTU (2017) Techniques for synthesis of nanomaterials (I). Available from https://www.ttu.ee/public/m/Mehaanikateaduskond/Instituudid/Materjalitehnika_instituut/MTX9100/Lecture11_Synthesis.pdf. Accessed on November 23, 2017.

Vadivambal R., Jayas D.S. (2016) *Bio-Imaging: Principles, Techniques, and Applications*. CRC Press, Taylor & Francis Group, LLC, Boca Raton, FL. p. 381.

Vengatesan M.R., Mittal, V. (2015) Surface modification of nanomaterials for application in polymer nanocomposites: An overview. *Surface Modification of Nanoparticle and Natural Fiber Fillers*, 1–28.

Weiss J., Takhistov P., McClements D.J. (2006) Functional materials in food nanotechnology. *Journal of Food Science* 71:R107–R116.

Whitesides G.M. (2005) Nanoscience, nanotechnology, and chemistry. *Small* 1(2):172–179.

Wood M.A. (2007) Colloidal lithography and current fabrication techniques producing in-plane nanotopography for biological applications. *Journal of the Royal Society Interface* 4(12):1–17.

Xu H., Aylott J.W., Kopelman R., Miller T.J., Philbert M.A. (2001) A real-time ratiometric method for the determination of molecular oxygen inside living cells using sol-gel-based spherical optical nanosensors with applications to rat C6 glioma. *Analytical Chemistry* 73(17):4124–4133.

Yam K.L., Takhistov P.T., Miltz J. (2005) Intelligent packaging: Concepts and applications. *Journal of Food Science* 70:R1–R10.

Yang Q., Sha J., Ma X., Yang D. (2005) Synthesis of NiO nanowires by a sol-gel process. *Materials Letters* 59(14–15):1967–1970.

Yokel R.A., MacPhail R.C. (2011) Engineered nanomaterials: Exposures, hazards, and risk prevention. *Journal of Occupational Medicine and Toxicology* 6(1):7.

Yusop S.M., O'Sullivan M.G., Preuß M., Weber H., Kerry J.F., Kerry J.P. (2012) Assessment of nanoparticle paprika oleoresin on marinating performance and sensory acceptance of poultry meat. *LWT-Food Science and Technology* 46:349–355.

2

Impact of Nanomaterials on Food Functionality

Introduction

Applications of nanoscience and nanotechnology are expanding quickly in food and beverages industry and nanotechnology is making revolutionary changes in food and beverages industry in the development and production of food and beverage products and their packaging and storage. Even though the nanotechnology deals with the materials in small scale, its impact in food and beverages manufacturing and packaging is huge (Wesley et al. 2014). Nanotechnology-based products and techniques have been applied in food and beverage industry using nanoengineered products, nanoencapsulation, and nanoemulisifiers. These applications have been as ingredients and additives (coloring, flavoring, and anticaking agents) and in pathogen detection and control, as well as packaging and enhancement of nutrient or bioavailability (Arend et al. 2017; He and Hwang 2016; Mogol et al. 2013; Santillán-Urquiza et al. 2017; Shimoni 2009; van Vlerken et al. 2007; Wajda et al. 2007). Recent developments in nanotechnology helped to produce various nanoengineered materials through different preparation techniques, and the physicochemical properties of the nanoparticles can be altered through these preparation methods. Application of nanoengineered products in food and beverage production alters the physical, chemical, textural, and sensory properties of the food. For example, use of nanosize food additives like coloring and flavoring agents can enhance the color and flavor of the food products. The functionality of the processed food or beverage, such as color, flavor, taste, antimicrobial effect, storability, and the quality of the food product, is modified or altered by the use of nanotechnology-based products or processes during food and beverage manufacturing processes (He and Hwang 2016; Oehlke et al. 2014). The list of potential nanoengineered materials that can be applied in food and beverages sector along with their functionality on food products is given in Table 2.1.

The major advantage of applying nanoengineered materials in food and beverages is an enhancement of bioavailability of nutrients. The higher surface area to mass/volume nature of nanosize particles increases the absorption of nutrients in the gastrointestinal tract (Oehlke et al. 2014). Nanotechnology-based encapsulation of the bioactive compounds like vitamins, minerals, and other nutritional compounds could help to reduce or arrest the degradation of bioactive compound due to the temperature, gases (oxygen, carbon dioxide), and moisture before consumption or reaching its targeted location (for example gastrointestinal system). The nanoencapsulation technique is also used for targeted nutrient delivery system in food, and targeted drug delivery system in pharmaceutical sectors (Astray et al. 2009; Lesmes et al. 2009; Livney 2010; McClements et al. 2009; Weiss et al. 2008; Zorilla et al. 2011). Nanoencapsulation can also help to add some bioactive components, which have great health benefits, but use in raw form or direct intake might be limited due to other properties like taste or flavor might be unpleasant to some people (Shimoni 2009). For example, the unpleasant flavor of the fish oil, which is an excellent source of omega-3 fatty acid, can be masked, and the nutrient can be delivered directly into intestine by using the nanoencapsulation technique (Cushen et al. 2012). Some of the bioactive components like antioxidants can be quickly degraded by oxidation process, and nanoencapsulation could help to minimize the loss of nutrient in food product until it reaches the targeted organ in the human digestion system.

Physical, chemical, textural, and sensory properties of foods and beverages can be altered by the use of nanotechnology-based products and processes. For example, application of food coloring and flavoring agents in the nanoscale level could lead to enhanced color and flavor of developed food product due to the greater penetration and absorption of nanoparticles than the regular-size food additives (Alfadul and Elneshwy 2010). The amount of food additives used for food product manufacturing can also be minimized by using nanoengineered food ingredients, which will economically help the food

Table 2.1 Potentially Useful Nanoengineered Products and Their Impact in Functionality of Food Products

NANOENGINEERED PRODUCTS	PREPARATION METHOD AND COMPOSITION	FUNCTIONALITY	STRUCTURE[a]	REFERENCES
LIPID AND SURFACTANT BASED PRODUCTS				
Micelles; microemulsions/swollen micelles	Self-assembly after dissolving of surface active compounds in ternary mixture of emulsifiers, oil, water	Increasing bioavailability by nanoencapsulation		McClements et al. (2009)
Nanoemulsions	High-pressure homogenization, ultrasound-assisted homogenization	Enhancing nutrient intake		McClements and Rao (2011)
Solid lipid nanoparticles (SLN)	Hot emulsification of high-melting lipids	Preventing chemical degradation of bioactive compounds and increase the nutrient absorption		Weiss et al. (2008)
Lipid nanocarriers (LNC), nanostructured lipid carriers (NLC)	Hot emulsification of high-melting lipids with certain proportion of low-melting lipids	Enhance lutein delivery		Liu and Wu (2010)

(Continued)

Table 2.1 (*Continued*) Potentially Useful Nanoengineered Products and Their Impact in Functionality of Food Products

NANOENGINEERED PRODUCTS	PREPARATION METHOD AND COMPOSITION	FUNCTIONALITY	STRUCTURE[a]	REFERENCES
Liposomes/vesicles	Mixture of phospholipids, evaporation of solvent under reduced pressure	Enhancing the structural stability, bioavailability, and shelf life of temperature- and oxidization-sensitive ingredients	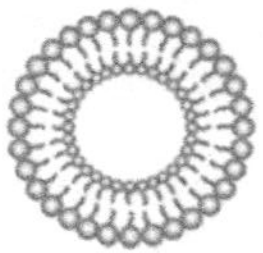	Mozafari et al. (2008)
POLYSACCHARIDE-BASED PRODUCTS				
Molecular complexes: cyclodextrin inclusion complexes, amylose complexes	Solubilization under appropriate conditions	Increasing stability and bioavailability of omega-3 polyunsaturated fatty acids; flavor stabilization and preventing browning; altering long-chain fatty acid delivery; enhance bioavailability and prevent degradation of sensitive nutrients	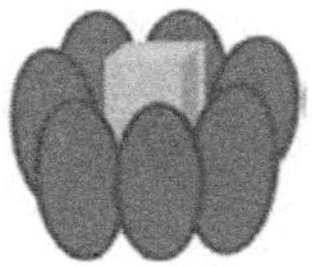	Zimet et al. (2011); Astray et al. (2009); Lesmes et al. (2009); Livney (2010)
Biopolymeric nanogels: chitosan particles, alginate gels	Ionic or covalent cross-linking of polymers	Enhancing Catechin oral bioavailability; encapsulation of active compounds		Dudhani and Kosaraju (2010); Morris et al. (2010)
PROTEIN-BASED PRODUCTS				
Protein inclusion complexes (e.g., with β-lactoglobulin)	Solubilization under appropriate conditions	Enhancing antioxidant (Epigallocatechin-3-gallate [EGCG]) activity through increasing bioavailability		Zorilla et al. (2011)

(*Continued*)

Table 2.1 (*Continued*) Potentially Useful Nanoengineered Products and Their Impact in Functionality of Food Products

NANOENGINEERED PRODUCTS	PREPARATION METHOD AND COMPOSITION	FUNCTIONALITY	STRUCTURE[a]	REFERENCES
Protein nanotubes	Self-assembly	Nanoencapsulation of bioactive compounds		Graveland-Bikker and de Kruif (2006)
Re-assembled casein micelles	Self-assembly	Enhance bioavailability of nutrients and prevent chemical degradation		Livney (2010)
PLANT AND INORGANIC MATERIALS, NANOCRYSTALS				
Plant materials, minerals	Comminution of larger materials	Improve calcium intake; Enhance the absorption of lignan glycosides from sesame meal		Chen et al. (2008); Liao et al. (2010)
Nanocrystals		Increasing the nutrient absorption		Müller et al. (2011)

Source: Oehlke, K. et al., *Food Funct.*, 5, 1341–1359, 2014.
[a] Schematic diagrams of the structures are not to the scale.

and beverage industry by reducing manufacturing costs (Yusop et al. 2012). The enhanced quality parameters like color, flavor, and antimicrobial properties of the processed food products using nanotechnology also help to keep the product safely for a long time, leading to a higher profit for the food and beverage manufacturer. Application of nanotechnology in the food and beverage industry affects the quality of food product, as well as the manufacturing process in many different ways, and the impact of applying nanotechnology-based products (nanoengineered materials) and processes on the functionality of the processed food product is discussed in brief in this chapter. The detailed discussion about the applications of nanoscience and nanotechnology in various sectors of the food and beverage industry and their effects on product and process quality is available in Chapters 5 through 9 of this book.

Impact on Bioavailability of Nutrient Compounds

One of the major advantages of using nanotechnology-based or nanoengineered products in food and beverages is increasing the bioavailability of the nutritional and other bioactive components of the food matrix. Application of nanotechnology to improve the bioavailability of the bioactive components like vitamins, calcium, iron, curcumin, and coenzyme Q10 (CoQ10) were studied in detail, and various nutrition delivery vehicles like nanofibers, nanoencapsulation, nanoemulsions, and nanolaminates were developed in these studies using nanoscience and nanotechnology (He and Hwang 2016; Zhou et al. 2014). Oehlke et al. (2014) stated that the nanotechnology-based nutrient delivery systems are increasing the bioavailability of the bioactive components of a food matrix using either one or a combination of the following four mechanisms:

1. Increasing the solubility of the bioactive components by reducing the particle size of the component to nanosize, which will increase the surface area
2. Improving the permeation or absorption of the bioactive component in the gastrointestinal system (e.g., coating and formulating with surfactants, which increase nutrient absorption)

3. Increasing the chemical stability of the bioactive component by encapsulation or emulsification
4. Targeted release with enhanced residence period in the gastrointestinal system

The bioavailability of the nutritional/bioactive component can be quantitatively calculated using the following formula (Salvia-Trujillo et al. 2016):

$$BA = B^* \times A^* \times T^*$$

where:

BA is the bioavailability of a nutritional/bioactive component of a food matrix.

B^* is the bioaccessibility of the component.

A^* is the absorption or permeation of bioactive component at the gastrointestinal system.

T^* is the transformation or alteration rate of bioactive component molecules.

Based on this equation, the increase in bioaccessibility can improve the bioavailability, and when the particle size is reduced to nano level, the surface-to-volume ratio increases, which results in increasing the bioaccessibility. The increase of permeation by using selective surfactants or coating agents can increase the bioavailability. The control of molecular transformation rate of the bioactive component also increases the bioavailability, for example, polyethylene glycol can be used as hydrophilic molecules in nutrient nano-delivery systems for controlled and targeted release of nutrients and other bioactive components (van Vlerken et al. 2007). A detailed list of nanotechnology-based methods used to enhance the bioavailability of nutrients and other bioactive compounds is given in Table 2.2.

Delivery of probiotics from dairy products directly to the gut ecosystem can improve the nutritional availability and encapsulation of probiotic bacterium, and targeted delivery into the gastrointestinal system has been tested by the researchers in order to improve the bioavailability of the nutrients (Huang et al. 2009; Yu and Huang 2010). The fortified yogurt prepared with calcium, zinc, and iron nanoparticles (average particle size of 50–80 nm) showed more stability in

Table 2.2 List of Nanotechnology-Based Materials Used in Food Products and Their Impact on Bioavailability of Bioactive Compounds

BIOACTIVE COMPONENT	NANOMATERIAL TESTED	CHANGE IN BIOAVAILABILITY	TESTING METHOD	REFERENCES
Curcumin	Suspension/lecithin mixture/liposomes (263 nm)	5-fold increase in area under the plasma level vs time curve (AUC) when compared with control	In vivo test using rats at an administration rate of 100 mg/kg body weight	Takahashi et al. (2009)
Curcumin	Lipid nanoparticles (135 nm) prepared with Tween 20 solution	39-fold increase in plasma levels	In vivo test using rats at an administration rate of 50 mg/kg body weight	Kakkar et al. (2011)
Quercetin	Lipid nanoparticles (155 nm)	5.7-fold increase in AUC	In situ test using rats at an administration rate of 50 mg/kg body weight	Li et al. (2009)
CoQ10	Nanoemulsion at a particle size of 60 nm	1.7-fold increase in AUC	In vivo test using rats at an administration rate of 60 mg/kg bodyweight	Hatanaka et al. (2008)
CoQ10	Nano micelles in gelatin capsules (30–60 nm)	5-fold increase in AUC	In vivo test with humans (100 mg/person)	Wajda et al. (2007)
Vitamin E	Nano micelles in gelatin capsules (50 nm)	5-fold increase in AUC	In vivo test with humans (120 mg/person)	Wajda et al. (2007)
Iron	Amorphous $FePO_4$ particles of different sizes: 11, 31, and 64 nm	0.96, 0.70, 0.61- fold increase in Hb repletion	In vivo test using Fe depleted rats (170 to 484 mg/day for 15 days)	Rohner et al. (2007)
Calcium	Pearl powder with a particle size of 84 nm	1.4-fold increase in calcium levels in serum	In vivo test with humans at a feed rate of 780 mg/person	Chen et al. (2008)
Calcium	Pearl powder with a particle size of 40–80 nm	2-fold increase in calcium levels in bones	In vivo test with rats at a feed rate of 10–70 mg/day for 4 weeks	Gao et al. (2008)
Chromium	Chromium picolinate with a particle size of 70 nm	2-fold increase in chromium levels in bones	In vivo test with rats at a feed rate of 300 μg/kg bodyweight/day for 18 days	Lien et al. (2009)

Source: Oehlke, K. et al., *Food Funct.*, 5, 1341–1359, 2014.

syneresis tests and had stronger gel structure than the regular yogurt (Santillán-Urquiza et al. 2017). The firmness and cohesiveness of nano Ca fortified samples were 0.93 N, and -0.65, respectively, but the control yogurt (with no supplement) had a firmness of 0.60 N, and a cohesiveness of -0.42. Santillán-Urquiza et al. (2017) also tested the textural properties of yogurt with microsize Ca, Zn, and Fe particles with the control and nanosize mineral fortified samples and found the yogurt with nanosize particles had higher firmness and cohesiveness. The sensory and digestibility tests also showed higher scores for Ca, Zn, and Fe nanoparticles fortified yogurt than the regular yogurt. The solubility of the yogurt fortified with nanosize Ca, Zn, and Fe particles was higher than that of control and microsize mineral fortified yogurt samples, which resulted in the increase in digestibility (Santillán-Urquiza et al. 2017). Consumer's preference is turning towards lower cholesterol dairy products due to the health benefits, but at the same time consumers do not want to lose the natural taste of dairy products. Chitosan, a derivative from the chitin nanoparticles, was tested to maintain the original taste and enhance the shelf life of low cholesterol yogurt by Seo et al. (2009) using chitosan nanoparticles (NPs) with an average particle size of 536 nm, and found the chitosan NPs mixed low cholesterol yogurt had higher shelf life than the regular yogurt. The pH of the chitosan NPs containing low cholesterol yogurt did not change significantly when stored at refrigeration temperature (4°C) for 20 days, but the pH value of the regular yogurt sample decreased from 4.21 to 3.94. There were no significant changes in the sensory scores between low cholesterol yogurt with or without Chitosan NPs (Seo et al. 2009). The nano Ca supplementation in milk resulted in the higher trabecular bone area and elongated trabeculae tissues when tested in rats than the no Ca supplementation and regular Ca supplement milk (Figures 2.1 and 2.2) (Park et al. 2007). Rats were fed with milk supplemented with carbonated Ca, ionized Ca, and nano Ca along with regular milk (no Ca supplementation), and the bone mineral density, trabecular area ratio, serum Ca, and tissue Ca were measured along with the microscopic imaging of trabecular bone tissues. The bone mineral density and trabecular area ratios were significantly higher in rats fed with nano Ca supplemented milk when compared with the other Ca supplementation and regular milk-fed rats. Park et al. (2007) attributed the supplementation with

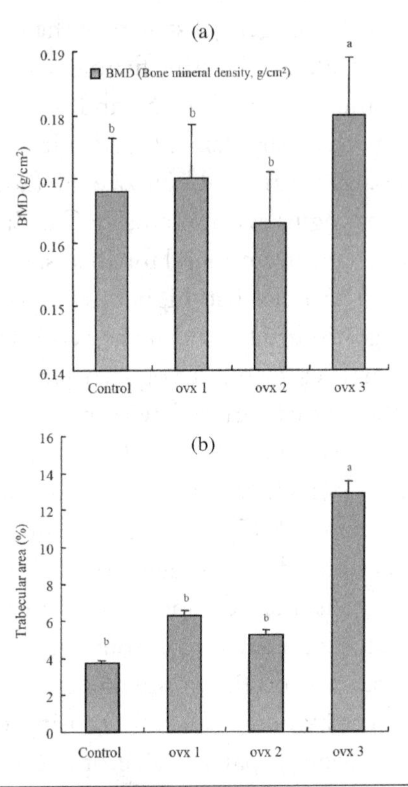

Figure 2.1 (a) Bone mineral density and (b) trabecular area ratio of rats fed with milk containing (control: no Ca supplementation regular milk), (ovx 1) carbonated Ca, (ovx 2) ionized Ca, and (ovx 3) nano Ca. (Reproduced from Park, H.S. et al., *Asian-Aust. J. Anim. Sci.*, 20, 1266–1271, 2007.)

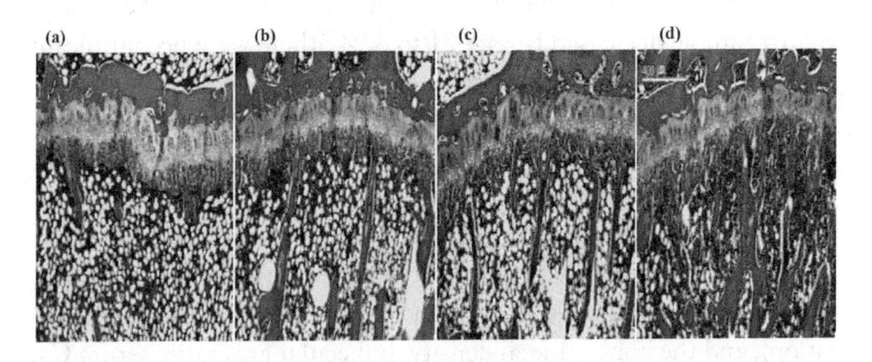

Figure 2.2 **(See color insert.)** Microscopic images of trabecular bone in tibia part of rats fed with milk containing (a) no Ca supplementation (regular milk), (b) carbonated Ca, (c) ionized Ca, and (d) nano Ca. (Reproduced from Park, H.S. et al., *Asian-Aust. J. Anim. Sci.*, 20, 1266–1271, 2007.)

nanosize Ca particles increased the bioavailability of Ca (Ca intake), resulted in elongated trabecular tissues, and increased bone mineral density and trabecular area ratio.

Curcumin is rich with health benefits ranging from anti-inflammatory and antioxidant properties to anticancer properties and is naturally found in turmeric roots. The turmeric is used as a food additive, mainly as a food colorant, and in recent times curcumin is used in food and pharmaceutical products due to its health benefits. The oral bioavailability of the curcumin is very low if it is used in raw format in food and pharmaceutical products, and the nanoemulsion technique has been successfully tested to increase the oral bioavailability of the curcumin (Onoue et al. 2010; Yu and Huang 2012). Organogel-based nanoemulsions were prepared with curcumin (9% curcumin concentration) and tested with the rats for determining oral bio-availability (Figure 2.3). The results showed the nanoemulsification of curcumin with organogel increased the bioavailability of the curcumin

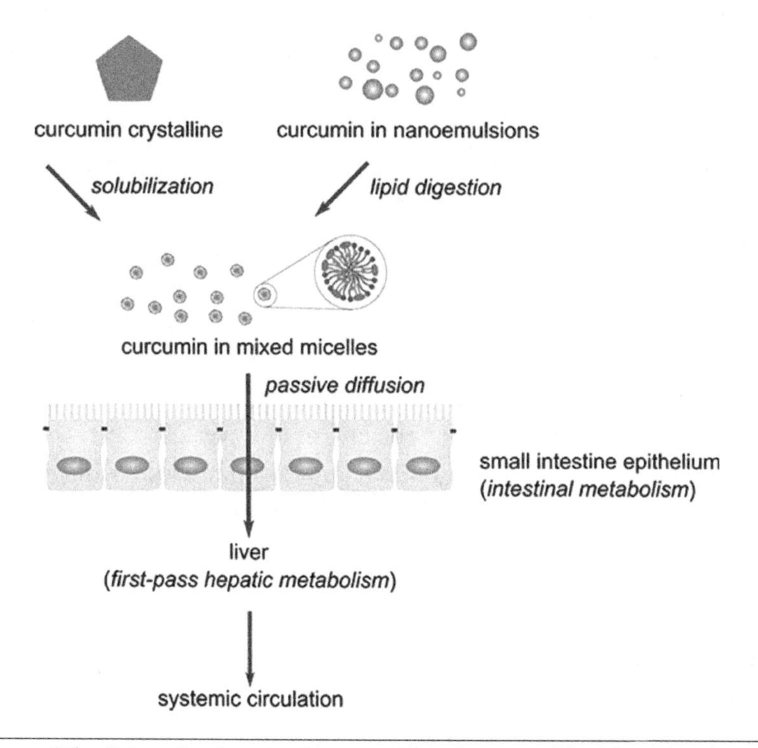

Figure 2.3 Schematic diagram of organogel-based nanoemulsions for enhancing oral bioavailability of curcumin. (Reproduced from Yu, H., and Huang, Q., *J. Agric. Food Chem.*, 60, 5373–5379, 2012.)

up to ninefold and had a higher rate and extent of digestion when compared with unformulated curcumin (Yu and Huang 2012).

Phenolic components are responsible for the antioxidant properties of fruit juices, and these phenolic components are very sensitive to heat. Thermal treatment of fruit juices for purification and sterilization might degrade the phenolic compounds resulting in poor antioxidant properties (Patras et al. 2010; Sui et al. 2014). Filtration of fruit juices using nanofiltration membranes can help to maintain the natural antioxidant properties of the fruit juices. Nanofiltration of strawberry juice helped to recover 97.0% and 97.8% of phenolic and anthocyanins compounds, and the antioxidant activity was also 99% of initial value even after processing (Arend et al. 2017). Membrane filtration using nanotechnology also retained 90% of total solid sugars and 100% of anthocyanins in blackberry juices.

Acetaldehyde (CH_3CHO) is a common compound of wine, and the acceptable acetaldehyde concentrations in red and white wine are 30 and 80 mg/L, respectively. The wine industry wants to control acetaldehyde below these concentration levels, because the elevated acetaldehyde results in nausea and headaches to more severe health issues, like liver fibrosis and pregnancy fetal injury in humans (Salaspuro 2011). The acetaldehyde content of the wine may increase due to the yeast and acetic acid bacteria reaction during fermentation and storage. Gold (Au) nanoparticles along with chitosan have been tested to keep the acetaldehyde level below acceptable level by Yu et al. (2010). The AuNPs of three different diameters (50, 30, and 10 nm) were prepared on the chitosan, mixed with the commercial white wine (initial acetaldehyde content 95 ppm), and kept at room temperature for seven days. At the end of one-week storage, the acetaldehyde content reduced to 68, 39, and 24 ppm while using 50, 30, and 10 nm size AuNPs with chitosan (Yu et al. 2010).

Impact on Physicochemical and Textural Properties

Silicon dioxide (SiO_2) is one of the coloring agents used in food products for a long time, and the development of nanotechnology helped to produce nanosize SiO_2 particles, which can be used as food color with enhanced properties. The nanosize SiO_2 particles have more surface-area-to-mass ratio, which increases the solubility and adsorption into the food matrix, which might result in enhanced food color with the use of less quantity of SiO_2. Titanium dioxide (TiO_2) is also approved as a food additive along with SiO_2, and Al_2O_3, and the European Union and US FDA have allowed food and beverages industry to use nanoscale TiO_2, SiO_2, and Al_2O_3 particles as food coloring agents in their products at a maximum limit of 2% w/w basis (He and Hwang 2016). SiO_2 is also used as a flavoring and anticaking agent in food products.

Marinating chicken breast fillets with nanosize (30 nm) paprika oleoresin improved the surface color of the chicken breast fillets (Yusop et al. 2012). The use of nanoparticle paprika oleoresin also increased the absorption of marinade and reduced the cooking losses while using homogenized milk as an ingredient carrier solution. The chicken fillet marinated with nanosize paprika oleoresin in homogenized milk had higher yellow and red color value while testing with CIELAB color scores, and the sensory evaluation of the cooked chicken fillets resulted in higher scores for nanoparticle marinated fillets than the regular paprika-oleoresin-marinated fillets (Yusop et al. 2012). In the meat industry, ginger is used as a natural tenderizer, and the use of nanosize ginger particles improved the penetration into the meat and resulted in tenderizing meat in shorter time with the lower amount of tenderizing material (nanosize ginger powder) compared to the use of normal particle size ginger (Zhao et al. 2009). Bioactive components like fatty acids and vitamins can be added into meat products (like cured meat and sausages) by using nanosized micelles (average size of 30 nm), and Aquanova (Darmstadt, Germany) is commercially producing this nanosize micelle with the commercial name Novasol. Using the Novasol micelles to encapsulate coloring agent enhanced the stability of the sausage color with a lower cost and processing time (Alfadul and Elneshwy 2010).

Encapsulation of nutrients and bioactive compounds are becoming popular in the food and beverage industry because the encapsulation of nutrients with specific carrier materials could help to reduce the degradation of nutrients such as enzymes due to temperature, pH, or other factors, since most of the bioactive compounds are sensitive to oxidation and temperature and have lower water solubility (Shimoni 2009). Microencapsulation of food ingredients such as flavoring and coloring agents and nutrients are successfully integrated into food and beverage manufacturing, and the nanoencapsulation of ingredients and nutrients is gaining more attention in the present time among researchers and industry due to their improved functionality than the microencapsulation (Cushen et al. 2012). Nanoencapsulation of fish oil, which is a good source of omega-3 fatty acid, can help to mask the odor and flavor of the fish oil (Chaudhry et al. 2008; Cushen et al. 2012). This nanoencapsulated fish oil can be used to produce omega-3 enriched bread, in which the consumers do not smell the unpleasant odor of the fish oil, and the omega-3 is released directly into the gastrointestinal system (Cushen et al. 2012).

A white bread was baked with omega-3 fatty acids derived from cold-pressed flax oil nanoencapsulated using high-amylose corn starch (HACS). The bread characteristics (bread density, color, and volume), oxidation, and thermal degradation of omega-3 fatty acids were measured (Mogol et al. 2013). The results showed nanoencapsulation significantly reduced the thermal degradation and oxidation of omega-3 fatty acids, and the sensory test proved the unpleasant flavor of omega-3 fatty acids was also masked by nanoencapsulation (Mogol et al. 2013). The textural properties (hardness and chewiness) of the bread increased with a higher nanoencapsulated HACS-omega-3-fatty-acid concentration, and nanoencapsulation also reduced the oxidation and thermal degradation of omega-3 fatty acid lipids (Gökmen et al. 2011). The SiO_2-galic acid nanoparticles developed by incorporating SiO_2 nanoparticles in galic acid showed excellent antioxidant properties, which can be used as a stable antioxidant in food and beverage production and storage at a low cost (Deligiannakis et al. 2012). The browning of fresh-cut apples and other fruits was controlled by using nano-ZnO-based coatings. The nano-ZnO acted as an antioxidant to control the reaction of phenolic components with the oxygen (Ghidelli and Pérez-Gago 2016; Rojas-Graü et al. 2009). Edible coating using this kind of antioxidants developed using nanotechnology could increase the shelf life of oxidation-sensitive food products like fresh-cut fruits and vegetables. The silver-montmorillonite (Ag-MMT) nanoparticles incorporated in calcium-alginate coating of fresh-cut carrots inhibited the development of Enterobacteriaceae, mesophilic, psychrotrophic bacteria as well as yeast (Costa et al. 2011). The ZnO nanoparticle incorporated in chitosan-edible coating for fresh-cut kiwifruit reduced the loss of water and the softening of fruit as well as controlled the production of carbon dioxide and ethylene (Meng et al. 2014). The nanoencapsulated rosemary oil using liposome was used in the edible coating of fresh-cut banana, which maintained the quality of banana for nine days without any reduction in weight of the product and sensory parameters (Alikhani-Koupaei 2014).

Wheat bran dietary fiber was ultra-finely ground to the particle size of 10–620 nm using a nano-ball-mill with multidimensional swinging ability and high-energy (Model: TJH-2-4L, Qinghuangdao Taiji Ring Nano-Products Co., Ltd., Hebei, China), it significantly decreased the swelling, water retention, and water-holding capacities of the dietary fiber (Zhu et al. 2010). The size reduction to nanoscale level ground wheat significantly increased the total phenolic content, reducing power, and ferrous chelating activity, but nanoscale size reduction reduced the radical scavenging activity of 1,1-diphenyl-2-picrylhydrazyl (DPPH) (Figure 2.4) (Zhu et al. 2010).

Impact on Antimicrobial Properties

The development of antimicrobial agents using nanotechnology showed promising results on inhibiting foodborne pathogen growth in food and beverages by which the quality of the food product can be kept at a high level and shelf life of the product can be improved (He and Hwang 2016; Hwang et al. 2012). The antimicrobial properties of silver and gold are well known, and the application of nanoparticles of silver and gold can improve the antimicrobial properties of food and beverages. Other metal oxides like zinc oxide (ZnO) and titanium dioxide (TiO_2) can also be used as antimicrobial agents in the food and beverage manufacturing industry. The metal oxide nanoparticles have been mostly used as a food-packaging material but sometimes are used as ingredient- or food-coating material (He and Hwang 2016).

Chitosan (chemical name: poly-β-[1-4]-D-glucosamine) is a polysaccharide made mainly from the crustacean shells and used in food and pharmaceutical sector due to its antioxidant and antimicrobial activity. Chitosan has been successfully tested against Gram-negative and Gram-positive bacteria, and this antimicrobial property of chitosan made this as one of the main potential natural food preservation agents (Prashanth and Tharanathan 2007). The nanoparticles of chitosan exhibit superior physicochemical properties than the regular chitosan particles.

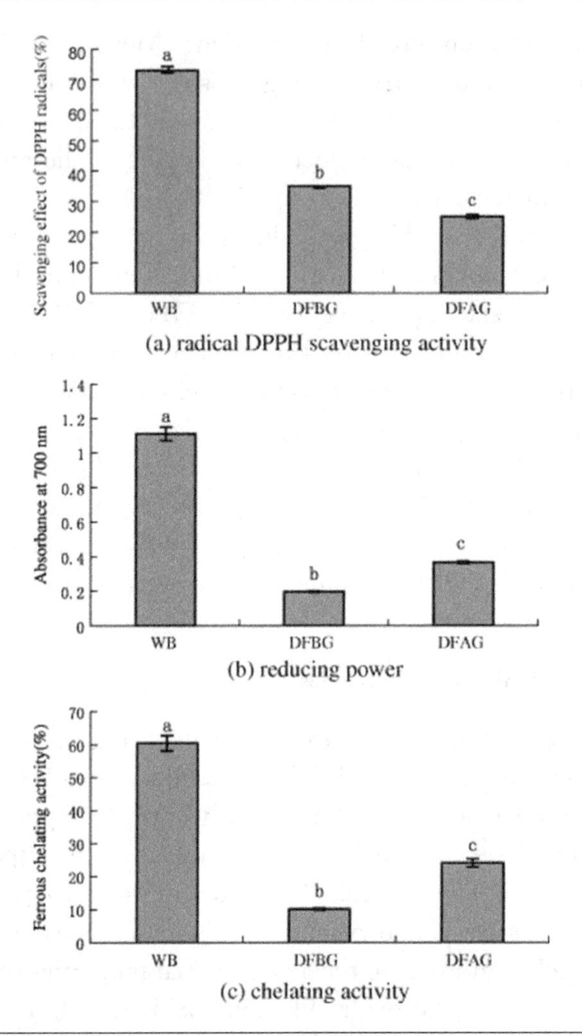

Figure 2.4 The DPPH radical scavenging activity (a), reducing power (b), and ferrous chelating activity (c) of wheat bran (WB), dietary fiber before size reduction to nanoscale (DFBG), and dietary fiber after size reduction to nanoscale (DFAG). (Reproduced from Zhu, K. et al., *Food Res. Int.*, 43, 943–948, 2010.)

Chitosan nanoparticles with sodium tripolyphosphate form chitosan tripolyphosphate nanoparticles through ionotropic gelation, which are used as a carrier material for targeted drug delivery in the pharmaceutical sector. Chitosan nanoparticles also have been tested in the meat industry to increase the shelf life of fish products by controlling bacteria and mold growth.

The coating of carp fish fingers with chitosan nanoparticles could improve the storability of the fish fingers by controlling the growth of bacterium, mold, and yeast (Abdou et al. 2012). The fish fingers coated with chitosan nanoparticles and regular edible coatings were stored for 6 months at -18°C and the total microbial count was determined. The results showed the total microbial count in fish fingers with regular edible coating was 5.27 log CFU/g, and chitosan nanoparticles-coated fish fingers had a total microbial count of 2.87 log CFU/g. The solution of lysozyme with chitosan nanoparticles also showed higher antimicrobial effects against major foodborne pathogens (*E. coli* O157: H7 and *Bacillus subtilis*), so it can be used in meat and other sectors of food industry as an active food coating agent or can be incorporated into food packaging materials (Wu et al. 2017). The development in the food packaging sector with the nanotechnology resulted in various nanocomposite materials (like polymers and plastics incorporated with nanoclay or nanoparticles of metal oxides) with increased physical, thermal, mechanical, gas and moisture transmission, and antimicrobial properties. These helped to maintain the quality of the meat and other products for a long time. The scientific studies on the application of the edible coating and packaging materials developed using the nanotechnology in food products showed enhanced shelf life and quality when compared with the products packed with regular packing materials. The meat and fresh-cut fruits packed in nanotechnology-based packaging systems remained fresh because the nanotechnology-based packaging materials reduced the gas and moisture permeation, and also increased antimicrobial activity. Detailed discussion on the impact of using nanotechnology-based materials in packaging process is available in Chapter 3, and the impacts on various food products like beverages, bakery, meat, and dairy are available in Chapters 5 through 9.

Development of oil-in-water nanoemulsions using nanotechnology is also getting increased attention in the food and beverage industry due to their improved functionality. A sunflower-oil-based nanoemulsion was tested to increase the shelf life of fish steaks, and processing of Indo-Pacific king mackerel fish steaks with this nanoemulsion inhibited the microbial growth and increased the shelf life of fish steaks upto 48 h when compared with control (Joe et al. 2012). The green tea particles were reduced to a nanosize level, and this nanotea showed enhanced antioxidant activity than the regular green tea (Ramachandraiah et al. 2015). The β-carotene, an oil-soluble pigment, is used as a coloring agent in food and beverages, and the nanoβ-carotene particles produced using calcium ions and alginic acid enhanced the color values of the food and beverages due to the higher penetration and absorbance of nanoparticles into the food matrix. Another advantage of this nanoβ-carotene-based coloring agent was the color of the food product can be easily modified from bright yellow to dark orange by just changing the concentration of β-carotene in the nanostructures (Astete et al. 2009).

Conclusion

The use of nanotechnology in food and beverage manufacturing can potentially lead to the development of new and novel food products with enhanced taste, flavor, color, texture, and nutrient content, as well as antimicrobial properties. Nanotechnology-based materials and processing operations can also increase the storability and bioavailability of the nutrients in the food product. Recent developments in nanoscience and nanotechnology can help the food and beverage industry to produce "healthier" products with "cheaper" production cost. The enhancement of bioavailability through various nanotechnology-based products like nanosized ingredients, nanoencapsulation, and nanoemulification can help the consumer to get more benefits from the product at the same time the food manufacturing industry can reduce the amount of ingredients used, since the nutritional benefits can be reached with the lower amount of ingredients due to increased absorption. It is evident that use of nanotechnology-based products as ingredient or additive help to improve the physicochemical, textural, and sensory properties, and also to enhance the bioavailability and antimicrobial properties of a food product; however, the impact of increased absorption of bioactive compounds in the human digestion system, higher penetration, ability of nanoscale bioactive, and other materials used in food production is not completely studied at this time. The toxicity of the nanoscale food additives and antimicrobial agents used in food and beverage manufacturing and packaging on human health, as well as on the environment, has also not been explored. The food industry and scientific community should focus more on studying these issues to help clarify the consumers' questions about the impact of modified functional characteristics of the food products due to the application of nanotechnology-based products and processes in food and beverage manufacturing. This will help the industry to improve their socio-economic ratings.

References

Abdou E.S., Osheba A., Sorour M. (2012) Effect of chitosan and chitosan-nanoparticles as active coating on microbiological characteristics of fish fingers. *International Journal of Applied Science and Technology* 2:158–169.

Alfadul S., Elneshwy A. (2010) Use of nanotechnology in food processing, packaging and safety–review. *African Journal of Food, Agriculture, Nutrition and Development* 10:2719–2739.

Alikhani-Koupaei M. (2014) Liposomal and edible coating as control release delivery systems for essential oils: Comparison of application on storage life of fresh-cut banana. *Quality Assurance and Safety of Crops and Foods* 7:175–185.

Arend G.D., Adorno W.T., Rezzadori K., Di Luccio M., Chaves V.C., Reginatto F.H., Petrus J.C.C. (2017) Concentration of phenolic compounds from strawberry (*Fragaria X ananassa* Duch) juice by nanofiltration membrane. *Journal of Food Engineering* 201:36–41.

Astete C.E., Sabliov C.M., Watanabe F., Biris A. (2009) Ca^{2+} cross-linked alginic acid nanoparticles for solubilization of lipophilic natural colorants. *Journal of Agricultural and Food Chemistry* 57:7505–7512.

Astray G., Gonzalez-Barreiro C., Mejuto J., Rial-Otero R., Simal-Gándara J. (2009) A review on the use of cyclodextrins in foods. *Food Hydrocolloids* 23:1631–1640.

Chaudhry Q., Scotter M., Blackburn J., Ross B., Boxall A., Castle L., Aitken R., Watkins R. (2008) Applications and implications of nanotechnologies for the food sector. *Food Additives and Contaminants: Part A* 25:241–258. doi:10.1080/02652030701744538.

Chen H., Chang J., Wu J. (2008) Calcium bioavailability of nanonized pearl powder for adults. *Journal of Food Science* 73(9):246–251.

Costa C., Conte A., Buonocore G., Del Nobile M. (2011) Antimicrobial silver-montmorillonite nanoparticles to prolong the shelf life of fresh fruit salad. *International Journal of Food Microbiology* 148:164–167.

Cushen M., Kerry J., Morris M., Cruz-Romero M., Cummins E. (2012) Nanotechnologies in the food industry—recent developments, risks and regulation. *Trends in Food Science and Technology* 24:30–46.

Deligiannakis Y., Sotiriou G.A., Pratsinis S.E. (2012) Antioxidant and antiradical SiO nanoparticles covalently functionalized with Gallic acid. *ACS Applied Materials and Interfaces* 4:6609–6617. doi:10.1021/am301751s.

Dudhani A.R., Kosaraju S.L. (2010) Bioadhesive chitosan nanoparticles: Preparation and characterization. *Carbohydrate Polymers* 81:243–251.

Gao H., Chen H., Chen W., Tao F., Zheng Y., Jiang Y., Ruan H. (2008) Effect of nanometer pearl powder on calcium absorption and utilization in rats. *Food Chemistry* 109:493–498. doi:10.1016/j.foodchem.2007.12.052.

Ghidelli C., Pérez-Gago M.B. (2016) Recent advances in modified atmosphere packaging and edible coatings to maintain quality of fresh-cut fruits and vegetables. *Critical Reviews in Food Science and Nutrition* 1–18. doi:10.1080/10408398.2016.1211087.

Gökmen V., Mogol B.A., Lumaga R.B., Fogliano V., Kaplun Z., Shimoni E. (2011) Development of functional bread containing nanoencapsulated omega-3 fatty acids. *Journal of Food Engineering* 105:585–591. doi:10.1016/j.jfoodeng.2011.03.021.

Graveland-Bikker J., De Kruif C. (2006) Unique milk protein-based nanotubes: Food and nanotechnology meet. *Trends in Food Science and Technology* 17:196–203.

He X., Hwang H.M. (2016) Nanotechnology in food science: Functionality, applicability, and safety assessment. *Journal of Food and Drug Analysis* 24:671–681.

Huang S., Chen J.C., Hsu C.W., Chang W.H. (2009) Effects of nano calcium carbonate and nano calcium citrate on toxicity in ICR mice and on bone mineral density in an ovariectomized mice model. *Nanotechnology* 20:375102.

Hwang H.M., Ray P., Yu H., He X. (2012) Toxicology of designer/engineered metallic nanoparticles. In: *Sustainable Preparation of Metal Nanoparticles.* Royal Society of Chemistry (RSC) Publishing, Cambridge, UK. pp. 190–212.

Joe M.M., Chauhan P.S., Bradeeba K., Shagol C., Sivakumaar P.K., Sa T. (2012) Influence of sunflower oil based nanoemulsion (AUSN-4) on the shelf life and quality of Indo-Pacific king mackerel (*Scomberomorus guttatus*) steaks stored at 20°C. *Food Control* 23:564–570.

Kakkar V., Singh S., Singla D., Kaur I.P. (2011) Exploring solid lipid nanoparticles to enhance the oral bioavailability of curcumin. *Molecular Nutrition and Food Research* 55:495–503.

Lesmes U., Cohen S.H., Shener Y., Shimoni E. (2009) Effects of long-chain fatty acid unsaturation on the structure and controlled release properties of amylose complexes. *Food Hydrocolloids* 23:667–675.

Li H., Zhao X., Ma Y., Zhai G., Li L., Lou H. (2009) Enhancement of gastrointestinal absorption of quercetin by solid lipid nanoparticles. *Journal of Controlled Release* 133:238–244.

Liao C.D., Hung W.L., Jan K.C., Yeh A.I., Ho C.T., Hwang L.S. (2010) Nano/sub–microsized lignan glycosides from sesame meal exhibit higher transport and absorption efficiency in Caco-2 cell monolayer. *Food Chemistry* 119:896–902.

Lien T.F., Yeh H.S., Lu F.Y., Fu C.M. (2009) Nanoparticles of chromium picolinate enhance chromium digestibility and absorption. *Journal of the Science of Food and Agriculture* 89:1164–1167.

Liu C.H., Wu C.T. (2010) Optimization of nanostructured lipid carriers for lutein delivery. *Colloids and Surfaces A: Physicochemical and Engineering Aspects* 353:149–156.

Livney Y.D. (2010) Milk proteins as vehicles for bioactives. *Current Opinion in Colloid and Interface Science* 15:73–83.

McClements D.J., Decker E.A., Park Y., Weiss J. (2009) Structural design principles for delivery of bioactive components in nutraceuticals and functional foods. *Critical Reviews in Food Science and Nutrition* 49:577–606.

McClements D.J., Rao J. (2011) Food-grade nanoemulsions: Formulation, fabrication, properties, performance, biological fate, and potential toxicity. *Critical Reviews in Food Science and Nutrition* 51:285–330.

Meng X., Zhang M., Adhikari B. (2014) The effects of ultrasound treatment and nano-zinc oxide coating on the physiological activities of fresh-cut kiwifruit. *Food and Bioprocess Technology* 7:126–132.

Mogol B.A., Gokmen V., Shimoni E. (2013) Nano-encapsulation improves thermal stability of bioactive compounds Omega fatty acids and silymarin in bread. *Agro Food Industry Hi-Tech* 24:62–65.

Morris G.A., Kök S.M., Harding S.E., Adams G.G. (2010) Polysaccharide drug delivery systems based on pectin and chitosan. *Biotechnology and Genetic Engineering Reviews* 27:257–284.

Mozafari M.R., Johnson C., Hatziantoniou S., Demetzos C. (2008) Nanoliposomes and their applications in food nanotechnology. *Journal of Liposome Research* 18:309–327.

Müller R.H., Gohla S., Keck C.M. (2011) State of the art of nanocrystals—Special features, production, nanotoxicology aspects and intracellular delivery. *European Journal of Pharmaceutics and Biopharmaceutics* 78:1–9.

Oehlke K., Adamiuk M., Behsnilian D., Gräf V., Mayer-Miebach E., Walz E., Greiner R. (2014) Potential bioavailability enhancement of bioactive compounds using food-grade engineered nanomaterials: A review of the existing evidence. *Food and Function* 5:1341–1359.

Onoue S., Takahashi H., Kawabata Y., Seto Y., Hatanaka J., Timmermann B., Yamada S. (2010) Formulation design and photochemical studies on nanocrystal solid dispersion of curcumin with improved oral bioavailability. *Journal of Pharmaceutical Sciences* 99:1871–1881.

Park H.S., Jeon B.J., Ahn J., Kwak H.S. (2007) Effects of nanocalcium-supplemented milk on bone calcium metabolism in ovariectomized rats. *Asian-Australasian Journal of Animal Sciences* 20:1266–1271. doi:10.5713/ajas.2007.1266.

Patras A., Brunton N.P., O'Donnell C., Tiwari B. (2010) Effect of thermal processing on anthocyanin stability in foods: Mechanisms and kinetics of degradation. *Trends in Food Science and Technology* 21:3–11.

Prashanth K.H., Tharanathan R. (2007) Chitin/chitosan: Modifications and their unlimited application potential—An overview. *Trends in Food Science and Technology* 18:117–131.

Ramachandraiah K., Han S.G., Chin K.B. (2015) Nanotechnology in meat processing and packaging: Potential applications—A review. *Asian-Australasian Journal of Animal Sciences* 28:290–302. doi:10.5713/ajas.14.0607.

Rohner F., Ernst F.O., Arnold M., Hilbe M., Biebinger R., Ehrensperger F., Pratsinis S.E., Langhans W., Hurrell R.F., Zimmermann M.B. (2007) Synthesis, characterization, and bioavailability in rats of ferric phosphate nanoparticles. *The Journal of Nutrition* 137:614–619.

Rojas-Graü M.A., Soliva-Fortuny R., Martín-Belloso O. (2009) Edible coatings to incorporate active ingredients to fresh-cut fruits: A review. *Trends in Food Science and Technology* 20:438–447.

Salaspuro M. (2011) Acetaldehyde and gastric cancer. *Journal of Digestive Diseases* 12:51–59.

Salvia-Trujillo L., Martín-Belloso O., McClements D. (2016) Excipient nanoemulsions for improving oral bioavailability of bioactives. *Nanomaterials* 6(17): 1–16.

Santillán-Urquiza E., Méndez-Rojas M.Á., Vélez-Ruiz J.F. (2017) Fortification of yogurt with nano and micro sized calcium, iron and zinc, effect on the physicochemical and rheological properties. *LWT-Food Science and Technology* 80:462–469. doi:10.1016/j.lwt.2017.03.025.

Seo M.H., Lee S.Y., Chang Y.H., Kwak H.S. (2009) Physicochemical, microbial, and sensory properties of yogurt supplemented with nanopowdered chitosan during storage. *Journal of Dairy Science* 92:5907–5916. doi:10.3168/jds.2009-2520.

Shimoni E. (2009) Nanotechnology for foods: delivery systems. In: B.C. Gustavo, M. Alan, L. David, S. Walter, B. Ken, & C. Paul (Eds.) *Global Issues in Food Science and Technology*. Academic Press, SanDiego, CA. pp. 411–424.

Sui X., Dong X., Zhou W. (2014) Combined effect of pH and high temperature on the stability and antioxidant capacity of two anthocyanins in aqueous solution. *Food Chemistry* 163:163–170.

Takahashi M., Uechi S., Takara K., Asikin Y., Wada K. (2009) Evaluation of an oral carrier system in rats: Bioavailability and antioxidant properties of liposome-encapsulated curcumin. *Journal of Agricultural and Food Chemistry* 57:9141–9146. doi:10.1021/jf9013923.

van Vlerken L.E., Vyas T.K., Amiji M.M. (2007) Poly (ethylene glycol)-modified nanocarriers for tumor-targeted and intracellular delivery. *Pharmaceutical Research* 24:1405–1414.

Wajda R., Zirkel J., Schaffer T. (2007) Increase of bioavailability of coenzyme Q10 and vitamin E. *Journal of Medicinal Food* 10:731–734.

Weiss J., Decker E.A., McClements D.J., Kristbergsson K., Helgason T., Awad T. (2008) Solid lipid nanoparticles as delivery systems for bioactive food components. *Food Biophysics* 3:146–154.

Wesley S.J., Raja P., Sunder-Raj A.A., Tiroutchelvamae D. (2014) Review on-Nanotechnology applications in food packaging and safety. *International Journal of Engineering Research* 3:645–651.

Wu T., Wu C., Fu S., Wang L., Yuan C., Chen S., Hu Y. (2017) Integration of lysozyme into chitosan nanoparticles for improving antibacterial activity. *Carbohydrate Polymers* 155:192–200.

Yu C.C., Yang K.H., Liu Y.C., Chen B.C. (2010) Photochemical fabrication of size-controllable gold nanoparticles on chitosan and their application on catalytic decomposition of acetaldehyde. *Materials Research Bulletin* 45:838–843.

Yu H., Huang Q. (2010) Enhanced in vitro anti-cancer activity of curcumin encapsulated in hydrophobically modified starch. *Food Chemistry* 119:669–674.

Yu H., Huang Q. (2012) Improving the oral bioavailability of curcumin using novel organogel-based nanoemulsions. *Journal of Agricultural and Food Chemistry* 60:5373–5379. doi:10.1021/jf300609p.

Yusop S.M., O'Sullivan M.G., Preuß M., Weber H., Kerry J.F., Kerry J.P. (2012) Assessment of nanoparticle paprika oleoresin on marinating performance and sensory acceptance of poultry meat. *LWT-Food Science and Technology* 46:349–355.

Zhao X., Yang Z., Gai G.,Yang Y. (2009). Effect of superfine grinding on properties of ginger powder. *Journal of Food Engineering* 91:217–222.

Zhou H., Liu G., Zhang J., Sun N., Duan M., Yan Z., Xia Q. (2014) Novel lipid-free nanoformulation for improving oral bioavailability of coenzyme Q10. *BioMed Research International* 2014.

Zhu K., Huang S., Peng W., Qian H., Zhou H. (2010) Effect of ultrafine grinding on hydration and antioxidant properties of wheat bran dietary fiber. *Food Research International* 43:943–948.

Zimet P., Rosenberg D., Livney Y.D. (2011) Re-assembled casein micelles and casein nanoparticles as nano-vehicles for ω-3 polyunsaturated fatty acids. *Food Hydrocolloids* 25:1270–1276.

Zorilla R., Liang L., Remondetto G., Subirade M. (2011) Interaction of epigallocatechin-3-gallate with β-lactoglobulin: Molecular characterization and biological implication. *Dairy Science and Technology* 91:629.

3

Impact of Incorporating Nanomaterials in Food Processing

Introduction

The food and beverages industry applies nanotehnology and nanoscience-based materials and methods in production, storage, and logistics sectors as an ingredient, packaging materials as well as in modified processes and tracking techniques. Nanotechnology-based materials and methods have been applied from raw material receipt to packed material delivery. The particle size of raw materials for food and beverages processing has been reduced to the nanoscale level to enhance the flavor, taste, bioavailability, texture, and sensory properties of food products and beverages. Application of nanotechnology-based products as an ingredient or food additive can impact the food processing technique by improving the intake by the food material and reducing the use of additives, which can result in a reduction in processing cost and energy requirement for processing. For example, food coloring agents like silicon dioxide (SiO_2) can be used as nanosize SiO_2 particles, which can improve the color of food with less quantity of food coloring agent by improving solubility and adsorption of nanosize food color into the food matrix. Similarly, the use of nanosize (30 nm) paprika oleoresin as a marinating agent for chicken breast fillets enhanced the surface color of the fillets during cooking (Yusop et al. 2012). Nanoencapsulation of active ingredients and nutrient compounds with the food products during processing of food and beverages helped to improve the nutritional quality and bioavailability of the nutrients (Shimoni 2009; Zhu et al. 2010). The impact of application of nanotechnology on food products' physical, chemical, microbial, nutritional, and sensory properties has been discussed in detail in Chapter 2.

Advancements in the nanoscience and nanotechnology have particularly impacted the packaging of food and beverages, food analysis and safety, as well as wastewater treatments. Nanotechnology-based packaging films and sensors make the simple, nonresponsive packaging system into an intelligent, smart, and active packaging. Incorporation of nanoparticles of clay, metal oxides, with the organic and inorganic polymeric materials improves the mechanical, tensile properties of the films, and also makes the film more resistant to heat. Nanotechnology-based packing films also have improved moisture and gas transmission properties due to the alternation in the path of moisture and gas migration through the packaging films. Nanotechnology-based sensors attached to the food packaging materials help to track the quality and condition of food products and their surrounding from the moment it is packed to the moment it reaches consumer's dining table. Application of nanotechnology-based materials for food packaging and its impact on packaging technology is discussed briefly in this chapter, and more detailed discussion on the impact on packaging as well as product quality is discussed in Chapters 4 through 9.

Nanoparticles of metal oxide have been used for microbial inactivation, and the nanosensors have been successfully tested for detection of major foodborne pathogens. Most of the nanotechnology-based sensors have the ability to detect pathogenic microorganisms in lesser number and also in a much shorter time when compared with the regular analytical methods. Nanosensors also have the ability to detect the food adulterants, chemical and antibiotic residuals in the food product. Summary of the impact of nanotechnology on food quality determination, detection of microbes, and adulterants has been discussed briefly in this chapter, and a complete discussion based on the type of food product is given in Chapters 4 through 9 of this book.

The development of nanofiltering membranes, nanotechnology-based nanosorbents, and photocatalysts have the ability to reduce the microbial load as well as hazardous chemical load in industrial wastewater from the food and beverages manufacturing units. Applications of nanotechnology to the water purification and wastewater treatment are discussed briefly here, and more detailed description is given in Chapter 9.

Impact on Packaging of Food and Beverages

Packaging technologies for food and beverages are improving day by day due to the development of new packaging materials or composites through the advancements in the fields of nanoscience and nanotechnology. The nanotechnology has helped transform the food packaging sector from "passive" to "active." In conventional packaging, the packaging technique or the material only offers the protection of the product from the biotic and abiotic factors, which cause the qualitative and quantitative damage to the food. But with the growth of nanotechnology-based packaging materials and sensors, the packing material has an active contact with the food material, and any changes (either qualitative or quantitative) can be monitored by the consumer or the processor through visible color or other changes. Nanomaterial-based packaging films have become a reality now, and some of the food manufacturers are even using nanotechnology-based sensors and radio-frequency identification (RFID) chips for real-time monitoring of internal and external characteristics (temperature, humidity, gas composition, etc.) of the packed food products. Incorporation of nanoparticles or nanotechnology-based materials with regular food packaging films, like plastic films, made the packaging films stronger, and at the same time made the materials lighter, which improved the performance of the packaging film. The addition of nanoparticles, which have excellent antimicrobial properties, like silver dioxide or titanium dioxide, helps to prevent or reduce the total microbial load on the food products, resulting in reduction of food spoilage and improved shelf life of the product. The packaging polymer with nanoclay improve the moisture and gas transmission properties of the packaging films. The addition of nanoclay with regular packaging polymer materials help to restrict the movement of oxygen and carbon dioxide in and out of the package, which helps to restrict and control the unwanted microbial growth in the food product. Incorporation of nanoclay particles also restricts the movement of moisture from product to surrounding or surrounding to the product, which keeps the product at optimal moisture content and also reduces spoilage. With the regular packaging plastic films, nanosize (<100 nm size) particles like the nanoclay, nanoparticles of metal and metal oxides, and nanocarbon tubes can be incorporated at a rate of 1%–2% by weight in order to produce polymeric nanocomposites, which have the improved mechanical strength and heat resistance, and also improved gas and moisture transmission properties (Sekhon 2010). The presence or incorporation of nanosize particles improves the antimicrobial capability of the packaging film and makes the packaging become a "smart package," in which the packaging material indicates to the consumer about possible contamination of pathogenic microorganisms inside the package. Some of the smart packaging materials alter or repair themselves based on the temperature, humidity, and the gas composition inside the packaging with the presence of nanoparticles (Ravichandran 2010). Not only the research institutions and government agencies but also the big global food and beverages manufacturing and packaging material manufacturing giants (e.g., Bayer, Kodak, Kraft) are investing millions of dollars in the development of nanotechnology-based food packaging materials. These materials have the ability to increase the shelf life of food product, alert the consumers if the food is spoiled or pathogenic microorganisms are present in the food, as well as absorb or restrict the movement of moisture and gas (especially oxygen) from product to ambient or ambient to product (Brody et al. 2008).

The nanocomposite polymer packaging film with the inclusion of nanoclay (or nanosize silicate) particles reduces or restricts the permeation of oxygen through the packing film, and also prevents the moisture migration from the product. The Bayer polymers (Pittsburg, PA, USA), one of the leading chemical manufacturers in the world, developed and is marketing the food packing film with the brand name "Durethan," which contains nanosize silicate particles with enhanced gas and moisture transmission properties in order to restrict spoilage of packed food products. The plastic bottles manufactured with nanocrystals by Nanocor (Arlington Heights, IL, USA) for packing beer keep the product safe and fresh by restricting the oxygen migration into the product and leak of the carbon dioxide from the product. Nanocor reported this new nanocomposite-based plastic bottle provided a shelf life of six months for the brewed beer (Ravichandran 2010).

Incorporation of the carbon nanotubes with the polymeric packaging films not only improved the mechanical strength and heat resistance, but it also improved the antimicrobial properties. Lanzon et al. (2009) reported that the pathogenic bacteria were eradicated in no time when it got direct contact with the carbon nanotubes. The nanotechnology-based nanofibers, nanotubes, and nanowheels have been tested widely throughout the world by scientists for enhanced thermal and barrier properties. Brody et al. (2008) projected the nanotechnology-based packing polymers and packing techniques have the enormous potential in the food and beverage packaging sector since they have the ability to restrict and control the growth of microbes, development of unfavorable volatile compounds, migration of moisture, and postponing oxidation process. The packaging polymers with silver and chitosan-based films exhibit excellent antimicrobial properties against all major foodborne pathogenic microorganisms (Rhim et al. 2006). Coating of packaging films with nanoparticles of silver and other metal oxides, or aqueous-based nanoparticles, acted as an oxygen scavenger in the food packaging, and the internal coating of nanoscale amorphous carbon inside the polyethylene terephthalate (PET) plastic bottles improved the gas transmission properties of the packaging system (McHugh 2008). An intelligent packaging technique has been developed by the scientists from the Netherlands that has the ability to release a food preservative directly on to the food products whenever it detects any spoilage or onset of spoilage inside the package (Sekhon 2010). The nanotechnology-based bioswitch has been used in these packaging systems as a sensing tool as well as the commanding tool for the release of preservatives. The smart packaging films and bottles, which have the ability to restrict oxygen movement, were used in packaging of dairy and the brewery products. Packing films manufactured or coated with metal oxide nanoparticles helped to inhibit the pathogenic microorganisms from the water, juices, beverages, and milk without regular thermal treatments (Maftoum 2017). Intelligent packaging with nanotechnology-based polymer with light-emitting diodes have the ability to store and display the product information as well as retrieve the information for food safety inspections.

Nanotechnology-based smart packaging films and containers have the ability to monitor the condition of the food product or the surrounding environment continuously using the nanosensors affixed on or in the package (Yam et al. 2005). These nanosensors measure or monitor the temperature, moisture, CO_2, and O_2 composition inside the package or the storage environment.

Researchers also developed nanosensors, which can detect the presence of adulterants, foodborne pathogens, and toxic substances in the food product (Silvestre et al. 2011). The product's quality and the condition of the food product can be monitored from the moment it is packed inside these smart packages to the time it reaches the consumer's hand without any damage. The package integrity is the main advantage of smart packaging using nanosensors (Ramachandraiah et al. 2015). As discussed earlier, incorporation of nanoclay with the regular packaging polymers enhance barrier properties for gases (O_2, CO_2) and controls the permeation of O_2 and CO_2 through the packaging films (Akbari et al. 2006). Incorporation of nanoclay with most of the regular polymeric materials (polyimides, PET, nylons, ethylene-vinylacetate [EVA] copolymer, polyamides [PA], polyurethane, and polystyrene [PS]) used for manufacturing food packaging films have been tested for production of nanoclay-polymer nanocomposites for food packaging. This hybrid nanocomposite packaging film had better thermal, mechanical, and gas transmission properties, and also these nanopolymers increased the shelf life of the food product and kept the product's quality throughout the storage period (Akbari et al. 2006). The oxygen concentration inside and outside the food package was monitored with the nanosensors with titanium oxide and methyl bromide ink. These sensors were placed in and out of the packaging film, irradiated using ultraviolet (UV) light to make them colorless. These sensors were colorless if the modified packaging contains no oxygen but turned into a blue color whenever the oxygen entered into the package (Mills 2005).

Specific applications of nanotechnology-based food and beverages packaging materials have been discussed in a more detail for different type of food products in Chapters 4 through 9. Summary of the manufacturing process of nanomaterials incorporated food packaging films, the effect of incorporation in packaging films' mechanical, thermal, and gas transmission properties are given in Table 3.1, and explained graphically in Figure 3.1.

Table 3.1 Effect of Nanocomposite Incorporation with Packaging Materials on Mechanical, Thermal, and Gas Transmission Properties

POLYMER	NANOMATERIAL	LEVEL OF INCORPORATION	METHOD OF PROCESSING	EFFECT ON MATERIAL PROPERTIES	REFERENCES
Poly (vinyl alcohol) (PVA)	Cellulose nanocrystals (CNC)	1%–5% (wt) dry basis	Solvent casting	Tensile strength (TS) increased by up to 45%	Fortunati et al. (2013)
Poly (e-caprolactone) (PCL)	CNC	0%–12% (wt)	Film casting/ evaporation technique	Water vapor permeability (WVP) decreased from 1,09,513 × 10^{-10} to 13,211 × 10^{-10} cm^3 (STP). cm.cm^{-2}.s^{-1}.$cmHg^{-1}$ when filler content increased from 3% to 12%	Follain et al. (2013)
Low-density polyethylene/ Linear Low-density polyethylene LDPE/LLDPE	Nanoclay	0%–50% (wt)	Co-rotating twin-screw extruding	Elastic modulus (EM) of LDPE/LLDPE blend enhanced by 58% after adding nanoclay	Hemati and Garmabi (2011)
Polyethylene (PE)	Layered silicate	5%–15% (wt)	Micro-extruding	Crystallization temperature increased from 102°C to 111°C with the addition of silicate	Chrissopoulou et al. (2005)
Polypropylene (PP)/ ethylene-propylene-diene rubber (EPDM) blend	Montmorillonite (MMT)-based organoclay	3%–7% (wt)	Melt extrusion	Adding 1.5 vol% organoclay increased both O_2 and CO_2 barrier properties by about twofold	Frounchi et al. (2006)
Maleated PE	Silicate	15% (wt)	Melt extrusion	The stiffness of the film increased with the addition of silicate filler	Hyun et al. (2003)
Ethylene-vinyl alcohol copolymer (EVOH)/ Poly(lactic acid) (PLA)	Clay	4%–5% (wt)	Melt blending	Barrier to oxygen enhanced with the addition of clay	Lagaron et al. (2005)
PP	Silica	0%–2.2% (vol)	Extruding	TS and toughness increased with the addition of nanosilica	Wu et al. (2005)

(*Continued*)

Table 3.1 (*Continued*) Effect of Nanocomposite Incorporation with Packaging Materials on Mechanical, Thermal, and Gas Transmission Properties

POLYMER	NANOMATERIAL	LEVEL OF INCORPORATION	METHOD OF PROCESSING	EFFECT ON MATERIAL PROPERTIES	REFERENCES
LDPE/LLDPE	MMT organoclay	3 to 7 parts per hundred polymer (phr).	Twin extruder and then film blown	Oxygen permeability (OP) decreased by 50% with the addition of organoclay	Dadbin et al. (2008)
LDPE	Clay	1%–7% (wt)	Melt-mixing and extruding	OP decreased by 24% with addition of 7% clay. Tensile modulus (TM) at 7 wt % of clay increased by 100% in transverse direction, and 17% in machine direction	Arunvisut et al. (2007)
LDPE	Potassium permanganate ($KMnO_4$)	2%–7% (wt)	Twin screw extruding	Oxygen barrier properties improved up to 24% after adding $KMnO_4$ nanoparticles	Khosravi et al. (2013)
LDPE	Organic montmorillonite (OMMT)	0%–4% (wt)	Extruding	Barrier to CO_2 and O_2 increased by 7 and 4 times, respectively, after adding 0.5 wt% of OMMT. WVP decreased by 2.5 times after adding 2 wt% OMMT. TS improved by 49.5% after adding 4 wt% OMMT	Xie et al. (2012)
Polyvinyl chloride (PVC)	Organoclay	3–10 phr	Melt intercalation	TS increased by adding 3 phr nanoclay	Mingliang and Demin (2008)
PVC/ethylene vinyl acetate copolymer (EVA)/	OMMT	0%–6% (wt)	Melt compounding	Impact strength improved up to 6.86 kJ m^2 by adding 100 PVC/5 EVA/2 OMMT	Chuayjuljit et al. (2008)
Poly (ethylene-co-vinyl acetate) (EVA)	Nanosilica	1–9 phr	Two-roll mixing	TS, hardness, and abrasion resistance increased by adding nanosilica	Dasan et al. (2010)

(*Continued*)

Table 3.1 (*Continued*) Effect of Nanocomposite Incorporation with Packaging Materials on Mechanical, Thermal, and Gas Transmission Properties

POLYMER	NANOMATERIAL	LEVEL OF INCORPORATION	METHOD OF PROCESSING	EFFECT ON MATERIAL PROPERTIES	REFERENCES
Unsaturated polyester (UP) and vinyl ester oligomer (VEO)	Organophilic MMT clay	0%–5% (wt)	Casting	TS of 10% VEO-toughened UP resin increased by 22% and 38% with the addition of 2.5% and 5% organoclay, respectively. Flexural strength increased by 19 and 41%, respectively, with the addition of the same amounts of organoclay.	Sharmila et al. (2010)
Poly (methyl methacrylate) (PMMA)/	Layered silicate	1%–5% (wt)	Injection molding	TM increased by 35% compared to neat PMMA	Mohanty and Nayak (2010)
PP	Clay	0.6%–7.4% (wt)	Twin screw extruder	Oxygen transmission rate (OTR) reduced by 21.4% with 4% nanoclay. Water vapor transmission rate (WVTR) decreased by 28.1% with 2% nanoclay	Manikantan and Varadharaju (2011)
PI	Clay	3%–9% (wt)	Casting	A 3 wt% clay reduced gas permeability and oxygen permeability to less than half compared with neat PI	Khayankarn et al. (2003)
PI	Clay (Na ion-exchanged clays Naþ-saponite (SPT), Naþ-mica (Mica), and Naþ-MMT	0%–1% (wt)	Casting	Highest TS of 101 MPa was observed at 0.5 wt% clay. The initial modulus increased from 2.62 to 3.45 GPa as nanoclay content increased from 0% to 1.0 wt%	Park and Chang (2009)
Polystyrene (PS)	Organomontmorillonite – Cloisite 20A (C20A))	0%–7% (wt)	Melt processing with a twin-screw extruding	Tensile, flexural, and impact strengths increased by 83%, 55%, and 74%, respectively by adding 5% clay	Nayak and Mohanty (2009)

(*Continued*)

Table 3.1 (*Continued*) Effect of Nanocomposite Incorporation with Packaging Materials on Mechanical, Thermal, and Gas Transmission Properties

POLYMER	NANOMATERIAL	LEVEL OF INCORPORATION	METHOD OF PROCESSING	EFFECT ON MATERIAL PROPERTIES	REFERENCES
PI	Clay	10% (wt)	Casting	Adding clays improved thermal stability and storage modulus	Gao et al. (2010)
Polyethylene terephthalate (PET)	Clay	0%–8% (wt)	Extruding	Increasing clay content from 2% to 8% increased yield stress from ~0 to 166 Pa	Ghanbari et al. (2013)
PVA/Poly (vinyl pyrrolidone) (PVP)	Sodium MMT	0%–8% (wt)	Solution mixing process	TS of PVA increased by 21.26% and 5.17% with the addition of 4% and 8% sodium MMT, respectively. TS of PVA/PVP blend increased by 14.97% and 3.19% after adding 4% and 8% sodium MMT, respectively. Moisture absorption of PVA decreased from 14.02% to 12.09% and 11.66% with the addition of 4% and 8% sodium MMT, respectively. While in PVA/PVP, moisture absorption decreased from 17.73% to 16.55% and 15.64%, respectively	Mondal et al. (2013)
PP	Corn zein	1%–7.5% (wt)	Solution intercalation method	OP reduced by four times with nanocoating of PP films. WVP reduced by 30% by adding 5 wt.% OMMT	Ozcalik and Tihminlioglu (2013)
Polyimide (PI)	Clay (Cloisite 30B)	10%–40% (wt)	Casting	OP of the film with 10 wt% clay decreased by 95% (42 cc/m^2/day) compared with neat PI film (768 cc/m^2/day)	Kim and Chang (2013)
HDPE	Clay (MMT)	2%–15% (wt)	Melt mixing	OP decreased by 30% when 15% wt of clay was added	Horst et al. (2014)

(*Continued*)

Table 3.1 (*Continued*) Effect of Nanocomposite Incorporation with Packaging Materials on Mechanical, Thermal, and Gas Transmission Properties

POLYMER	NANOMATERIAL	LEVEL OF INCORPORATION	METHOD OF PROCESSING	EFFECT ON MATERIAL PROPERTIES	REFERENCES
EVOH	Organically modified MMT	1%–7% (wt)	Casting	Oxygen and water vapor barrier properties increased by 59% and 90%, respectively with the addition of 3 wt % clay	Kim and Cha (2014)
Poly (ethylene glycol) grafted polypropylene (PP-g-PEG), Mt/polypropylene (PP) The Mt/PP-g-PEG/PP	MMT	1%–5% (wt)	Melt blending	Impact strength and elongation at break increased by 148% and 43%, respectively compared with neat PP	Zhu et al. (2014)
Soy protein isolate	MMT	5%–15%	Melt extrusion	TS increased from 2.26 ± 0.48 MPa to 15.60 ± 1.69 MPa as the MMT content increased from 0% to 15%. WVP reduced by 22.1% with 5% MMT	Kumar et al. (2010)
Cellulose acetate	OMMT	5%	Solvent casting	OTR and water vapor transmission rate reduced by 50% and 10%, respectively	Rodríguez et al. (2012)
Poly (lactic acid) (PLA)	CNC and silver (Ag) nanoparticles	5% CNC 1% Ag nanoparticles	Solvent casting	Adding 1% Ag reduced OTR by 22% compared with pure PLA OTR of PLA/CNC (5%)/Ag(1%) was reduced by 60% compared with pure PLA. WVP was reduced by 46% in PLA/Ag (1%) film. WVP was reduced by 59% in PLA/CNC (5%)/Ag(1%) film	Fortunati et al. (2013)

(*Continued*)

Table 3.1 (*Continued*) Effect of Nanocomposite Incorporation with Packaging Materials on Mechanical, Thermal, and Gas Transmission Properties

POLYMER	NANOMATERIAL	LEVEL OF INCORPORATION	METHOD OF PROCESSING	EFFECT ON MATERIAL PROPERTIES	REFERENCES
Chitosan and clove essential oil	Magnesium Oxide (MgO)	10% MgO	Solution casting	TS increased from 30 MPa to 63 MPa. Elongation (E) increased from 7.2% to 15.3%. Water solubility (WS) decreased from 78.3% to 28.99%.	Sanuja et al. (2014)
Soy protein	Clay (Cloisite® 30B)	1%–8%	Solution intercalation	OP reduced by 6 times with 8% clay	Swain et al. (2012)
Alginate	MMT and cellulose nanoparticles (CNP)	1%–5%	Solvent casting	Adding 5% MMT reduced WS from 99% to 61%. Adding 5% CNP reduced E from 99% to 77%. TS and E of the nanocomposite films were improved with an increase in CNP content and reached a maximum at the highest loading level (5 wt%). WVP decreased by 20% and 18% in alginate/MMT alginate/CNP films, respectively	Abdollahi et al. (2013)
Starch	Silicon carbide (SiC)	0%–10%	Solution technique	OP decreased by 3.5 times with the addition of 10% SiC	Dash and Swain (2013)
Pectin/polyethylene glycol 20,000 (PEG) blend	Halloysite nanotubes (HNTs)	0%–50%	Casting	Water uptake decreased from 8.2% to 4.7% at 75% RH condition. EM increased up to 50% with the addition of 30% filler	Cavallaro et al. (2013)
Thermoplastics starch	Clay MMT	1%–5%	Combination of intercalation and melt-processing	Addition of 5% clay reduced water absorbing capacity of the starch matrix around 30%	Aouada et al. (2013)

(*Continued*)

Table 3.1 (*Continued*) Effect of Nanocomposite Incorporation with Packaging Materials on Mechanical, Thermal, and Gas Transmission Properties

POLYMER	NANOMATERIAL	LEVEL OF INCORPORATION	METHOD OF PROCESSING	EFFECT ON MATERIAL PROPERTIES	REFERENCES
Agar	Silver nanoparticles (AgNPs)	13%	Solvent casting	WVP decreased by 10%	Rhim et al. (2014)
PLLA (poly(L-lactic acid))	Clay MMT (SMMTC18 and NMMTC18) and one fluoro-mica	5%	Melt extrusion	Oxygen barrier properties of SMMTC18- and NMMTC18-based composites were higher than that of the MC18-based composites. OP reduced from 803 (PLLA) to 371 mL-mil m^{-2} day^{-1} with NMMTC18	Chaudhry et al. (2008)
PLA and PLA/PEG	Organoclay (Cloisite 30B)	0%–5%	Melt intercalation	Incorporation of 3% improved modulus by 25%	Ozkoc and Kemaloglu (2009)
Chitosan/ poly(vinylalcohol) (PVA) matrix	Sepiolite (SP)	1%–5%	Casting	TS increased approximately by 35% after adding SP. Moisture uptake was reduced nearly 25%	Huang et al. (2012)
PLA/SiO_2 hybrids		0%–78.3%	Bar coater and casting	The transparency and optical transmittance of films were over 92%, and oxygen and water vapor barrier properties were increased by 69.7% and 45.7%, respectively, over those of neat PLA film	Bang and Kim (2012)
Sago starch and bovine gelatin	Zinc oxide nanorods (ZnO-nr)	1%–5%	Solvent casting	OP decreased by 55% in 5% of ZnO-nr/starch/ gelatin films. Mechanical and heat seal properties increased more than 20% by adding ZnO-nr	Nafchi et al. (2013)

Source: Mihindukulasuriya, S.D.F., and Lim, L.T., *Trends Food Sci. Technol.*, 40, 149–167, 2014.

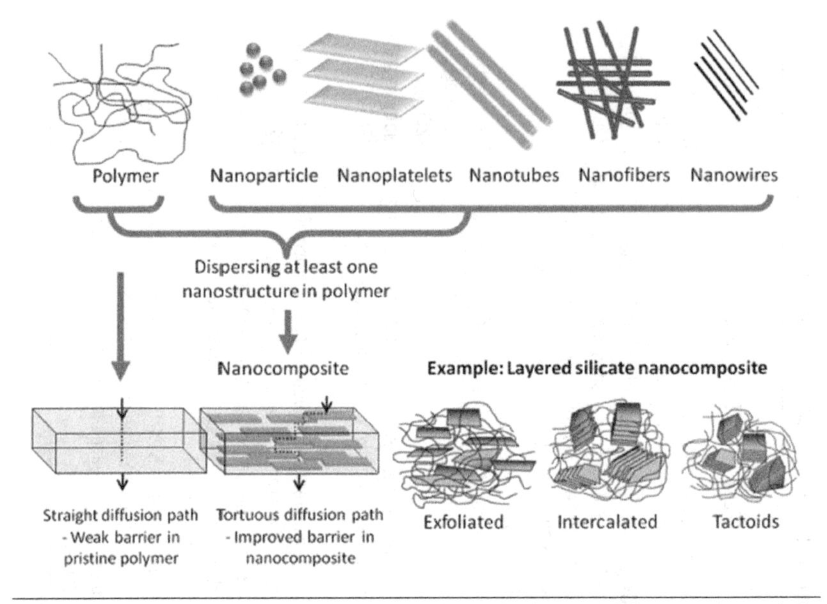

Figure 3.1 Schematic diagram of nanocomposite packaging material preparation and the enhancement of the gas and moisture barrier properties. (Reproduced from Mihindukulasuriya, S.D.F., and Lim, L.T., *Trends Food Sci. Technol.*, 40, 149–167, 2014.)

Food Safety and Quality Analysis

Another sector in the food and beverages industry that benefitted from the developments in nanoscience and nanotechnology is food safety. The spoilage and pathogenic microorganisms cause a huge amount of quality and quantity losses in food products, and the food industry loses millions of dollars due to these microorganisms. Incorporation of nanoparticles with antimicrobial properties as an ingredient or additive or a preservative can help to inhibit the growth of microorganisms in the food. Nanotechnology-based sensors have been developed for detection and identification of microorganisms in the food products. Some of the developed nanosensors or nanobiosensors have the ability to quantify the amount of microbial load in the food product, which might be helpful for quality control personnel to take necessary control measures. Some of the nanosensors contain an array of nanoparticles dispersed in the food packaging films and containers, and when the array comes in contact with any type of pathogenic microorganisms it emits light in various colors or changes the color based on the type or level of pathogenic microorganism infection (Bhattacharya and Gupta 2005). Nanosensors have been successfully tested for detection of foodborne pathogens in food and beverages using the volatile gases released by spoilage and pathogenic microbes when the sensors were directly placed on or incorporated into the packaging materials. These "electronic nose" and "electronic tongue" nanosensors have the ability to detect the foodborne pathogens quickly and can be used as rapid and online food safety assessment tools (García et al. 2006; Lange and Wittman 2002).

Nanotechnology-based "nanoelectromechanical systems" (NEMS) have been tested for detecting not only the foodborne pathogens but also the chemical constituents of the food product. Polychoromix (Wilmington, MA, USA) developed a digital transform spectrometer with NEMS, and it has the ability to find the level of trans fat in the food product (Ritter 2005). The "micro and nanotechnology" (MNT)-based sensors have been tested for identifying as well as continuous monitoring of entry of common food adulterants into food during processing, packaging, logistics, and storage (Canel 2006). Nanotechnology-based cantilever sensors (nanocantilevers) have also been successfully tested as a food safety tool for detection of pathogenic microorganisms.

The polyvinyl chloride (PVC) packaging film incorporated with silver nanoparticles has been tested for inhibiting the growth of *Escherichia coli* and *Staphylococcus aureus* in minced beef and showed a promising result of elevated inhibition up to seven days of storage at refrigeration temperature (4°C) (Mahdi et al. 2012). In the same study, the minced meat packed in regular packaging had a shelf life of two days, whereas meat products packaged with the nanotechnology-based packing films had the enhanced shelf life (up to seven days). The freshness and the condition of the meat product inside the package was monitored using the electronic tongue with nanosensors placed in the package with the analyses of odor and gases released during spoilage of meat product in the package (Bowles and Lu 2014).

The fresh vegetables, fruits, and meat products release some moisture, and the meat products release exudate fluids inside the packaging. This moisture and fluids provide a favorable condition for the development and growth of foodborne pathogens, and also results in loss of freshness of the product. The absorbent pads are used to absorb this unwanted moisture and fluids from the products. In general, cellulose-based materials are used for manufacturing absorbent pads. Researchers developed cellulose-based absorbent pads with silver nanoparticles to inhibit the growth of pathogenic microorganisms like *E. coli* and *S. aureus* (Fernández et al. 2009, 2010a; Lloret et al. 2012). The results of these studies showed a decreased microbial count with increased silver nanoparticle concentration, kept the microbial load 1 log CFU/g lower than the packages with regular absorbent pads under modified atmospheric storage for 11 days at refrigeration temperature (4°C ± 0.2°C), and 90% less microbial load than the packages with regular absorbent pads when storing minimally processed meat products, at 4°C for 10 days (Fernández et al. 2009, 2010a; Lloret et al. 2012).

The *E. coli* O157: H7 pathogen was successfully detected using a DNA sensor (with quartz crystal microbalance) with magnetic nanoparticles (MNPs) (Mao et al. 2006), and a circulating-flow piezoelectric biosensor with gold nanoparticles successfully detected *E. coli* O157: H7 as low as 1.20×10^2 CFU/mL (Chen et al. 2008). These nanosensors can be used as food safety analysis tools for detection and quantification of pathogenic microorganisms. Nanotechnology-based sensors are not only used for detection of spoilage and pathogenic microorganisms, but these can also be used to detect adulteration in food products. Nanosensors containing gold nanoparticles (AuNPs) successfully detected meat adulteration in chicken and beef meatballs (Ali et al. 2011, 2012, 2014; Inbaraj and Chen 2016; Sonawane et al. 2014). These nanosensors containing gold nanoparticles with an average size of 20 nm coated with citrate and dispersed in deionized water remained at its original red color when it incubated with the material having single-stranded DNA (like chicken or beef meat alone). But, when an adulterant meat (swine meat) was mixed with chicken or beef meat, this double-stranded DNA (swine DNA with beef or chicken DNA) created a color change (red to purple grey) after a three-minute incubation (Ali et al. 2012).

In the dairy industry, nanotechnology-based sensors and sensing methods have been successfully tested for detection of adulterants like melamine, pathogenic microorganisms, mycotoxins, and unwanted chemical and antibiotics residues (Ai et al. 2009; Basheer and Lee 2004; Font et al. 2008; Giovannozzi et al. 2014; Guo et al. 2014; Inoue et al. 2000; Moreno et al. 2011; Ni et al. 2014; Song et al. 2015; Xin et al. 2015; Zhou et al. 2014). Melamine is commonly used in plastic and dishware production, and in recent times it has been mixed with diluted milk by the milk producers and processors in order to artificially boost the protein content of the milk to get a better price. Melamine in the milk can cause serious health issues to the consumers, and the standard analytical testing methods used for detection of melamine in milk like high-performance liquid chromatography (HPLC), mass spectroscopy (MS), and surface-enhanced Raman spectroscopy (SERS) need longer detection time and higher initial cost. Nanotechnology-based sensors have been tested for detection of melamine, and the results showed that the melamine presence in the milk can be easily detected by the simple color change using these nanotechnology-based sensing methods (Ai et al. 2009; Cai et al. 2014; Giovannozzi et al. 2014; Guo et al. 2014; Kumar et al. 2014; Ni et al. 2014; Song et al. 2015; Xin et al. 2015).

Gold nanoparticle-based nanosensors have been developed and tested for melamine detection in raw milk, and these colorimetric nanosensors changed their color from wine red to blue based on the amount of melamine present in the milk (Ai et al. 2009; Xin et al. 2015). This simple color change can be easily identified by the naked eye. Similar to these AuNP-based nanosensors, Song et al. (2015) developed a nanosensor for melamine detection using silver nanoparticles (AgNPs) modified with sulfanilic acid (SAA). Guo et al. (2014) and Kumar et al. (2014) developed nanosensors for detection and quantification of the melamine presence in milk using the hollow gold chip with SERS and citrate-stabilized nanosensors, respectively. Ni et al. (2014) developed a colorimetric nanosensor using bare mixed with 3,3′,5,5′-tetramethlybenzidine to detect melamine in milk and milk products. The AuNPs nanosensor with SERS developed by Giovannozzi et al. (2014) rapidly detected the melamine mixed with milk up to a concentration level of 0.17 mg/L. The melamine-sensing probe with label-free AgNPs developed by Ping et al. (2012) had the ability to detect melamine in the milk with a concentration level of 2.32 mM.

The chemicals, steroids, and antibiotics enter into the cow's body through feed and medicine, and these can end up in the milk as a residue. These chemical and antibiotic residues can also cause health issues, and regular analytical methods like HPLC, gas chromatography coupled with mass spectrometry (GC/MS), and enzyme-linked immunosorbent assay (ELISA) methods need a long time for analysis and higher capital cost. Nanotechnology-based sensing methods facilitate a simple and rapid way for detection of chemical and antibody residues in milk and other dairy products. An ELISA-based detection system with MNPs was successfully tested for detecting antibiotic (sulfonamide) residues with concentrations as low as 0.5 μg/L in raw milk samples. This sensor can be easily employed for real-time detection of sulfonamide in milk since the European Union's maximum allowable residue limit in milk is 100 μg/L (Font et al. 2008). Estrogen is a steroid hormone commonly found in raw milk since most of the milk at dairy farms is produced from pregnant cows. Estrogen may pose health risks like carcinogenesis cancers in humans. A nanosensor with polypyrrole-coated magnetic nanoparticles (PPy/MNP) coupled with liquid chromatography/mass spectrometry was developed and successfully tested for detecting estrogen levels in raw milk, which had a detection limit of 5.1–66.7 ng/L, and detection time of 3 min (Gao et al. 2011).

Bisphenol A (BPA) is a commonly used component in food and beverage container manufacturing, and it can cause reproductive disorders to heart disease in humans if these get into the human body through food and beverages (Basheer and Lee 2004; Inoue et al. 2000; Moreno et al. 2011). An aptamer nanobiosensor with AuNPs was able to detect the BPA residues in milk samples (Zhou et al. 2014). All the major foodborne pathogens affecting dairy products like *Bifidobacterium lactis* (Bb12), *Streptococcus thermophiles*, *Lactobacillus bulgaricus*, *Bifidobacterium longum*, and *L. acidophilus* can be detected using nanosensors developed with AgNPs coupled with Matrix-assisted laser desorption/ionization mass spectrometry (MALDI-MS) (Lee et al. 2012).

Impact on Filtration and Wastewater Treatment from the Food and Beverages Industry

Another major area where the current advancements in nanoscience and nanotechnology have had major impact is water purification and wastewater treatment from the food and beverages manufacturing industry. Nanotechnology-based nanofilters, nanosorbents, and photocatalysis agents are commonly used for water purification and wastewater treatment (Bora and Dutta 2014). Membranes with nanoscale pores are used in nanofiltration technology, and the properties like pore size and applied pressure lie in between reverse osmosis and ultrafiltration technologies (Baker 2004; Hong et al. 2006; Salehi et al. 2011). The membranes with a pore size of 1–5 nm are used for removing hazardous organic and inorganic solutes in the wastewater by the nanofilter membranes (Salehi et al. 2011). Wastewater from the food and beverage manufacturing plants contain organic and inorganic pollutants, as well as microbial components, which may cause land, water, and air pollution if these are released from the food plants without any treatment. The recent developments in nanofiltration technology have helped the food and beverages industry to treat the wastewater in a more efficient way with low cost (Lipnizki 2010; Madaeni et al. 2011; Salehi 2014; Thanuttamavong et al. 2001, 2002; Vandanjon et al. 2002). The nanofiltering membranes are generally manufactured with nanoclay, carbon nanotubes, metal oxides, zinc oxide (ZnO), titanium dioxide (TiO_2), zirconium dioxide (ZrO_2), and silica (SiO_2) (Bora and Dutta 2014). Incorporating nanoclay with the conventional inorganic filter membrane manufacturing materials can reduce the manufacturing cost and extend the flexibility of its application (Van der Bruggen et al. 2008). The carbon nanotube (CNT)-based nanofilter membrane has higher chemical and thermal resistance, as well as extended flexibility (Wan et al. 2009). The integrated wastewater treatment facility with isoelectric precipitation, nanofiltration, and anaerobic fermentation techniques in a dairy processing plant showed excellent results in recovery of economically beneficial nutrients like caseins, acetate, butyrate, and hydrogen, as well as filtered out the pollutants and hazardous microorganisms (Chen et al. 2016). A ceramic nanofilter with CNT was able to remove the microorganisms and heavy metals from wastewater (Parham et al. 2013). The nanofilters were also used to recover up to 75% commercial single-phase detergents used for clean-in-place systems (Jayas et al. 2000) in food and beverage manufacturing units for reuse (Fernández et al. 2010b). The wastewater treatment systems with nanostructured photocatalytic medium (commonly metal oxide semiconductors) were successfully tested for purification of water and removal of pollutants from the wastewater (Bora and Dutta 2014). The photocatalytic nanostructured semiconductors with ZnO and TiO_2 nanoparticles also have the ability to inactivate and destroy the harmful bacteria like *E. coli* and *S. aureus* present in the food and beverage industry wastewater (Baruah et al. 2012a, 2012b; Jaisai et al. 2012; Sapkota et al. 2011).

Nanosorbents for water purification and wastewater treatment are generally made with activated carbon nanoparticles, CNTs, MNPs, metal oxide-based nanomaterials (titanium dioxide, alumina, tungsten oxide, iron oxide), and silica (Çeçen and Aktaş 2011; Hu et al. 2010; Kuo et al. 2008; Li et al. 2007; Luo et al. 2012a; Zhang et al. 2010, 2013). The nanosorbent material with copper and silver nanoparticles has been developed by Mostafa and Darwich (2014) and tested for meat industry wastewater treatment. The results showed this hybrid nanocomposite material inactivated 100% foodborne pathogens present in the meat industry wastewater. The heavy metal ions present in the industrial wastewater were successfully removed by absorption using the ferrous-oxide-based nanosorbents (Hu et al. 2010; Iram et al. 2010; Jeon and Yong 2010; Nassar 2010). A detailed discussion about water purification and wastewater treatment using nanotechnology-based nanofilters, nanosorbents, and photocatalytic material is given in Chapter 9.

Impact on Food Processing Operations When Nanoparticles Are Incorporated as an Ingredient, Additive, or Preservative

Application of nanoscience and nanotechnology in the food and beverages industry is growing at an exponential scale in recent times, and the effect of nanotechnology-based products and processes on food and beverage products' physical, chemical, mechanical, rheological, physico-chemical, nutritional, and functional properties has been studied extensively by the researchers throughout the world (discussed in detail in Chapter 2 of this book). Researchers are also working throughout the globe to check the effect of nanotechnology in food and beverages on the consumers' health. But very few studies have been conducted to study the effects of application of nanotechnology-based products on the process properties/parameters in food and beverages. Most of the studies just analyzed the effect of nanotechnology application in efficiency of the process. For example, in filtration of juices, water, and wastewater from the food and beverages industry, application of nanotechnology-based filtration membranes (nanofiltration) increased the efficiency and permeate flux of the process (Acosta et al. 2017; Arend et al. 2017; Berovic et al. 2014; Luo et al. 2012b). Arend et al. (2017) applied microfiltration and nanofiltration techniques for phenolic component recovery from the juices, and found the polyvinylidene difluoride nanofiltration membrane had higher permeate flux (4.0 L h^{-1} m^{-2}) than the microfiltration membrane (3.0 L h^{-1} m^{-2}) at same operating temperature (20°C ± 2°C) and pressure conditions (600 kPa). Similar results were obtained by Acosta et al. (2017), Andrade et al. (2014), and Luo et al. (2012b) when the nanofiltration was applied for different beverage processing operations. The fouling rate was also reduced when the nanotechnology-based membranes were used for the filtration process (Acosta et al. 2017; Andrade et al. 2014; Chen et al. 2016; Luo and Ding 2011). Packaging is the major unit operation in the food and beverages manufacturing and in where the nanotechnology is extensively applied. The effect of incorporation of nanomaterials with the packaging polymers on the mechanical, thermal, and gas and moisture barrier properties has been studied extensively (Table 3.1). The effect of incorporation of nanoclay, metal, and metal oxide nanoparticles with the food packaging film polymers on gas and moisture permeation properties has been predominantly studied with respect to food and beverages sector since the oxygen, carbon dioxide, and moisture transmission from product to ambient or ambient to product through the packaging film plays a major role in keeping the food and beverage product fresh and safe for a long time (Akbari et al. 2006; Brody et al. 2008; Ramachandraiah et al. 2015; Ravichandran 2010; Sekhon 2010). But not so much research has been carried out in other sectors like the baking and meat industry to check the effect of nanotechnology-based material application on the process parameters. In the baking industry, nanoencapsulated omega-3 fatty acids were incorporated with the dough during the baking process to improve the nutritive value and bioavailability of the nutrients, and researchers mainly concentrated on the effect of nanoencapsulation on reduction in degradation of nutrient and improvement in bioavailability, and also studied the effect of inclusion of nanoencapsulated products on color, texture, and other physico-chemical properties of the product but not much on the process parameters (Gökmen et al. 2011; Mogol et al. 2013). More research is needed to check the effect of application of nanotechnology-based products on process temperature, processing time, flowability, and other rheological properties of the ingredients in order to understand the whole picture of changes or variations due to nanomaterial incorporation. Understanding the influence of nanomaterials on the process parameters will not only help to improve the food and beverage product quality, it will also help the process engineers to make modifications in processing equipment. Optimization of processes and processing equipment will be beneficial to the food and beverages industry quantitatively (reduction in processing cost and time) as well as qualitatively (will yield better quality products).

Conclusion

Nanotechnology-based materials and processes not only make the food product tastier and nutritious but also make the processing operations simple and more efficient. Size reduction of raw materials into the nanoscale level makes the ingredients, additives, preservatives, and food-coloring agents more efficient by increasing the solubility of the products and increasing surface contact area. The nanotechnology-based packaging materials alter the mechanical and thermal properties of the packaging materials in a positive manner and also restrict the moisture and gas entry and exit from the product. Nanotechnology-based smart packaging, intelligent packaging, and active packaging allow the packaging material to interact with the product and environment, and also helps to monitor the quality of the product and storage environment's condition throughout the logistics and storage period. Nanotechnology also helps to detect the pathogenic microorganisms in a rapid manner, and also inactivates the microorganisms more efficiently even after the packaging. Nanotechnology-based membranes and nanosorbents have the ability to effectively remove the chemicals and microorganisms that are hazardous to the environment and humans from the food and beverage industry wastewater. Even though nanotechnology provides enormous benefits to the food and beverage industry during processing, it may have some ill effects, like leaching of nanomaterials into food products from the packaging films, increasing intake by consumers. More detailed research is needed to study these issues in order to get consumer acceptance of this new emerging technology in the food and beverages sector.

References

Abdollahi M., Alboofetileh M., Rezaei M., Behrooz R. (2013) Comparing physico-mechanical and thermal properties of alginate nanocomposite films reinforced with organic and/or inorganic nanofillers. *Food Hydrocolloids* 32(2):416–424.

Acosta O., Vaillant F., Pérez A.M., Dornier M. (2017) Concentration of polyphenolic compounds in blackberry (*Rubus adenotrichos* Schltdl.) juice by nanofiltration. *Journal of Food Process Engineering* 40:e12343.

Ai K., Liu Y., Lu L. (2009) Hydrogen-bonding recognition-induced color change of gold nanoparticles for visual detection of melamine in raw milk and infant formula. *Journal of the American Chemical Society* 131:9496–9497. doi:10.1021/ja9037017.

Akbari Z., Ghomashchi T., Aroujalian A. (2006) Potential of nanotechnology for food packaging industry. In: *Nano and Micro Technologies in the Food and HealthFood Industries*, Amsterdam, the Netherlands. pp. 25–26.

Ali M.E., Hashim U., Mustafa S., Man Y.B.C., Adam T., Humayun Q. (2014) Nanobiosensor for the detection and quantification of pork adulteration in meatball formulation. *Journal of Experimental Nanoscience* 9:152–160.

Ali M.E., Hashim U., Mustafa S., Man Y.B.C., Islam K.N. (2012) Gold nanoparticlesensor for the visual detection of pork adulteration in meatball formulation. *Journal of Nanomaterials* 1–7. doi:10.1155/2012/103607.

Ali M.E., Hashim U., Mustafa S., Man Y.B.C., Yusop M.H.M., Bari M.F., Islam K.N., Hasan M.F. (2011) Nanoparticle sensor for label free detection of swine DNA in mixed biological samples. *Nanotechnology* 22:195503.

Andrade L., Mendes F., Espindola J., Amaral M. (2014) Nanofiltration as tertiary treatment for the reuse of dairy wastewater treated by membrane bioreactor. *Separation and Purification Technology* 126:21–29.

Aouada F.A., Mattoso L.H., Longo E. (2013) Enhanced bulk and superficial hydrophobicities of starch-based bionanocomposites by addition of clay. *Industrial Crops and Products* 50:449–455.

Arend G.D., Adorno W.T., Rezzadori K., Di Luccio M., Chaves V.C., Reginatto F.H., Petrus J.C.C. (2017) Concentration of phenolic compounds from strawberry (*Fragaria X ananassa* Duch) juice by nanofiltration membrane. *Journal of Food Engineering* 201:36–41.

Arunvisut S., Phummanee S., Somwangthanaroj A. (2007) Effect of clay on mechanical and gas barrier properties of blown film LDPE/clay nanocomposites. *Journal of Applied Polymer Science* 106(4):2210–2217.

Baker R.W. (2004) *Membrane Technology and Applications*. John Wiley & Sons, London, UK. pp. 96–103.

Bang G., Kim S.W. (2012) Biodegradable poly (lactic acid)-based hybrid coating materials for food packaging films with gas barrier properties. *Journal of Industrial and Engineering Chemistry* 18(3):1063–1068.

Baruah S., Pal S.K., Dutta J. (2012a) Nanostructured zinc oxide for water treatment. *Nanoscience and Nanotechnology-Asia* 2:90–102.

Baruah S., Pal S.K., Dutta J. (2012b) Nanostructured zinc oxide for water treatment. *Nanoscience and Nanotechnology-Asia* 2:90–102

Basheer C., Lee H.K. (2004) Analysis of endocrine disrupting alkylphenols, chlorophenols and bisphenol-A using hollow fiber-protected liquid-phase microextraction coupled with injection port-derivatization gas chromatography–mass spectrometry. *Journal of Chromatography A* 1057:163–169.

Berovic M., Berlot M., Kralj S., Makovec D. (2014) A new method for the rapid separation of magnetized yeast in sparkling wine. *Biochemical Engineering Journal* 88:77–84.

Bhattacharya D., Gupta R.K. (2005) Nanotechnology and potential of microorganisms. *Critical Reviews in Biotechnology* 25:199–204.

Bora T., Dutta J. (2014) Applications of nanotechnology in wastewater treatment—A review. *Journal of Nanoscience and Nanotechnology* 14:613–626.

Bowles M., Lu J.J. (2014) Removing the blinders: A literature review on the potential of nanoscale technologies for the management of supply chains. *Technological Forecasting and Social Change* 82:190–198. doi:10.1016/j.techfore.2013.10.017.

Brody A.L., Bugusu B., Han J.H., Sand C.K., Mchugh T.H. (2008) Innovative food packaging solutions. *Journal of Food Science* 73:R107–R116.

Cai H.H., Yu X., Dong H., Cai J., Yang P.H. (2014) Visual and absorption spectroscopic detections of melamine with 3-mercaptopriopionic acid-functionalized gold nanoparticles: A synergistic strategy induced nanoparticle aggregates. *Journal of Food Engineering* 142:163–169. doi:10.1016/j.jfoodeng.2014.04.018.

Canel C. (2006) Micro and nanotechnologies for food safety and quality applications. *Micro-Nano-Engineering* 6:219–225.

Cavallaro G., Lazzara G., Milioto S. (2013) Sustainable nanocomposites based on halloysite nanotubes and pectin/polyethylene glycol blend. *Polymer Degradation and Stability* 98(12):2529–2536.

Çeçen F., Aktaş Ö. (2011) Water and wastewater treatment: Historical perspective of activated carbon adsorption and its integration with biological processes. *Activated Carbon for Water and Wastewater Treatment: Integration of Adsorption and Biological Treatment* 1–11.

Chaudhry Q., Scotter M., Blackburn J., Ross B., Boxall A., Castle L., Aitken R., Watkins R. (2008) Applications and implications of nanotechnologies for the food sector. *Food Additives and Contaminants* 25(3):241–258.

Chen S.H., Wu V.C., Chuang Y.C., Lin C.-S. (2008) Using oligonucleotide-functionalized Au nanoparticles to rapidly detect foodborne pathogens on a piezoelectric biosensor. *Journal of Microbiological Methods* 73:7–17.

Chen Z., Luo J., Chen X., Hang X., Shen F., Wan Y. (2016) Fully recycling dairy wastewater by an integrated isoelectric precipitation–nanofiltration–anaerobic fermentation process. *Chemical Engineering Journal* 283:476–485.

Chrissopoulou K., Altintzi I., Anastasiadis S.H., Giannelis E.P., Pitsikalis M., Hadjichristidis N., Theophilou N. (2005) Controlling the miscibility of polyethylene/layered silicate nanocomposites by altering the polymer/surface interactions. *Polymer* 46(26):12440–12451.

Chuayjuljit S., Thongraar R., Saravari O. (2008) Preparation and properties of PVC/EVA/organomodified montmorillonite nanocomposites. *Journal of Reinforced Plastics and Composites* 27(4):431–442.

Dadbin S., Noferesti M., Frounchi M. (2008) Oxygen barrier LDPE/LLDPE/Organoclay nano-composite films for food packaging. *Macromolecular Symposia* 274(1):22–27.

Dasan P.K., Purushothaman E., Unnikrishnan G. (2010) Analysis of physical and solvent transport behavior of poly (ethylene-co-vinyl acetate)/silica composites. *Journal of Reinforced Plastics and Composites* 29(2):238–246.

Dash S., Swain S.K. (2013) Synthesis of thermal and chemical resistant oxygen barrier starch with reinforcement of nano silicon carbide. *Carbohydrate Polymers* 97(2):758–763.

Fernández A., Picouet P., Lloret E. (2010a) Cellulose-silver nanoparticle hybrid materials to control spoilage-related microflora in absorbent pads located in trays of fresh-cut melon. *International Journal of Food Microbiology* 142:222–228.

Fernández P., Riera F.A., Álvarez R., Álvarez S. (2010b) Nanofiltration regeneration of contaminated single-phase detergents used in the dairy industry. *Journal of Food Engineering* 97:319–328.

Fernández A., Soriano E., López-Carballo G., Picouet P., Lloret E., Gavara R., Hernández-Muñoz P. (2009) Preservation of aseptic conditions in absorbent pads by using silver nanotechnology. *Food Research International* 42:1105–1112.

Follain N., Belbekhouche S., Bras J., Siqueira G., Marais S., Dufresne A. (2013) Water transport properties of bio-nanocomposites reinforced by *Luffa cylindrica* cellulose nanocrystals. *Journal of Membrane Science* 427:218–229.

Font H., Adrian J., Galve R., Estévez M.C., Castellari M., Gratacós-Cubarsí M., Sánchez-Baeza F., Marco M.P. (2008) Immunochemical assays for direct sulfonamide antibiotic detection in milk and hair samples using antibody derivatized magnetic nanoparticles. *Journal of Agricultural and Food Chemistry* 56:736–743. doi:10.1021/jf072550n.

Fortunati E., Puglia D., Luzi F., Santulli C., Kenny J.M., Torre L. (2013) Binary PVA bio-nanocomposites containing cellulose nanocrystals extracted from different natural sources: Part I. *Carbohydrate Polymers* 97(2):825–836.

Frounchi M., Dadbin S., Salehpour Z., Noferesti M. (2006) Gas barrier properties of PP/EPDM blend nanocomposites. *Journal of Membrane Science* 282(1–2):142–148.

Gao Y., Choudhury N.R., Dutta N.K. (2010) Systematic study of interfacial interactions between clays and an ionomer. *Journal of Applied Polymer Science* 117(6):3395–3405.

Gao Q., Luo D., Bai M., Chen Z.W., Feng Y.Q. (2011) Rapid determination of estrogens in milk samples based on magnetite nanoparticles/polypyrrole magnetic solid-phase extraction coupled with liquid chromatography–tandem mass spectrometry. *Journal of Agricultural and Food Chemistry* 59:8543–8549. doi:10.1021/jf201372r.

García M., Aleixandre M., Gutiérrez J., Horrillo M.C. (2006) Electronic nose for wine discrimination. *Sensors and Actuators B: Chemical* 113(2):911–916.

Ghanbari A., Heuzey M.C., Carreau P.J., Ton-That M.T. (2013) Morphological and rheological properties of PET/clay nanocomposites. *Rheologica Acta* 52(1):59–74.

Giovannozzi A.M., Rolle F., Sega M., Abete M.C., Marchis D., Rossi A.M. (2014) Rapid and sensitive detection of melamine in milk with gold nanoparticles by surface enhanced Raman scattering. *Food Chemistry* 159:250–256. doi:10.1016/j.foodchem.2014.03.013.

Gökmen V., Mogol B.A., Lumaga R.B., Fogliano V., Kaplun Z., Shimoni E. (2011) Development of functional bread containing nanoencapsulated Omega-3 fatty acids. *Journal of Food Engineering* 105:585–591. doi:10.1016/j.jfoodeng.2011.03.021.

Guo Z., Cheng Z., Li R., Chen L., Lv H., Zhao B., Choo J. (2014) Onestep detection of melamine in milk by hollow gold chip based on surface-enhanced Raman scattering. *Talanta* 122:80–84. doi:10.1016/j.talanta.2014.01.043.

Hemati F., Garmabi H. (2011) Compatibilised LDPE/LLDPE/nanoclay nanocomposites: I. Structural, mechanical, and thermal properties. *The Canadian Journal of Chemical Engineering* 89(1):187–196.

Hong S.U., Miller M.D., Bruening M.L. (2006) Removal of dyes, sugars, and amino acids from NaCl solutions using multilayer polyelectrolyte nanofiltration membranes. *Industrial & Engineering Chemistry Research* 45:6284–6288.

Horst M.F., Quinzani L.M., Failla, M.D. (2014) Rheological and barrier properties of nanocomposites of HDPE and exfoliated montmorillonite. *Journal of Thermoplastic Composite Materials* 27(1):106–125.

Huang D., Mu B., Wang A. (2012) Preparation and properties of chitosan/poly (vinyl alcohol) nanocomposite films reinforced with rod-like sepiolite. *Materials Letters* 86:69–72.

Hu H., Wang Z., Pan L. (2010) Synthesis of monodisperse Fe_3O_4@ silica core–shell microspheres and their application for removal of heavy metal ions from water. *Journal of Alloys and Compounds* 492:656–661.

Hyun K., Chong W., Koo M., Chung I.J. (2003) Physical properties of polyethylene/silicate nanocomposite blown films. *Journal of Applied Polymer Science* 89(8):2131–2136.

Inbaraj B.S., Chen B. (2016) Nanomaterial-based sensors for detection of foodborne bacterial pathogens and toxins as well as pork adulteration in meat products. *Journal of Food and Drug Analysis* 24:15–28.

Inoue K., Kato K., Yoshimura Y., Makino T., Nakazawa H. (2000) Determination of bisphenol A in human serum by high-performance liquid chromatography with multi-electrode electrochemical detection. *Journal of Chromatography B: Biomedical Sciences and Applications* 749:17–23.

Iram M., Guo C., Guan Y., Ishfaq A., Liu H. (2010) Adsorption and magnetic removal of neutral red dye from aqueous solution using Fe_3O_4 hollow nanospheres. *Journal of Hazardous Materials* 181:1039–1050.

Jaisai M., Baruah S., Dutta J. (2012) Paper modified with ZnO nanorods–antimicrobial studies. *Beilstein Journal of Nanotechnology* 3:684.

Jayas D.S., Shatadal P., Cenkowski S. (2000) Cleaning-in-place (CIP). In: Y.H. Hui (Ed.) *Encyclopaedia of Food Science and Technology*, 2nd edition. John Wiley & Sons, New York. pp. 351–353.

Jeon S., Yong K. (2010) Morphology-controlled synthesis of highly adsorptive tungsten oxide nanostructures and their application to water treatment. *Journal of Materials Chemistry* 20:10146–10151.

Khayankarn O., Magaraphan R., Schwank J.W. (2003) Adhesion and permeability of polyimide–clay nanocomposite films for protective coatings. *Journal of Applied Polymer Science* 89(11):2875–2881.

Khosravi R., Hashemi S.A., Sabet S.A., Rezadoust A.M. (2013) Thermal, dynamic mechanical, and barrier studies of potassium permanganate-LDPE nanocomposites. *Polymer-Plastics Technology and Engineering* 52(2):126–132.

Kim S.W., Cha S.H. (2014) Thermal, mechanical, and gas barrier properties of ethylene–vinyl alcohol copolymer-based nanocomposites for food packaging films: Effects of nanoclay loading. *Journal of Applied Polymer Science* 131(11):40289.

Kim Y., Chang J.H. (2013) Colorless and transparent polyimide nanocomposites: Thermo-optical properties, morphology, and gas permeation. *Macromolecular Research* 21(2):228–233.

Kumar P., Sandeep K.P., Alavi S., Truong V.D., Gorga R.E. (2010) Preparation and characterization of bio-nanocomposite films based on soy protein isolate and montmorillonite using melt extrusion. *Journal of Food Engineering* 100(3):480–489.

Kumar N., Seth R., Kumar H. (2014) Colorimetric detection of melamine in milk by citrate-stabilized gold nanoparticles. *Analytical Biochemistry* 456:43–49. doi:10.1016/j.ab.2014.04.002.

Kuo C.Y., Wu C.H., Wu J.Y. (2008) Adsorption of direct dyes from aqueous solutions by carbon nanotubes: Determination of equilibrium, kinetics and thermodynamics parameters. *Journal of Colloid and Interface Science* 327:308–315.

Lagaron J.M., Cabedo L., Cava D., Feijoo J.L., Gavara R., Gimenez E. (2005) Improving packaged food quality and safety. Part 2: Nanocomposites. *Food Additives and Contaminants* 22(10):994–998.

Lange J., Wittmann C. (2002) Enzyme sensor array for the determination of biogenic amines in food samples. *Analytical and Bioanalytical Chemistry* 372(2):276–283.

Lanzon N., Kahl J., Ploeger A. (2009) Nanotechnology in the context of organic food processing. Wissenschaftstagung Ökologischer Landbau; Zürich. February 11–13, 2009. Available from http://orgprints.org/14180/1/Lanzon_14180.pdf.

Lee C.H., Gopal J., Wu H.F. (2012) Ionic solution and nanoparticle assisted MALDI-MS as bacterial biosensors for rapid analysis of yogurt. *Biosensors and Bioelectronics* 31:77–83. doi:10.1016/j.bios.2011.09.041.

Li Y., Zhao Y., Hu W., Ahmad I., Zhu Y., Peng X., Luan Z. (2007) Carbon nanotubes-the promising adsorbent in wastewater treatment. *Journal of Physics: Conference Series* IOP Publishing: 698.

Lipnizki F. (2010) *Membrane Technology, Volume 3: Membranes for Food Applications*. Wiley-VchVerlag GmbH & Co. KGaA, Weinheim, Germany.

Lloret E., Picouet P., Fernández A. (2012) Matrix effects on the antimicrobial capacity of silver based nanocomposite absorbing materials. *LWT-Food Science and Technology* 49(2):333–338.

Luo J., Ding L. (2011) Influence of pH on treatment of dairy wastewater by nanofiltration using shear-enhanced filtration system. *Desalination* 278:150–156.

Luo J., Ding L., Wan Y., Jaffrin M.Y. (2012a) Threshold flux for shear-enhanced nanofiltration: Experimental observation in dairy wastewater treatment. *Journal of Membrane Science* 409:276–284.

Luo J., Ding L., Wan Y., Paullier P., Jaffrin M.Y. (2012b) Fouling behavior of dairy wastewater treatment by nanofiltration under shear-enhanced extreme hydraulic conditions. *Separation and Purification Technology* 88:79–86.

Madaeni S., Yasemi M., Delpisheh A. (2011) Milk sterilization using membranes. *Journal of Food Process Engineering* 34:1071–1085.

Maftoum N. (2017) Healthylicious—the hypnotizing world of nano foods. Available from http://nicolemaftoum.blogspot.com/2009_04_01_archive.html. Accessed June 26, 2017.

Mahdi S., Vadood R., Nourdahr R. (2012) Study on the antimicrobial effect of nanosilver tray packaging of minced beef at refrigerator temperature. *Global Veterinaria* 9:284–289.

Manikantan M.R., Varadharaju N. (2011) Preparation and properties of polypropylene-based nanocomposite films for food packaging. *Packaging Technology and Science* 24(4):191–209.

Mao X.L., Yang L.J., Su X.L., Li Y.B. (2006) A nanoparticle amplification based quartz crystal microbalance DNA sensor for detection of *Escherichia coli* O157: H7. *Biosensors & Bioelectronics* 21:1178–1185.

McHugh T.H. (2008) Food nanotechnology–Food packaging applications. *The World of Food Science* 4:1–3.

Mihindukulasuriya S.D.F., Lim L.T. (2014) Nanotechnology development in food packaging: A review. *Trends in Food Science & Technology* 40(2):149–167.

Mills A. (2005) Oxygen indicators and intelligent inks for packaging food. *Chemical Society Reviews* 34:1003–1011. doi:10.1039/b503997p.

Mingliang G.E., Demin J. (2008) Influence of organoclay prepared by solid state method on the morphology and properties of polyvinyl chloride/organoclay nanocomposites. *Journal of Elastomers & Plastics* 40(3):223–235.

Mogol B.A., Gokmen V., Shimoni E. (2013) Nano-encapsulation improves thermal stability of bioactive compounds Omega fatty acids and silymarin in bread. *Agro Food Industry Hi-Tech* 24:62–65.

Mohanty S., Nayak S.K. (2010) Effect of organo-modified layered silicates on the properties of poly (methyl methacrylate) nanocomposites. *Journal of Thermoplastic Composite Materials* 23(5):623–645.

Mondal D., Mollick M.M.R., Bhowmick B., Maity D., Bain M.K., Rana D., Chattopadhyay D. (2013) Effect of poly (vinyl pyrrolidone) on the morphology and physical properties of poly (vinyl alcohol)/sodium montmorillonite nanocomposite films. *Progress in Natural Science: Materials International* 23(6):579–587.

Moreno M.J., D'Arienzo P., Manclús J.J., Montoya Á. (2011) Development of monoclonal antibody-based immunoassays for the analysis of bisphenol A in canned vegetables. *Journal of Environmental Science and Health Part B* 46:509–517.

Mostafa T.B., Darwish A.S. (2014) An approach toward construction of tune chitosan/polyaniline/metal hybrid nanocomposites for treatment of meat industry wastewater. *Chemical Engineering Journal* 243:326–339. doi:10.1016/j.cej.2014.01.006.

Nafchi A.M., Nassiri R., Sheibani S., Ariffin F., Karim A.A. (2013) Preparation and characterization of bionanocomposite films filled with nanorod-rich zinc oxide. *Carbohydrate Polymers* 96(1):233–239.

Nassar N.N. (2010) Rapid removal and recovery of Pb (II) from wastewater by magnetic nanoadsorbents. *Journal of Hazardous Materials* 184:538–546.

Nayak S.K., Mohanty S. (2009) Dynamic mechanical, rheological, and thermal properties of intercalated polystyrene/organomontmorillonite nanocomposites: Effect of clay modification on the mechanical and morphological behaviors. *Journal of Applied Polymer Science* 112(2):778–787.

Ni P., Dai H., Wang Y., Sun Y., Shi Y., Hu J., Li Z. (2014) Visual detection of melamine based on the peroxidase-like activity enhancement of bare gold nanoparticles. *Biosensors and Bioelectronics* 60:286–291. doi:10.1016/j.bios.2014.04.029.

Ozkoc G., Kemaloglu S. (2009) Morphology, biodegradability, mechanical, and thermal properties of nanocomposite films based on PLA and plasticized PLA. *Journal of Applied Polymer Science* 114(4):2481–2487.

Ozcalik O., Tihminlioglu F. (2013) Barrier properties of corn zein nanocomposite coated polypropylene films for food packaging applications. *Journal of Food Engineering* 114(4):505–513.

Parham H., Bates S., Xia Y., Zhu Y. (2013) A highly efficient and versatile carbon nanotube/ceramic composite filter. *Carbon* 54:215–223. doi:10.1016/j.carbon.2012.11.032.

Park J.S., Chang J.H. (2009) Colorless polyimide nanocomposite films with pristine clay: Thermal behavior, mechanical property, morphology, and optical transparency. *Polymer Engineering & Science* 49(7):1357–1365.

Ping H., Zhang M., Li H., Li S., Chen Q., Sun C., Zhang T. (2012) Visual detection of melamine in raw milk by label-free silver nanoparticles. *Food Control* 23:191–197. doi:10.1016/j.foodcont.2011.07.009.

Ramachandraiah K., Han S.G., Chin K.B. (2015) Nanotechnology in meat processing and packaging: Potential applications—A review. *Asian-Australasian Journal of Animal Sciences* 28:290–302. doi:10.5713/ajas.14.0607.

Ravichandran R. (2010) Nanotechnology applications in food and food processing: Innovative green approaches, opportunities and uncertainties for global market. *International Journal of Green Nanotechnology: Physics and Chemistry* 1(2):72–96.

Rhim J.W., Hong S.I., Park H.M., Ng P.K. (2006) Preparation and characterization of chitosan-based nanocomposite films with antimicrobial activity. *Journal of Agricultural and Food Chemistry* 54(16):5814–5822.

Rhim J.W., Wang L.F., Lee Y., Hong S.I. (2014) Preparation and characterization of bio-nanocomposite films of agar and silver nanoparticles: Laser ablation method. *Carbohydrate Polymers* 103:456–465.

Ritter S.K. (2005) An eye on food. *Chemical Engineering News* 83:28–34.

Rodríguez F.J., Galotto M.J., Guarda A., Bruna J.E. (2012) Modification of cellulose acetate films using nanofillers based on organoclays. *Journal of Food Engineering* 110(2):262–268.

Salehi F. (2014) Current and future applications for nanofiltration technology in the food processing. *Food and Bioproducts Processing* 92:161–177.

Salehi F., Razavi S.M., Elahi, M. (2011) Purifying anion exchange resin regeneration effluent using polyamide nanofiltration membrane. *Desalination*, 278(1–3):31–35.

Sanuja S., Agalya A., Umapathy M.J. (2014) Studies on magnesium oxide reinforced chitosan bionanocomposite incorporated with clove oil for active food packaging application. *International Journal of Polymeric Materials and Polymeric Biomaterials* 63(14):733–740.

Sapkota A., Anceno A.J., Baruah S., Shipin O.V., Dutta J. (2011) Zinc oxide nanorod mediated visible light photoinactivation of model microbes in water. *Nanotechnology* 22(21):215703.

Sekhon B.S. (2010) Food nanotechnology—An overview. *Nanotechnology, Science and Applications* 3:1–15.

Sharmila R.J., Premkumar S., Alagar M. (2010) Preparation and characterization of organoclay-filled, vinyl ester-modified unsaturated polyester nanocomposites. *High Performance Polymers* 22(1):16–27.

Shimoni E. (2009) Nanotechnology for foods: Delivery systems. In: B.C. Gustavo, M. Alan, L. David, S. Walter, B. Ken, & C. Paul (Eds.) *Global Issues in Food Science and Technology*. Academic Press, SanDiego, CA. pp. 411–424.

Silvestre C., Duraccio D., Cimmino S. (2011) Food packaging based on polymer nanomaterials. *Progress in Polymer Science* 36(12):1766–1782.

Sonawane S.K., Arya S.S., LeBlanc J.G., Jha N. (2014) Use of nanomaterials in the detection of food contaminants. *European Journal of Food Research & Review* 4:301.

Song J., Wu F., Wan Y., Ma L. (2015) Colorimetric detection of melamine in pretreated milk using silver nanoparticles functionalized with sulfanilic acid. *Food Control* 50:356–361. doi:10.1016/j.foodcont.2014.08.049.

Swain S.K., Priyadarshini P.P., Patra S.K. (2012) Soy protein/clay bionanocomposites as ideal packaging materials. *Polymer-Plastics Technology and Engineering* 51(12):1282–1287.

Thanuttamavong M., Oh J., Yamamoto K., Urase T. (2001) Comparison between rejection characteristics of natural organic matter and inorganic salts in ultra-low-pressure nanofiltration for drinking water production. *Water Science and Technology: Water Supply* 1:77–90.

Thanuttamavong M., Yamamoto K., Oh J.I., Choo K.H., Choi S.J. (2002) Rejection characteristics of organic and inorganic pollutants by ultra-low-pressure nanofiltration of surface water for drinking water treatment. *Desalination* 145:257–264.

Vandanjon L., Cros S., Jaouen P., Quéméneur F., Bourseau P. (2002) Recovery by nanofiltration and reverse osmosis of marine flavours from seafood cooking waters. *Desalination* 144:379–385.

Van der Bruggen B., Mänttäri M., Nyström M. (2008) Drawbacks of applying nanofiltration and how to avoid them: A review. *Separation and Purification Technology* 63:251–263.

Wan R., Lu H., Li J., Bao J., Hu J., Fang H. (2009) Concerted orientation induced unidirectional water transport through nanochannels. *Physical Chemistry Chemical Physics* 11:9898–9902.

Wu C.L., Zhang M.Q., Rong M.Z., Friedrich K. (2005) Silica nanoparticles filled polypropylene: Effects of particle surface treatment, matrix ductility and particle species on mechanical performance of the composites. *Composites Science and Technology* 65(3–4):635–645.

Xie L., Lv X.Y., Han Z.J., Ci J.H., Fang C.Q., Ren P.G. (2012) Preparation and performance of high-barrier low density polyethylene/organic montmorillonite nanocomposite. *Polymer-Plastics Technology and Engineering* 51(12):1251–1257.

Xin J.Y., Zhang L.X., Chen D.D., Lin K., Fan H.C., Wang Y., Xia C.G. (2015) Colorimetric detection of melamine based on methanobactin-mediated synthesis of gold nanoparticles. *Food Chemistry* 174:473–479. doi:10.1016/j.foodchem.2014.11.098.

Yam K.L., Takhistov P.T. Miltz J. (2005) Intelligent packaging: Concepts and applications. *Journal of Food Science* 70(1):R1–R10.

Yusop S.M., O'Sullivan M.G., Preuß M., Weber H., Kerry J.F., Kerry J.P. (2012) Assessment of nanoparticle paprika oleoresin on marinating performance and sensory acceptance of poultry meat. *LWT-Food Science and Technology* 46:349–355.

Zhang S., Niu H., Hu Z., Cai Y., Shi Y. (2010) Preparation of carbon coated Fe_3O_4 nanoparticles and their application for solid-phase extraction of polycyclic aromatic hydrocarbons from environmental water samples. *Journal of Chromatography A* 1217:4757–4764.

Zhang S., Xu W., Zeng M., Li J., Li J., Xu J., Wang X. (2013) Superior adsorption capacity of hierarchical iron oxide@magnesium silicate magnetic nanorods for fast removal of organic pollutants from aqueous solution. *Journal of Materials Chemistry A* 1:11691–11697.

Zhou L., Wang J., Li D., Li Y. (2014) An electrochemical aptasensor based on gold nanoparticles dotted graphene modified glassy carbon electrode for label-free detection of bisphenol A in milk samples. *Food Chemistry* 162:34–40. doi:10.1016/j.foodchem.2014.04.058.

Zhu S., Chen J., Li H., Cao Y., Yang Y., Feng Z. (2014) Preparation and properties of montmorillonite/poly (ethylene glycol) grafted polypropylene/polypropylene nanocomposites. *Applied Clay Science* 87:303–310.

Zhu K., Huang S., Peng W., Qian H., Zhou H. (2010) Effect of ultrafine grinding on hydration and antioxidant properties of wheat bran dietary fiber. *Food Research International* 43:943–948.

4
Nanosensors for the Food Industry

Introduction

Nanosensors are the sensors built with at least one material in nanoscale level or the sensors used to detect components of a material in nanoscale level. The advancements in nanoscience and nanotechnology are leading to rapid improvements in the design and fabrication of nanosensors and also are widening the applications of nanosensors in various industrial sectors including the food industry. Nanosensors have the major advantage of sensing some components present in nanolevel in a complex food matrix quickly with high accuracy and simple operational procedure. Most of the common analytical techniques used in food industry to detect minor or major components of food matrix, like liquid chromatography (LC), gas chromatography/mass spectroscopy (GCMS), high-performance liquid chromatography (HPLC), and immunosensing techniques like enzyme-linked immunosorbent assay (ELISA), have some disadvantages, such as needing trained, qualified personnel and requiring a long analysis time or destruction of the sample. Some of the advance techniques like spectroscopy and hyperspectral imaging are either expensive or complex operationally. Nanosensors overcome these obstacles, and some of the recently developed nanosensors can be directly used in production line for identifying chemical components and for detecting foodborne pathogens (Augustin and Sanguansri 2009; Avella et al. 2011; Baeumner 2004; Bernardes et al. 2014; Gomes et al. 2015; Ramachandraiah et al. 2015). These sensors help to improve the productivity of the food industry by eliminating time wasted for laboratory analytical tests, and they also improve product quality since most of these nanosensors can detect components present in low level in the material, which cannot be detected by the common analytical tests.

The major applications of nanosensors in the food industry are for the detection and measurement of chemical components of raw and processed food materials, detection of food adulterants, detection and quantification of foodborne pathogens, identification and quantification of chemical residues like pesticides in food and beverages, and determination of quality of packed food from packaging line to the consumer's kitchen table. The quality of the packed food depends on the integrity of the package, which controls the permeation of major parameters (oxygen [O_2], carbon dioxide [CO_2], and moisture) affecting the rate of spoilage of food. Nanosensors integrated with food packaging materials and packages measure the O_2, CO_2, or moisture level from the manufacturing unit to the consumer's hand and help to identify the quality of the food with the naked eye (Akbari et al. 2006; Brody et al. 2008; De Azeredo 2009; Duncan 2011; Mills 2005).

Pesticides, herbicides, and insecticides are widely used in the agricultural and horticultural sectors to control pests in the field or in the storage. Residues of these pesticides end up down the stream in the food products through these agricultural products or water used for food manufacturing. Sensors developed using nanotechnology have the potential of detecting these pesticide residues quickly and with high accuracy well below the maximum allowable limits set by regulatory agencies (Gobi et al. 2005, 2007; Jiang et al. 2009; Liang et al. 2013; Navrátilová and Skládal 2004). Nanosensors can also be used to detect food adulterants (like melamine in milk, Sudan I dye in paprika powder or ketchup) present in the food matrix in very low levels. The major application of nanosensors in the food industry is in the detection of foodborne pathogens. Sensors developed by nanotechnology with the combination of florescence, spectroscopy, immune assay, or imaging (like spatial resonance) can be used to detect and also quantify the common foodborne pathogens presence in raw and processed food products. Most of these nanosensors and sensing techniques can detect foodborne pathogens quickly, and some of these sensors have the ability to detect pathogens in early stages (Akbari et al. 2015; Alfadul and Elneshwy 2010; Duncan 2011; Maurer et al. 2012; Yamada et al. 2014, 2016; Yang et al. 2015; Zhao et al. 2014). The application of nanosensors for rapid detection of foodborne pathogens at early stages and quantification saves billions of dollars every year to the food industry (Ramachandraiah et al. 2015).

Types of Nanosensors

Nanosensors can be classified based on the material used for fabrication, sensing, or detection technique used, and type of application. Based on the sensing/detection method used or coupled with the sensors, nanosensors can be broadly classified into following three categories (Augustin and Sanguansri 2009; Bogue 2006; Duran and Marcato 2013; Gomes et al. 2015; Hu et al. 2014; Ramachandraiah et al. 2015):

1. Optical nanosensors: These nanosensors use the optical signal developed by the interaction between a functionalized nanomaterial and a composition or adulterant or a microorganism in a food product for detection of a component or pathogen or adulterant.
2. Electrochemical nanosensors: In these nanosensors, a conductive nanomaterial (like TiO_2 nanobundle, carbon nanotube) is allowed to bind with selective antibodies, and then the conductivity of the nanomaterial is measured to identify food pathogens and components (protein, antioxidants, and flavor) in a food product.
3. Nanoparticle-based nanosensors: In these types of nanosensors, metal (nano-Cu, nano-Ag) or metal-oxide (nano-ZnO, nano-TiO_2)-based nanoparticles are used in the sensors to detect components, gases, microorganisms, and moisture of food product.

The nanoparticle sensors can be further divided into three categories: magnetic nanoparticle (MNP) sensors, nanoclay particle sensors, and metal (silver or gold) nanoparticle sensors. Most of the optical nanosensors use florescence, spectroscopy, magnetic resonance imaging, or scanning electroscopic microscopy imaging methods for detection. Some of the nanosensors use the colorimetric technique for detection and quantification of target materials (e.g., gas [O_2/CO_2] or moisture sensors used in food packaging). Electrochemical nanosensors use electrical conductivity, resistance, or impedance value generated at the sensor due to antigen–antibody reaction to detect and measure the target material (pathogen, chemical component/residue) in the food matrix.

Applications of Nanosensors in the Food Industry

Food Packaging

Applications of nanotechnology in food packaging are becoming popular, and in recent times, most of the food producers and processors are moving towards "active packaging" of food products. The food and beverages industry around the globe used about $6.5 billion worth of nanotechnology-enabled packaging materials in 2013, and it is estimated to increase to $15.0 billion in the year of 2020 (Ramachandraiah et al. 2015). Nanosensors are the major component of this active packaging concept, and these sensors measure or monitor the quality of food from the processing plant to the consumer's table. Nanosensors, embedded with packaging films, measure the gases developed from spoilage of food, and alarm the consumers by a change in color of the sensor. Some of the nanosensors detect the microbial activity and foodborne pathogens (like *Salmonella*, *E. coli*), and some sensors detect pesticide strains on the fruits and vegetables.

A nanoparticle-based nanosensor (technically an O_2 indicator film) was developed (Imran et al. 2016; Mills and Hafazy 2009; Mills 2005; Mills et al. 2011) with TiO_2 nanoparticles and methyl bromide indicator ink. One sensor was placed inside the packaging film, another sensor was placed on top of the packaging film, and then the package was flushed with CO_2 and sealed (Figure 4.1). After sealing, both the sensors were blue in color, and then the package was irradiated with UV light for 2 min. After UV irradiation, both the films were colorless, and after a few minutes, the sensor at the outside of the film turned into its original color (blue) because of the reaction with the oxygen in the ambient air. The sensor inside the film stayed colorless due to the absence of oxygen, and once the pack was opened, the inside sensor also changed to a blue color within a few minutes. Even for small quantities of the oxygen present inside the package due to leakage and perforation, the inside sensor turned to blue color. This sensor is helpful to consumers as well as food processors to visually check the integrity of modified atmospheric packaging of food products.

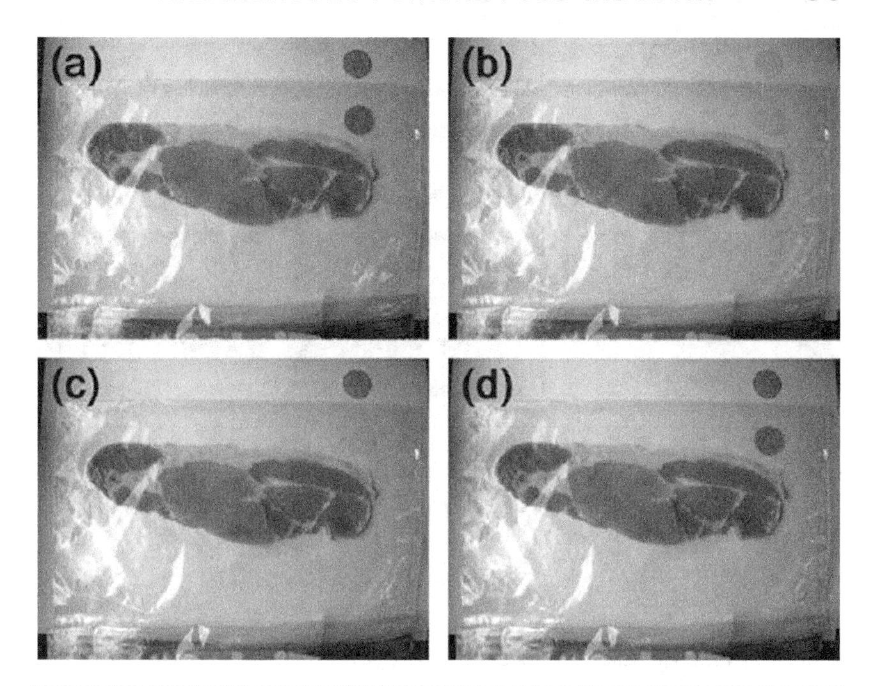

Figure 4.1 **(See color insert.)** Oxygen nanosensors for detection of O_2 levels inside a packed meat (one sensor at each side [inside and outside] of the packaging film) (a) right after sealing, (b) immediately after activation of sensors with UVA light, (c) few minutes after activation of sensors, (d) after opening the package. (Reproduced from Mills, A., *Chem. Soc. Rev.*, 34, 1003–1011, 2005.)

Moisture is another main parameter, which can be used to detect the freshness or spoilage of the food product. Nanosensor films have been developed using carbon-coated copper nanoparticles (Luechinger et al. 2007), which can be used with packing films to monitor moisture content changes inside the food packages using the color changes of the film. Metallic copper films were made by depositing carbon-coated copper nanoparticles (25–50 nm size) on a glass substrate. This metallic film was the color of copper, and exposure of this film to steam of ethanol vapor resulted in rapid and opalescent coloration of the film. When it was exposed to water vapor at room temperature, the film swelled, and it resulted in an intense green coloration.

When the film was kept in the water vapor at room temperature for three, six, and nine seconds, the color changed mostly to pink, blue, and orange, respectively (Figure 4.2). These visible color

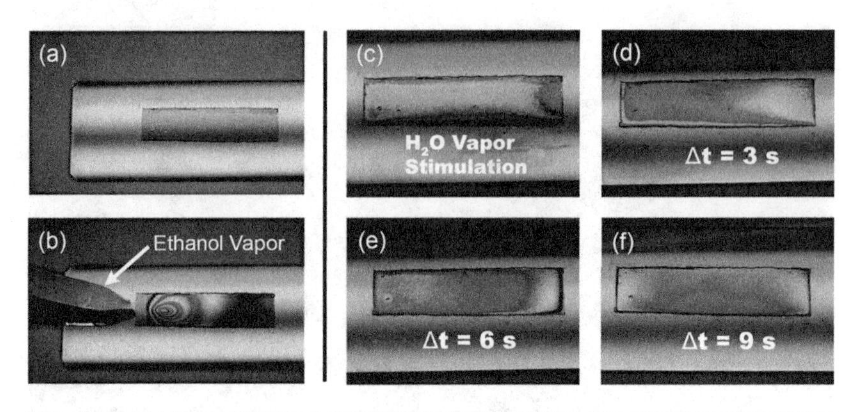

Figure 4.2 (See color insert.) Nanosensor for moisture measurement (a) moisture sensor with carbon-coated copper nanoparticles at the start; (b) when exposed to ethanol vapor; (c) when exposed to water vapor; (d–f) moisture dry out by time (3, 6, and 9 s). (Reproduced from Luechinger, N.A. et al., *Langmuir*, 23, 3473–3477, 2007.)

changes can be used to monitor the moisture levels of a packed food product, which can be useful to determine the freshness of the food without damaging the integrity of packaging films. They tested this sensor at different relative humidity (RH) levels (50%–80%) and observed significant changes in visible light even for 1% change in RH.

Detection of Chemical Components of Food and Food Adulterants

Antioxidants are the substances that inhibit the oxidation or remove the chemicals and agents causing oxidation in living organisms. Intake of fresh fruits and vegetables, which have antioxidants, helps to maintain the antioxidant level in a human body. The lower antioxidant level can cause multiple health issues, including cardiovascular diseases and cancers. Detection of antioxidant levels in food is gaining more attention nowadays due to their significant link to the health. Hu et al. (2014) developed a spectroscopic sensing method with gold nanoclusters (Au-NC) to analytically determine the antioxidants in fruit juices. The Au-NCs were prepared by mixing 2.5 mL of aqueous chloroauric acid ($HAuCl_4$) with 2.5 mL glutathione (GSH) solution (8 nM) at room temperature and stirred vigorously for 5 min, and kept at 70°C for 24 h for reaction. Then the solution was filtered with dialysis bags with 1000 Da molecular weight cut-off (MWCO) membrane. The commercial mango, apple, and orange juices were diluted to 250-, 300-, and 1000-fold, respectively with 10 mM phosphate-buffered saline (PBS) buffer solution, and 2.8 mL of diluted juices were mixed with 100 μL Au-NC solution. Then 50 μL Fe^{2+} (3 mM), H_2O_2 (3 mM) were added to the solution and incubated at room temperature. After a 5 min incubation period, the solution was scanned using a fluorescence spectrophotometer at 375 nm, and the observed florescence spectra were used for measuring antioxidant levels in commercial juices. This spectroscopic nanosensing technique can also be used to collect fluorescence microscopy images of antioxidants in food products. Hu et al. (2014) proved that antioxidants in fresh fruit juices can be easily detected by imaging in short time using the developed sensors compared with the regular spectroscopic imaging techniques, and also these sensors had good biocompatibility.

Jamali et al. (2014) synthesized nanosensors with Pt:Co nanoalloy at a surface of an ionic liquid (n-hexyl-3-methylimidazolium-hexafluoro phosphate) to determine vitamin B9 in food products. A nanoalloy carbon paste electrode (Pt:Co/CPE) was fabricated by mixing graphite powder (0.85 g), and Pt:Co (0.15 g) with paraffin at 70:30 (w/w) ratio for 40 min and packing this mixed paste into a glass tube. Then copper wires were inserted into the tube to make the electrical connection. The electrode with ionic liquid (Pt:Co/IL/CPE) was fabricated by mixing 0.3 g of n-hexyl-3-methylimidazolium hexafluoro phosphate with 0.15 g Pt:Co, 0.85 g graphite powder, and 0.70 g liquid paraffin for 40 min, and packed into the glass tube. The voltammetric response of this sensor to vitamin B9 in a test product was used to calibrate the sensor. The developed sensor was demonstrated for determining the vitamin B9 content in real products like mint vegetable, vitamin B9 tablet, apple juice, and orange juice.

Gallic acid (GA) is one of the major phenolic compounds available in fruits (blueberries, grapes, banana, cantaloupes, and many other fruits), and GA has strong antioxidant properties. GA also has anti-mutagenic and anti-carcinogenic properties. The food industry extensively uses GA as main additive during processing. Traditionally, chromatographic (GC and GCMS) and spectrophotometric techniques were used for quantification of GA, and now electrochemical sensors are the most commonly used technique for GA quantification by the analytical chemists around the globe. The major concern with the electrochemical sensor during GA quantification is direct oxidation of GA on the electrodes of the sensor, which affects the sensitivity and accuracy of the electrochemical sensor. Ghaani et al. (2016) developed a modified electrochemical sensor using glassy carbon electrode (GCE) with AgNP for GA quantification. Delphinidin coating was added on the GCE/AgNP electrode for selective oxidation of GA. First, the GCE electrode was polished with alumina abrasive slurry followed by rinsing with double-distilled water. Then the GCE was introduced to a continuous potential cycling in an $AgNO_3$ (1 mmol/L) nitric acid (100 mmol/L) solution from 0.7 to 1.9 V with the sweep rate of 80 mV/s for 8 cycles. In the second step of sensor fabrication, delphinidin coating was added by rinsing the GCE/AgNP electrode with double-distilled water, followed by continuous potential cycling at a sweep rate of 20 mV/s from 100 to 400 mV for 8 cycles in delphinidin (1.0 mmol/L) solution in a phosphate buffer solution (0.1 mmol/L, 7.0 pH). Ghaani et al. (2016) tested this GCE/AgNP/Delph sensor to measure GA levels in 5 different commercial juices (apple, peach, lemon, orange, and green tea) by measuring the GA oxidation current and found the final recovery rate >99.0%, which demonstrated the feasibility of using this sensor for GA quantification in real food samples.

Detection of adulterant materials in food products is important to ensure the quality of the food as well as ensuring consumers' health in case the adulterant is an allergen. Nanosensors have been developed to detect food adulterant even if present in very small quantities in food products. Melamine is the most commonly used adulterant material in infant formulas and pet foods in order to artificially inflate protein content. Cyanuric acid with gold nanoparticles binds with this melamine and produces a reproducible color change, which depends on the concentration of the melamine content (blue to red) (Ai et al. 2009). This color change is easily noticeable by the naked eye (Figure 4.3), and the nanosensors based on this concept can be used to measure the melamine content in dairy products and pet foods even in small quantities (up to 2.5 ppb). Ping et al. (2012) also developed a probe using label-free AgNPs to detect the melamine concentration in raw milk. Melamine binds with the AgNP solution and produces concentration dependant colorimetric changes, which can be noticed by the naked eye or monitored by a Vis-NIR spectrophotometer (Figure 4.4). Melamine concentration of 2.32 mM can be detected by this nanoparticle-based probe.

Most of the fruit and fruit juices naturally contain ascorbic acid (AA), which is known as vitamin C, and processors also add AA to fruit juices and soft drinks during manufacturing to increase vitamin C content. Sudan I is a toxic synthetic azo dye used as

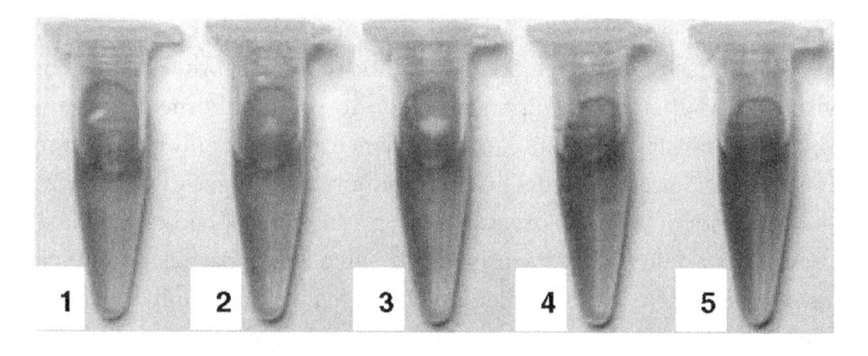

Figure 4.3 (See color insert.) Visual color change of the optimized sensor: (1) without any addition of melamine; (2) with the addition of the extract from blank raw milk; (3) with the addition of the extract containing 1 ppm (final concentration: 8 ppb); (4) with the addition of the extract containing 2.5 ppm (final concentration: 20 ppb); (5) with the addition of the extract containing 5 ppm (final concentration: 40 ppb). (Reproduced from Ai, K. et al., *J. Am. Chem. Soc.*, 131, 9496–9497, 2009.)

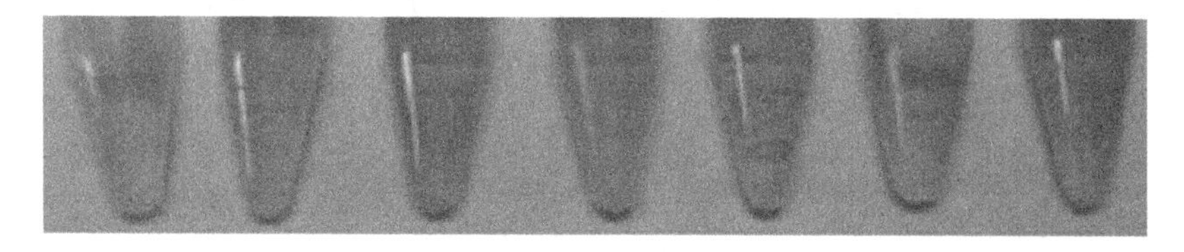

Figure 4.4 (See color insert.) Test tubes contain 800 mL Ag NPs solution mixed with 600 mL melamine solution. The concentrations of spiked melamine (from left to right) are 0, 0.002, 0.004, 0.008, 0.017, 0.085, 0.17 mM, respectively. (Reproduced from Ping, H. et al., *Food Control*, 23, 191–197, 2012.)

an adulterant in some food products (paprika, sausage, pie, and ketchup) in order to increase color and stability of the food products, even though it is strictly prohibited from use in food because it has carcinogenic and mutagenic effects on humans. Karimi-Maleh et al. (2014) developed a highly sensitive voltammetric nanosensor with modified nickel oxide nanoparticles (NiO/NPs) (9,10-dihydro-9,10-ethanoanthracene-11,12-dicarboximido)-4-ethylbenzene-1,2-diol (DEDED) carbon-paste electrode (NiO/NPs/DEDED/CPE) for simultaneous determination of AA and Sudan I in food samples. This nanosensor had detection limits of 0.006 and 0.2 μM for AA and Sudan I, respectively.

Detection of Food Pathogens

Since food is a complex matrix, detection of the target pathogen is a difficult task because food matrix also contains other pathogens, viruses, and chemical components. Due to the foodborne pathogens, the food industry loses millions of dollars every year, and nanosensors for rapid and accurate detection of foodborne pathogens will improve food industries' productivity as well as profitability. Zhao et al. (2004) developed a fluorescence sensor using silicon nanoparticles to detect *Escherichia coli* from food materials. The *E. coli* was immobilized against the monoclonal antibodies (mAbs) onto the surface of silica nanoparticles (RuBpy-doped silica). The reaction between antigen and antibody creates an emission of light, which can be captured using a florescence spectrometer (Figure 4.5). The probe contains silica nanoparticles with florescence dye molecules bonded with the mAbs and amplifies the florescence developed by the antigen–antibody reaction. Using this sensor, *E. coli* contamination in food can be detected within 20 min.

A single-walled carbon nanotube (SWCNT) biosensor was used to detect the *Salmonella* presence in the food materials (Jain et al. 2012). Due to the antibody–antigen reaction, a self-assembled monolayer of antibody–antigen complex was formed on the electrode of SWCNT, and the bacteria cell wall's insulation properties increased the electrical impedance. *Salmonella* presence can be detected by measuring these changes in electrical impedance before and after the antigen–antibody reaction. This sensor can be used to detect *Salmonella* concentrations as low as 1.6×10^4 cfu/mL.

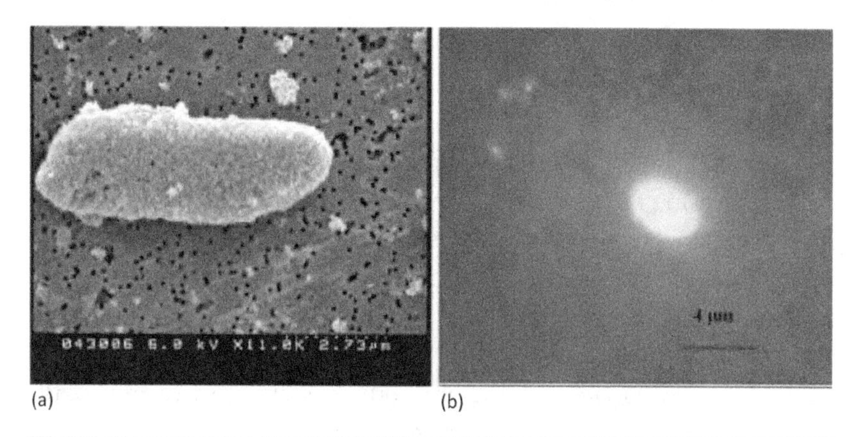

(a) (b)

Figure 4.5 (a) Scanning electron microscopy image of an *E. coli* bacterium; (b) the fluorescence image of *E. coli* cell following incubation with antibodies with silica nanoparticles. (Reproduced from Zhao, X. et al., *Proc. Natl. Acad. Sci. USA*, 101, 15027–15032, 2004.)

A multi-step sensor was developed using carbon nanotubes (CNT) with gold nanoparticles for rapid detection of *E. coli* (Maurer et al. 2012). They coated gold nanoparticles with thoilated RNA and found *E. coli* binding enhanced 189% with the RNA-coated gold nanoparticles compared to gold nanoparticles alone in the CNT. A new technique to detect *E. coli* from complex food mixture, called "reflective interferometry," was developed by Horner et al. (2006). It uses silicon nanoparticles for monitoring light scattering by the *E. coli* cells. This technique works on the concept that when *E. coli* bacteria in food mixture binds with the protein of known bacterium present in the silicon chip it produces a nanoscale light scattering effect. By capturing and analyzing this light scattering effect using imaging techniques, *E. coli* presence in the food can be detected.

Yamada et al. (2016) developed a multi-junction (2 × 2 junction array) SWCNT biosensor for rapid detection of *E. coli* and *Staphylococcus aureus* bacterium. Streptavidin and biotinylated antibodies (5 μL each) were applied on the SWCNT junctions and dried for 5 min for pathogen detection. This sensor was connected to a picoammeter to measure electric current in the circuit before and after the sensor was in contact with the pathogen strains. They tested pure and mixed *E. coli* and *S. aureus* samples and found that this sensor detected *E. coli* and *S. aureus* in small concentrations (10^2 CFU/mL). They also found that the electric current had negative correlation with the pathogen concentration in the testing media (with R^2 values more than 0.978, and 0.998 for *S. aureus* and *E. coli*, respectively).

Fluit et al. (1993) used an immunomagnetic separation (IMS) technique using nanomagnetic particles to separate the target bacteria from the complex food samples and quantifying them using standard techniques (PCR analysis, fluorescence). Yang et al. (2007) developed a nanosensor with magnetic iron oxide nanoparticles coupled with PCR to separate and quantify *Listeriosis monocytogenes* presence in the milk. *L. monocytogenes* specific antibody was attached to iron oxide nanoparticles to separate *L. monocytogenes* in the milk and then quantified by PCR analysis.

Wang et al. (2008) developed an immunosensor with titanium dioxide (TiO_2) nanowire bundle to detect *L. monocytogenes* using electrical impedance in the circuit. A TiO_2 nanowire bundle was placed between gold electrodes, and the electrodes were coated with 2-methyl-2-propanethiol ($SH–nCH_2)_3–CH_3$) in order to avoid cross binding with antibodies. Then monoclonal *L. monocytogenes* antibodies were placed on TiO_2 nanowire bundle, which captured the *L. monocytogenes* from the samples (Figure 4.6A). Due to the antibody–antigen reaction, increase in electrical impedance was noticed in the circuit, by which *L. monocytogenes* presence in the sample was detected (Figure 4.6B). Using this sensor, *L. monocytogenes* can be detected in 50 min, at a concentration level of 4.7×10^2 cfu/mL. Wang et al. (2008) also found there was no interference from other foodborne pathogens (*E. coli O157:H7, S. aureus,* and *S. typhimurium*) on *L. monocytogenes* detection using this immunosensor.

Su and Li (2004) developed a nanosensor with CdSe–ZnS nanocrystals quantum dots (QDs) coupled with IMS for detection of *E. coli*, which was separated from other foodborne pathogens and components using *E. coli* antibody-coated magnetic beads by IMS technique. The QDs acted as fluorescent markers for quantification of *E. coli* concentration using a charge-coupled device (CCD) array spectrometer. The samples with *E. coli* had a peak emission at 609 nm, and the detection limit and time of this sensor were $10^3–10^7$ CFU/mL, and 2 h, respectively. Standard bacterial plating techniques take 18–25 h for incubation and detection of foodborne pathogens. El-Boubbou et al. (2007) also developed a sensor based on IMS technique with

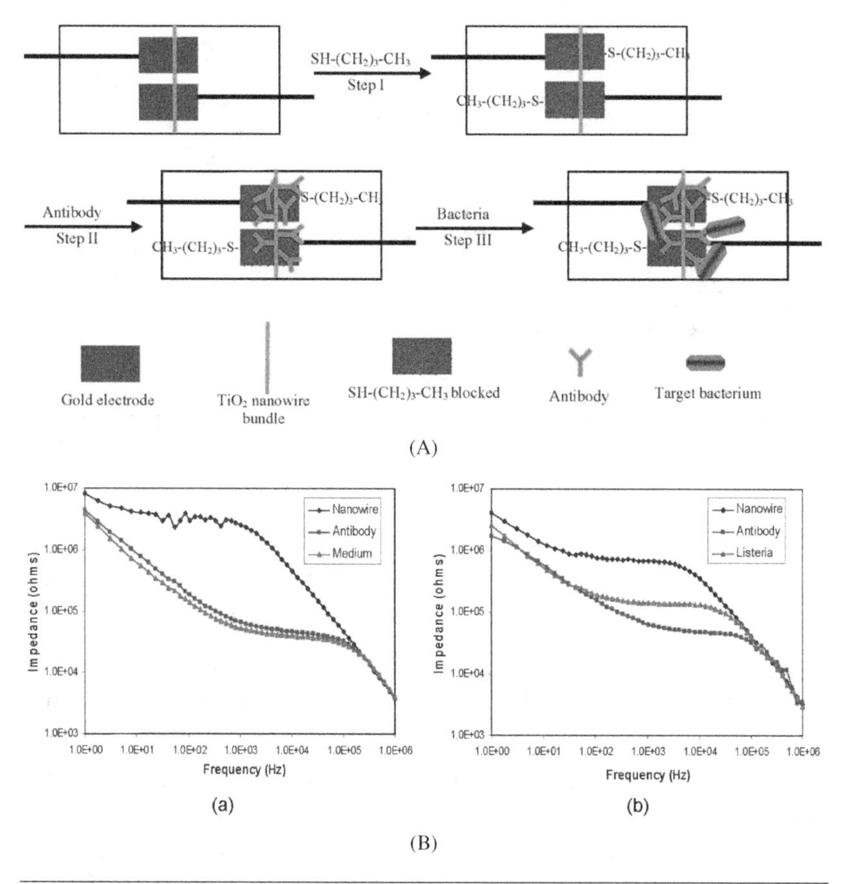

Figure 4.6 (A) Procedure of *L. monocytogenes* detection using TiO_2 nanowire bundle impedance immunosensor; (B) impedance spectra from TiO_2 nanowire bundle impedance immunosensor for (a) growth medium without *L. monocytogenes* and (b) growth medium with *L. monocytogenes* at a concentration of 4.65×10^3 CFU/mL. (Reproduced from Wang, R. et al., *Nano Lett.*, 8, 2625–2631, 2008.)

sugar-coated nanomagnetic (Fe_3O_4) particles (magnetic glyconanoparticle [MGNP]) coupled with a fluorescence microscope to separate and detect *E. coli* from the medium in 5 min. They demonstrated that this nondestructive method was not only able to detect *E. coli* but also can remove 88% of *E. coli* from the medium quickly (Figure 4.7). Ravindranath et al. (2009) developed a sensor with nanomagnetic (Fe_3O_4) particles activated using anti-*E. coli* and anti-*Salmonella typhimurium* antibodies coupled with benchtop Fourier transform infrared (FT-IR) spectrometer and mid-IR spectrometer to separate and detect *E. coli* and *S. typhimurium* contamination in 2%

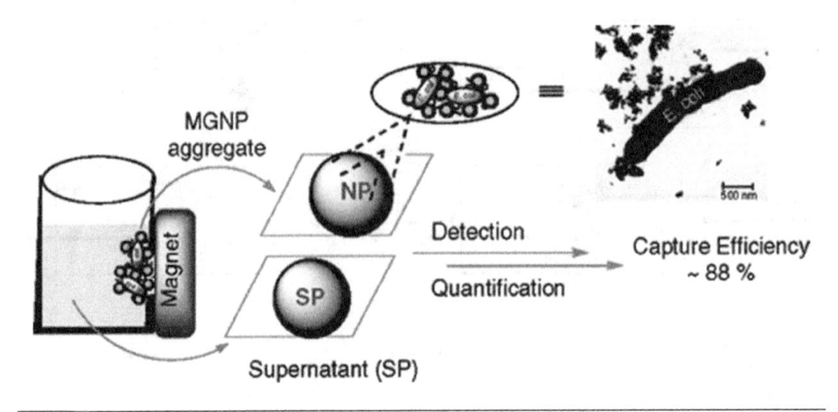

Figure 4.7 Detection of *E. coli* using MGNP. (Reproduced from El-Boubbou, K. et al., *J. Am. Chem. Soc.*, 129, 13392–13393, 2007.)

milk and spinach extract samples in 30 min with a detection limit of 10^4–10^5 CFU/mL. Wang and Irudayaraj (2008) developed a nano-surface plasmon resonance (nano-SPR) probe with gold nanorods (AuNRs) functionalized with *E. coli* and *S. typhimurium* antibodies to detect simultaneously *E. coli* and *S. typhimurium* at low concentrations ($<10^{-2}$ CFU/mL) in food complex in a short time (<30 min). Wang and Irudayaraj (2008) also developed magnetic-optical nanoprobes (Fe_3O_4-Au_{rod} necklace-like probes) for simultaneous detection of *E. coli* and *S. typhimurium* from a single sample at a concentration range of 10^2 CFU/mL.

Pal et al. (2007) developed a direct-charge transfer (DCT) biosensor using polyaniline nanowires to detect *Bacillus cereus* presence in the food samples. This sensor used the resistance signal developed during antibody–antigen reaction to identify *B. cereus*. The developed biosensor had four sections (application, conjugate, capture, and adsorption) with one membrane pad in each section. The sample was introduced in application section and flowed to the conjugate pad where it bound with primary antibodies, and then this mixture moved to the capture section, where it bound with secondary antibodies and created a sandwich complex. An electric circuit was formed between the silver electrodes at the capture section by the polyaniline nanowires bound with this mixture and created an electrical resistance signal due to the antibody–antigen reaction, which was measured by an ammeter and used for detection of *B. cereus* contamination. This sensor can detect *B. cereus* presence in the food sample at a concentration range of 10^1 to 10^2 CFU/mL in 6 min.

Nanoparticle-based test strip kits are commercially available for rapid detection of *E. coli* (RapidChek®-E. coli O157, Iquum®, MaxSignal®-E. coli O157:H47, Watersafe®), *S. typhimurium* (MaxSignal®-Salmonella, RapidChek® SELECT), and *L. monocytogenes* (RapidChek® Listeria, and Listeria NextDay) from food samples (Cushen et al. 2012; Duran and Marcato 2013; Kaittanis et al. 2010). Most of these sensors are based on the ELISA method and use the antibody–antigen reaction for detection of foodborne pathogens. For example, the RapidChek-E. coli O157 test kit contains *E. coli* specific antibodies (immobilized) in test line, and gold nanoparticles and secondary antibodies in the upstream of the test line. When the food sample containing *E. coli* passes through test line, the *E. coli* binds with the antibody and moves upstream by capillary action creating an antibody–antigen sandwich when it binds with the secondary antibody–gold mixture. This antibody–antigen reaction develops a red color line in the test line, by which the presence of *E. coli* in the food samples is identified. Based on the sample size, RapidChek-E. coli O157 takes 8–12 h for detection of *E. coli* from food samples, and 50 samples can be tested with one test kit. The detection level of this nanoparticle-based test kit is 10^5 CFU/mL. The Maxsignal O157:H47 test kit also uses the same testing principle and with one test kit 96 samples can be tested with a detection limit of 10^5CFU/mL and incubation time of 75 min.

Mycotoxin Detection

Mycotoxins are the toxic substances produced by secondary metabolism of fungal species (mainly *Aspergillus* and *Penicillium*) in grain and food samples. Some of the mycotoxins can cause serious health problems because these are mutagenic (can cause genetic problems), carcinogenic (can cause cancer), and teratogenic (can cause developmental problems) in nature. Aflatoxin and ochratoxin A are the most common mycotoxins produced by fungi in food and grain products. Rapid and early detection of mycotoxin will help to reduce potential health and recall issues for consumers and food producers, respectively. Immunoassay methods (ELISA test) and HPLC are the most commonly used techniques to detect aflatoxin and ochratoxin in food. Scientists across the world are developing nanotechnology-based sensors for accurate detection of mycotoxins from food samples quickly.

Hosseini et al. (2015) developed an aptamer (DNA, RNA, or peptides bind with selective proteins and peptides) based nanosensor with AuNPs to detect aflatoxin B1. They mixed an AuNP solution with a salt medium (NaCl), and an aflatoxin-specific aptamer (oligonucleotide(5-GTTGGGCACGTGTTGTCTCTCTGT GTCTCGTGCCCTTCGCTAGGCCCACA-3)) was added in to the solution. This aptamer stopped the aggregation of AuNP induced by the salt mixture, and when the aflatoxin-containing sample was introduced with this mixture, aptamer bonded with aflatoxin and salt-induced aggregation of AuNP triggered again. This reaction changed the color of AuNPs from red to purple. With this color change, the presence of aflatoxin B1 (AFB1) can be identified from the food sample. They successfully tested this sensor with rice and peanut samples with AFB1 concentrations of 80–200 nM (linear range) and found the detection limit of this colorimetric aptamer-based nanosensor was 7 nM.

Luan et al. (2015) developed a similar aptamer-based nanosensor with AuNPs to detect aflatoxin B2 (AFB2) in food and beverages. When the AFB2 aptamer (ssDNA (5′-GTTGGGCA CGTGTTGTCTCTCTGTGTCTCGTGCCCTTCGCTA GGCCCACA-3′) was mixed with AuNPs, the aptamer was absorbed by AuNPs due to the electrostatic interaction between ssDNA and AuNPs and kept AuNPs in red color in the presence of NaCl. Whenever the AFB2-containing sample was introduced into the mixture, aptamer bonded with AFB2 and freed up AuNPs, which induced the AuNP-NaCl aggregation and changed the color of AuNPs from red to blue-purple. A beer sample containing AFB2 was tested with this sensor, and AFB2 was detected in 15 min as low as 0.4 ng/mL.

Xiao et al. (2015) developed aptamer-based nanosensors with AuNPs to detect ochratoxin A from food samples. The OTA-specific aptamer acted as a linker to form an AuNP dimer between probe 1 and 2 in the absence of OTA. When the OTA-containing sample was introduced, OTA aptamers from probe 1 and 2 competed to bind with

OTA and disassembled the AuNP dimers. This reaction created a colorimetric change of the solution from blue to red. They successfully tested this sensor to detect OTA contamination in wine and found the detection limit was 0.05 nM. Soh et al. (2015) also developed a similar kind of aptamer-based colorimetric nanosensor with AuNPs to detect OTA, and in this system, the morphology and color of AuNPs were controlled by aptamer–OTA interactions. A low amount of aptamer coverage on the surface of AuNPs produced spherical-shaped NPs, which created a red color solution, and high surface coverage of aptamers on AuNPs produced branched NPs and blue color solutions. When they tested this method with red wine containing OTA, the solution color changed from blue to red, and they found detection limit of this sensor was 1.3 nM.

Wang et al. (2014) developed a sensing method to detect AFB1 using the ELISA technique with magnetic beads and AuNPs. Magnetic beads (Fe_3O_4) were modified as capture probes when AFB1-bovine serum albumin conjugates were applied on them and bound with anti-AFB1 antibodies labeled AuNPs. When AFB1 was introduced into the solution, the bond was broken and the supernatant solution was collected after magnetic separation. The absorption intensity of this solution was tested using UV-Vis spectroscopy, and it was found that the AFB concentration was directly proportional to absorption intensity. This sensor had a detection range of 20–800 ng/L, and the detection limit of 12 ng/L. Urusov et al. (2014) developed an immunoassay (ELISA) method combined with MNPs to detect AFB1 in barley and corn, and found this new technique can detect AFB1 in food solutions in 20 min of assay time and a detection limit of 20 ng/L. Hayat et al. (2015) developed a fluorescent-based aptasensor with MNPs to detect OTA. In this sensor, the carboxy-modified fluorescent particles functioned as signal generating probe, and the streptavidin MNPs functioned as solid separation support. The carboxy-modified fluorescent particles detached from the magnetic beads when OTA entered into the binding medium resulting in the emission of a fluorescent signal. By analyzing this fluorescent signal, OTA level in the samples was identified. Beer samples with OTA were tested with this sensor, and the OTA level was detected as low as 0.4 nm/L.

Pesticide Residue Detection

Application of pesticides to control insects and pests in the preharvest and postharvest agriculture for grains, oil seeds, vegetables, and fruits has been practiced for a long time. Pesticide application helps to reduce quality and quantity losses of agricultural products, but extensive use of pesticides poses serious hazards to humans and the environment (Jiang et al. 2009). Residues from these pesticides are ending up in the food products manufactured from these agricultural products and causing serious health issues to the consumers. Organophosphate (OP) is the most common toxic substance in the insecticides, herbicides, and pesticides used on agricultural products, and it has a neurotoxic effect on humans (Liang et al. 2013). A survey done by Stout et al. (2009) estimated around 0.5 million tonnes of OPs are released every year through pesticide and insecticide application into the environment and food streams. The common techniques used to detect OPs are LC/MS, GC, GC/MS, enzyme inhibition techniques, ELISA, and other immune assay techniques (Blesa et al. 2004; Sporty et al. 2010; Szinicz et al. 2007). Most of these techniques are expensive and not suitable for rapid infield analysis. Liang et al. (2013) developed a sensing technique with Fe_3O_4 MNPs for rapid detection of OPs. The peroxidase mimetic-based colorimetric detection assay was prepared with 50 μg Fe_3O_4 MNPs, 10 μM acetylcholinesterase (AChE), 5 μM acetylcholine chloride (Ach), and 1.2 mg/mL choline oxidase (CHO) in a 500 μL acetate buffer. A colorimetric substrate *(3,3,5,5-tetramethylbenzidine (TMB) was used along with this detection assay and incubated at room temperature (22°C) for 15 min in dark. The enzymes present in the assay, AChE and CHO, increased the rate of formation of H_2O_2 due to acetylcholine's reaction with water and oxygen during the incubation period. The MNPs in the solution got activated by the H_2O_2 and induced the oxidation of the TMB, which resulted in a color change in the solution (Figure 4.8). When OP compounds were mixed with the solution, the H_2O_2 production was less due to the inhibition of AChE's enzymatic activity. The lower H_2O_2 production resulted in a substantial reduction in oxidation of TMB, which caused a reduction in color intensity of the solution (Liang et al. 2013). This color intensity difference was noticed by the human eye and easily measured

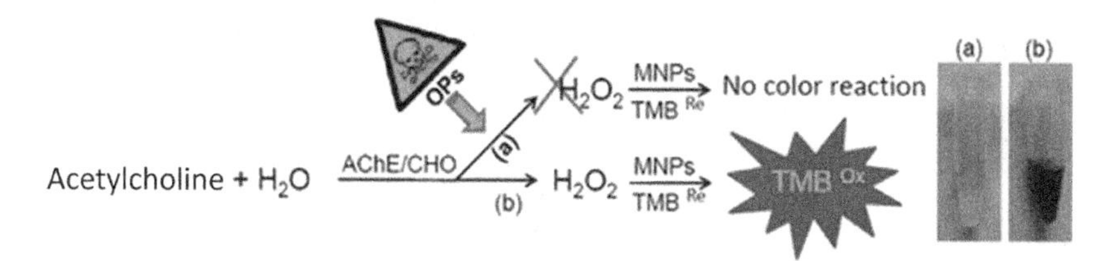

Figure 4.8 Basic principle of MNP peroxidase mimetic detection assay for detection of OP compounds (a) detection assay with OP compound after incubation, (b) control detection assay after incubation. (Reproduced from Liang, M. et al., *Analyt. Chem.*, 85, 308–312, 2013.)

with a spectrophotometer. Using this difference in color intensity of the solution after incubation, the presence of OP compounds in food and beverages can be detected. Liang et al. (2013) tested this MNP peroxidase mimetic sensing technique to identify acephate and methyl-paraoxon pesticides, as well as nerve agent Sarin, and found out this new sensor was more rapid and sensitive than the common enzyme incubation and immune assay techniques. Various concentration of OP-containing pesticides (acephate and methyl-paraoxon) and Sarin were incubated with AChE enzyme for 15 min and mixed with the MNP peroxidase detection assay solution. The color intensity of the solution after the reaction was measured using a microplate spectrophotometer at 450 nm. The absorbance curves for all three OP-containing solutions (Figure 4.9) indicated that increase in OP pesticide concentration had a negative correlation with the color intensity of the solution. This MNP peroxidase mimetic sensor had detection limits of 5 μM, 10 nM, and 1 nM for acephate, methyl-paraoxon, and Sarin, respectively.

Luckham and Brennan (2010) developed a paper dipstick sensor with gold nanoparticles to detect paraoxon. This dipstick sensor was made with silica-based Sol-gel, AChE enzyme, and AuNP composites. A 2.9 g of sodium silicate was mixed with 10 mL of double-distilled water (ddH_2O), and then 5 g of Dowex cation exchange resin was added and stirred for 30 s followed by vacuum filtration to prepare sodium silicate sols. The prepared sol solution was mixed with AChE and 5″-triphosphate disodium salt-AuNP at 1:1 volume ratio at 20°C and vortex mixed for 10 s (Luckham and Brennan 2010). Then the mixture was cast on thin paper strips to form a dipstick sensor and cured at room temperature for 24 h. A 10 μL of paraoxon was placed

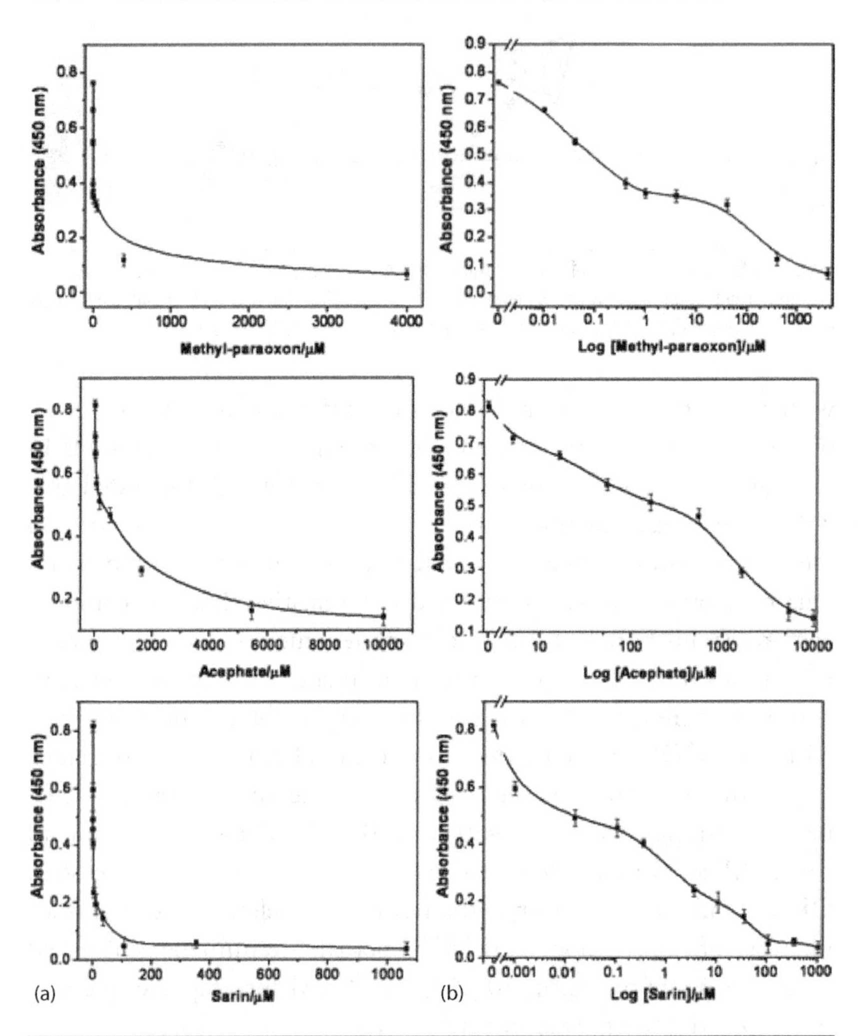

Figure 4.9 (a) Absorbance, and (b) semi-log curves of methyl-paraoxon, acepahate, and Sarin at 450 nm. (Reproduced from Liang, M. et al., *Analyt. Chem.*, 85, 308–312, 2013.)

on the dipstick sensor and incubated for 20 min at room temperature and then 20 μL of acetylthiocholine iodide (ATCh) + Au (III) salt solution was added. The color intensity of the dipstick paper was measured by analyzing the images taken by the digital camera (Canon Powershot A630) using ImageJ software. Luckham and Brennan (2010) also noticed a decrease in color intensity (Figure 4.10) with an increase in paraoxon concentration in the solution, which aligns with the results obtained by Liang et al. (2013). Luckham and Brennan

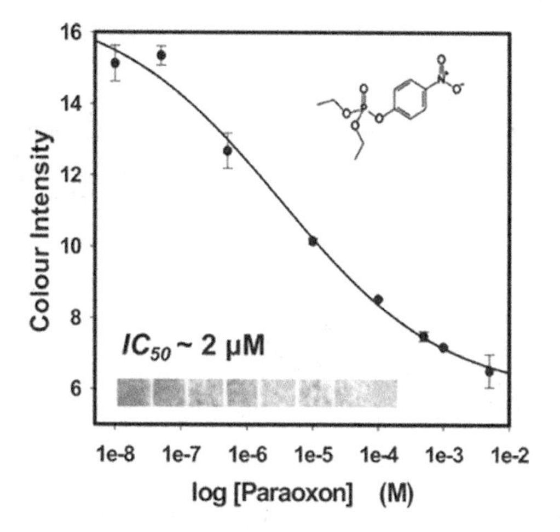

Figure 4.10 **(See color insert.)** Semi-log color intensity curve of dipstick paper sensor with gold nanoparticles tested against paraoxon. (Reproduced from Luckham, R.E., and Brennan, J.D., *Analyst*, 135, 2028–2035, 2010.)

(2010) found the detection limit of this dipstick paper sensor was 0.5 μM, which is well below the lethal dose (LD_{50}) of paraoxon (6.5 μM). Since the use of this dipstick-paper-based sensor is so easy, this sensor has great potential to apply for rapid and in-field detection of pesticide residues in food and beverages.

Navrátilová and Skládal (2004) developed an immunosensor based on electrical impedance measurement to detect 2,4-dichlorophenoxyacetic acid (2,4-D), which is one of the commonly used pesticides in agriculture. They tested three types of sensors (disc-shaped gold electrode, one integrated array electrode, and finger-shaped sensor with two electrodes), and monoclonal antibody (MAb) against 2,4-D was immobilized on the surface of all three sensors (Figure 4.11). The antibody-immobilized sensors were connected with an electrochemical impedance spectrometer (EIS) and immersed in a phosphate (50 mM, pH 7) + sodium chloride (145 mM) solution (PBS). The 2,4-D was slowly added to the solution, and the change in electrical impedance was measured using EIS. The antigen–antibody reaction created change in impedance in the electrical circuit with which the presence of 2,4-D in the solution was measured. Navrátilová and Skládal (2004) showed that the detection limit of this EIS-sensing technique was 45 nM/L.

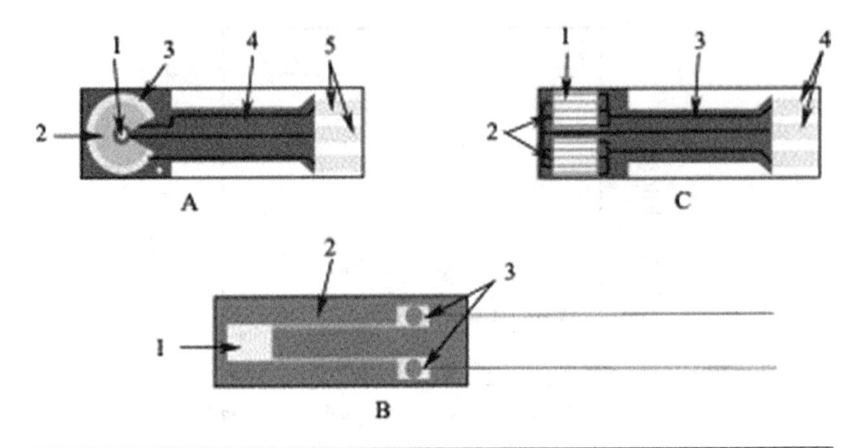

Figure 4.11 Schematic view of immunosensors for detection of 2,4-dichlorophenoxyacetic acid (2,4-D), (A) Strip sensor: 1–working electrode, 2–reference electrode, 3–auxiliary electrode, 4–alumina support, 5–contacts; (B) IDA electrode sensor: 1–Pt working interdigitated electrode, 2–connection (interdigitated array contacts), 3–contacts; (C) Finger strip sensor: 1–working electrode, 2–electrode sets, 3–alumina support, 4–contacts. (Reproduced from Navrátilová, I., and Skládal, P., *Bioelectrochemistry*, 62, 11–18, 2004.)

An immunosensor using a surface plasmon resonance (SPR) technique was developed by Gobi et al. (2007) to detect 2,4-D levels in solutions using SPR angles. A thin gold plate sensor was developed using microscopic cover glass plate (18 × 18 × 0.15 mm with a reflective index of 1.515). The glass plate was ultra-cleaned by dipping the plate in diluted H_2O_2 solution at 80°C for 10 min, and ammonia solution was added with the same temperature kept for another 10 min. Then the plate was washed with deionized distilled water and dried by clean compressed air. Then glass plate was coated with 5 nm-thick chromium and then gold was sputtered over at a thickness of 50 nm in a high-vacuum sputtering unit. The 2,4-D analyte was conjugated with bovine serum albumin (BSA) protein and then immobilized on the prepared gold plate sensor. When the monoclonal 2,4-D-BSA antibody (2,4-D Ab) was introduced onto the chip, it immobilized firmly with the 2,4-D-BSA. When a solution with free 2,4-D flowed over the sensor chip, 2,4-D-BSA conjugate and free 2,4-D competed to bond with 2,4-D Ab and reduced the binding ability of 2,4-D-BSA on the chip. This reaction created the change in SPR angle, which can be measured using an SPR analyzer, and with the changes in SPR angle concentration of 2,4-D in the solution can be measured

(Figure 4.12). Gobi et al. (2007) found that this sensor measured 2,4-D in the solution from 0.5 ng/mL to 1 μg/mL within 20 min. This sensor can be used for rapid detection of 2,4-D using simple a SPR analyzer, and another major advantage is the reusability of this sensor. The sensor chip can be regenerated when washing with pepsin solution for 2 min after every use.

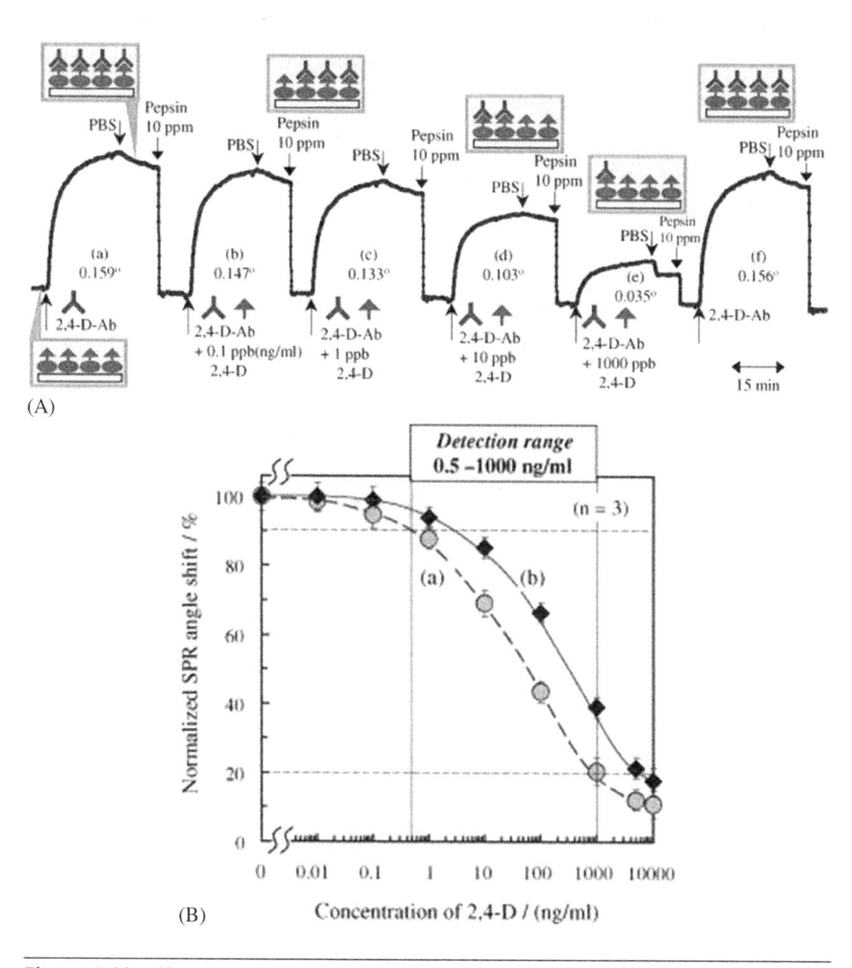

Figure 4.12 **(See color insert.)** (A) Surface plasmon resonance (SPR) sensorgram of flow 0.1, 1, 10 and 1000 ng/mL 2,4-D and fixed concentration (20 ng/mL) and 2,4-D-antibody solutions over the immunosensor chip immobilized with of 2,4-D-BSA; (B) normalized SPR angle shift observed through the immunosensor with various concentrations of 2,4-D. (Reproduced from Gobi K.V. et al., *Sensor. Actuat. B-Chem.*, 111–112:562–571, 2005.)

Alvarez et al. (2003) developed a nanomechanical sensor for detecting dichlorodiphenyltrichloroethane (DDT), the main compound of insecticides. This sensor consists of the micro-cantilever beam (width of 200 μm × 40 μm × 600 nm) coated with chromium (5 nm thickness) and gold (25 nm thickness) on one side. The DDT-BSA conjugate was immobilized on the gold-coated side of the sensor, and when the DDT-specific antibody was introduced on the sensor, it created a nanoscale-level deflection (bending) of the micro-cantilever due to the change in surface stress (Figure 4.13). This bending was measured using a laser beam and a four-quadrant photodetector. Alvarez et al. (2003) noticed the downward bending of the micro-cantilever was directly proportional to the DDT levels, and they proved this sensor can be regenerated by washing the cantilever beam with 100 μL of 100 mM HCl after every use.

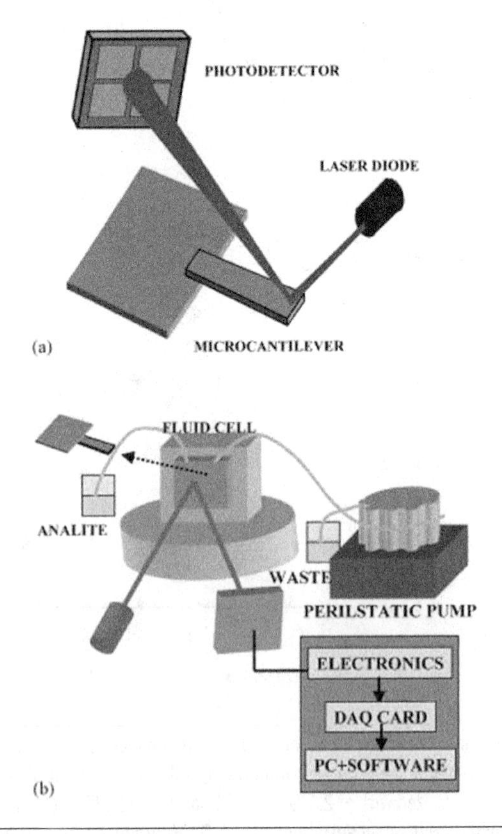

Figure 4.13 Schematic view of nanomechanical sensor for detection of DDT (a) optical measurement (b) experimental setup. (Reproduced from Alvarez, M. et al., *Biosens. Bioelectron.*, 18, 649–653, 2003.)

Valera et al. (2007) developed an immunosensor based on impedometric measurement for detection of atrazine, a selective herbicide used in agricultural operations, and its residues ended up in agricultural products and water, which has health risks to humans. The sensor was made by patterning thin interdigitized microelectrodes (IDμE's) made up of gold and chromium (Au/Cr, approximately 200 nm thickness) with an electrode gap of 6.8 μm on a 0.7 mm thick Pyrex 7740 glass substrate (Figure 4.14a). The developed sensor was immobilized

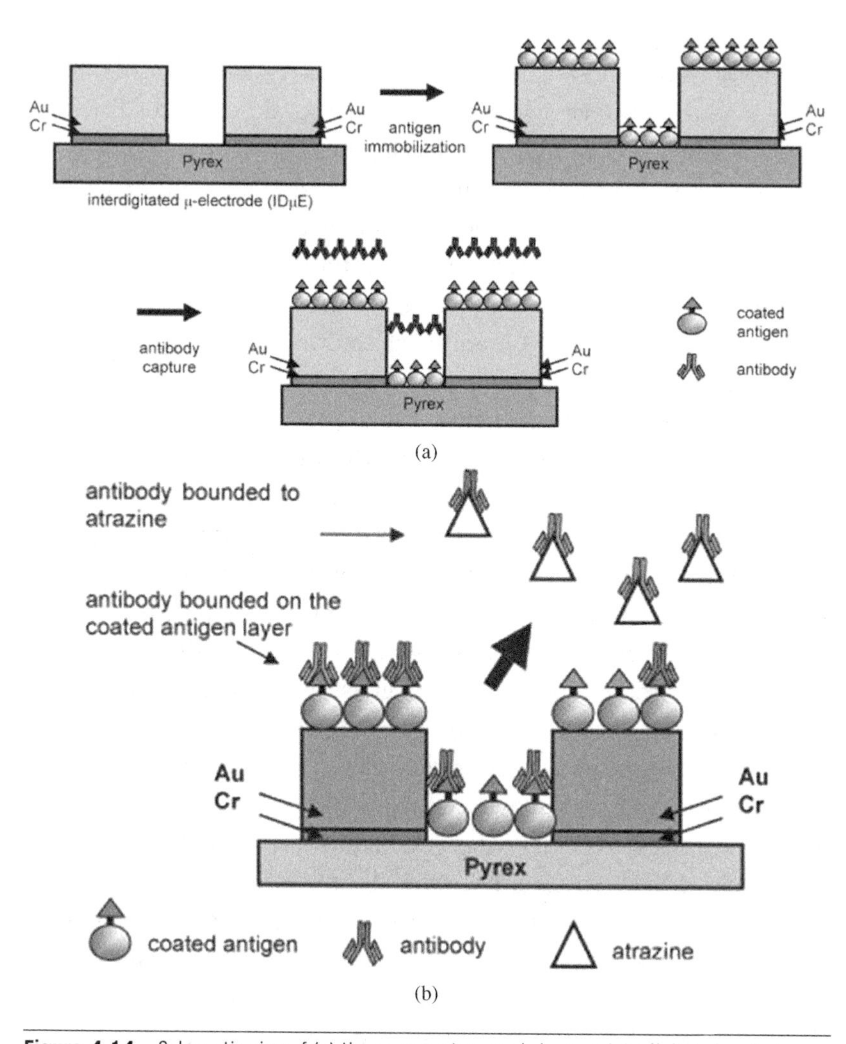

Figure 4.14 Schematic view of (a) the assay system carried out on interdigitated μ-electrodes, and (b) reaction of impedometric immunosensor. (Reproduced from Valera, E. et al., *Sens. Actuators B Chem.*, 125, 526–537, 2007.)

with 2d-BSA antigen by incubating at 25°C for overnight. When the atrazine-specific antibody and solution with atrazine was introduced on the sensor, a change in electrical impedance was noticed between the electrodes due to the antigen–antibody binding (Figure 4.14b). The change in impedance was based on the concentrations of atrazine, and using this impedance amount of atrazine was quantified (Valera et al. 2007).

Summary

Applications of nanosensors in the food industry are increasing day by day due to the advance developments in sensor design and fabrication through scientific research and development. Food packaging is the main sector using nanosensors for monitoring integrity of the package by which the quality of the food product is being monitored at each stage of the logistic line (processing facility, storage, transportation, retail facility, and consumer's kitchen) of the food material. The temperature, moisture, and gas monitoring sensors are integrated with the packaging film, and the food quality is continuously monitored. Food packaging polymers with nanosensors are commercially available throughout the world, and nowadays the food industry is regularly using these sensors. Another major application of nanosensors in the food industry is the detection of foodborne pathogens like *E. coli.* The food industry spends considerable funds for the detection of foodborne pathogens in raw materials and in their processing lines, as well as in their finished products, and still loses a huge amount due to the contamination of these foodborne pathogens in their products. Recently developed nanosensors help the food industry to rapidly detect and quantify foodborne pathogens in their raw or finished products, which will improve their productivity by reducing the detection time when compared with traditional pathogen identification methods. Some of the developed nanosensors have the ability to simultaneously detect all major types of foodborne pathogens, which reduces detection time significantly. Application of nanosensors to detect and quantify food adulterants, pesticide residues, and chemical components of raw material is gaining the food industry's interest in recent times. Most of the nanosensors for pathogen and chemical component (including adulterant and residue) detection have been tested only at the lab or in vitro conditions at present with promising results. Ongoing research and development in nanosensor design and fabrication will help the commercialization of these sensors in the near future for applications in the food industry.

References

Ai K., Liu Y., Lu L. (2009) Hydrogen-bonding recognition-induced color change of gold nanoparticles for visual detection of melamine in raw milk and infant formula. *Journal of the American Chemical Society* 131:9496–9497. doi:10.1021/ja9037017.

Akbari E., Buntat Z., Afroozeh A., Zeinalinezhad A., Nikoukar A. (2015) *Escherichia coli* bacteria detection by using graphene-based biosensor. *IET Nanobiotechnology* 9:273–279.

Akbari Z., Ghomashchi T., Aroujalian A. (2006) Potential of nanotechnology for food packaging industry. In: *Nano and Micro Technologies in the Food and HealthFood Industries*, Amsterdam, the Netherlands. pp. 25–26.

Alfadul S., Elneshwy A. (2010) Use of nanotechnology in food processing, packaging and safety–review. *African Journal of Food, Agriculture, Nutrition and Development* 10:2719–2739.

Alvarez M., Calle A., Tamayo J., Lechuga L.M., Abad A., Montoya A. (2003) Development of nanomechanical biosensors for detection of the pesticide DDT. *Biosensors and Bioelectronics* 18:649–653. doi:10.1016/S0956-5663(03)00035-6.

Augustin M.A., Sanguansri P. (2009) Nanostructured materials in the food industry. *Advances in Food and Nutrition Research* 58:183–213. doi:10.1016/S1043-4526(09)58005-9.

Avella M., Errico M.E., Gentile G., Volpe M.G. (2011) Nanocomposite sensors for food packaging. In: Reithmaier J., Paunovic P., Kulisch W., Popov C., Petkov P. (Eds.) *Nanotechnological Basis for Advanced Sensors*. NATO Science for Peace and Security Series B: Physics and Biophysics. Springer, Dordrecht, the Netherlands. pp. 501–510. doi:10.1007/978-94-007-0903-4_53.

Baeumner A. (2004) Nanosensors identify pathogens in food. *Food Technology* 58:51–55.

Bernardes P.C., de Andrade N.J., Soares N.D.F. (2014) Nanotechnology in the food industry. *Bioscience Journal* 30:1919–1932.

Blesa J., Soriano J., Moltó J., Marin R., Mañes J. (2004) Determination of aflatoxins in foods from local markets and supermarkets in Valencia, Spain. *Food Additives and Contaminants* 21(2):165–171.

Bogue R. (2016) Nanosensors and MEMS: Connecting the nanoscale with the macro world with microscale technology. *Sensor Review* 36:1–6.

Brody A.L., Bugusu B., Han J.H., Sand C.K., Mchugh T.H. (2008) Innovative food packaging solutions. *Journal of Food Science* 73:R107–R116. doi:10.1111/j.1750-3841.2008.00933.x.

Cushen M., Kerry J., Morris M., Cruz-Romero M., Cummins E. (2012) Nanotechnologies in the food industry—Recent developments, risks and regulation. *Trends in Food Science & Technology* 24:30–46.

De Azeredo H.M. (2009) Nanocomposites for food packaging applications. *Food Research International* 42:1240–1253.

Duncan T.V. (2011) Applications of nanotechnology in food packaging and food safety: Barrier materials, antimicrobials and sensors. *Journal of Colloid and Interface Science* 363:1–24. doi:10.1016/j.jcis.2011.07.017.

Duran N., Marcato P.D. (2013) Nanobiotechnology perspectives. Role of nanotechnology in the food industry: A review. *International Journal of Food Science and Technology* 48:1127–1134. doi:10.1111/ijfs.12027.

El-Boubbou K., Gruden C., Huang X. (2007) Magnetic glyco-nanoparticles: A unique tool for rapid pathogen detection, decontamination, and strain differentiation. *Journal of the American Chemical Society* 129:13392–13393.

Fluit A., Torensma R., Visser M., Aarsman C., Poppelier M., Keller B., Klapwijk P., Verhoef J. (1993) Detection of *Listeria monocytogenes* in cheese with the magnetic immuno-polymerase chain reaction assay. *Applied and Environmental Microbiology* 59:1289–1293.

Ghaani M., Nasirizadeh N., Ardakani S.A.Y., Mehrjardi F.Z., Scampicchio M., Farris S. (2016) Development of an electrochemical nanosensor for the determination of gallic acid in food. *Analytical Methods* 8:1103–1110. doi:10.1039/c5ay02747k.

Gobi K.V., Kim S.J., Tanaka H., Shoyama Y., Miura, N. (2007) Novel surface plasmon resonance (SPR) immunosensor based on monomolecular layer of physically-adsorbed ovalbumin conjugate for detection of 2, 4-dichlorophenoxyacetic acid and atomic force microscopy study. *Sensors and Actuators B: Chemical* 123(1):583–593.

Gobi K.V., Tanaka H., Shoyama Y., Miura N. (2005) Highly sensitive regenerable immunosensor for label-free detection of 2,4-dichlorophenoxyacetic acid at ppb levels by using surface plasmon resonance imaging. *Sensors and Actuators B: Chemical* 111–112:562–571. doi:10.1016/j.snb.2005.03.118.

Gomes R.C., Pastore V.A.A., Martins O.A., Biondi G.F. (2015) Nanotechnology applications in the food industry. A review. *Brazilian Journal of Hygiene and Animal Sanity* 9:1–8. doi:10.5935/1981-2965.20150001.

Hayat A., Mishra R.K., Catanante G., Marty J.L. (2015) Development of an aptasensor based on a fluorescent particles-modified aptamer for ochratoxin A detection. *Analytical and Bioanalytical Chemistry* 407:7815–7822.

Horner S.R., Mace C.R., Rothberg L.J., Miller B.L. (2006) A proteomic biosensor for enteropathogenic *E. coli*. *Biosensors and Bioelectronics* 21:1659–1663. doi:10.1016/j.bios.2005.07.019.

Hosseini M., Khabbaz H., Dadmehr M., Ganjali M.R., Mohamadnejad J. (2015) Aptamer-based colorimetric and chemiluminescence detection of aflatoxin B1 in foods samples. *Acta Chimica Slovenica* 62:721–728.

Hu L.Z., Deng L., Alsaiari S., Zhang D.Y., Khashab N.M. (2014) "Light-on" sensing of antioxidants using gold nanoclusters. *Analytical Chemistry* 86:4989–4994.

Imran M., Yousaf A.B., Zhou X., Liang K., Jiang Y.F., Xu A.W. (2016) Oxygen-deficient TiO_2–x/methylene blue colloids: Highly efficient photoreversible intelligent ink. *Langmuir* 32:8980–8987. doi:10.1021/acs.langmuir.6b02676.

Jain S., Singh S., Horn D., Davis V., Ram M., Pillai S. (2012) Development of an antibody functionalized carbon nanotube biosensor for foodborne bacterial pathogens. *Journal of Biosensors & Bioelectronics* 11:1–2.

Jamali T., Karimi-Maleh H., Khalilzadeh M.A. (2014) A novel nanosensor based on Pt:Co nanoalloy ionic liquid carbon paste electrode for voltammetric determination of vitamin B9 in food samples. *LWT-Food Science and Technology* 57:679–685. doi:10.1016/j.lwt.2014.01.023.

Jiang Y.F., Wang X.T., Jia Y., Wang F., Wu M.H., Sheng G.Y., Fu J.M. (2009) Occurrence, distribution and possible sources of organochlorine pesticides in agricultural soil of Shanghai, China. *Journal of Hazardous Materials* 170(2–3):989–997.

Kaittanis C., Santra S., Perez J.M. (2010) Emerging nanotechnology-based strategies for the identification of microbial pathogenesis. *Advanced Drug Delivery Reviews* 62:408–423.

Karimi-Maleh H., Moazampour M., Yoosefian M., Sanati A.L., Tahernejad-Javazmi F., Mahani M. (2014) An electrochemical nanosensor for simultaneous voltammetric determination of ascorbic acid and Sudan I in food samples. *Food Analytical Methods* 7:2169–2176. doi:10.1007/s12161-014-9867-x.

Liang M., Fan K., Pan Y., Jiang H., Wang F., Yang D., Lu D., Feng J., Zhao J., Yang L., Yan X. (2013) Fe_3O_4 magnetic nanoparticle peroxidase mimetic-based colorimetric assay for the rapid detection of organophosphorus pesticide and nerve agent. *Analytical Chemistry* 85:308–312. doi:10.1021/ac302781r.

Luan Y., Chen J., Xie G., Li C., Ping H., Ma Z., Lu A. (2015) Visual and microplate detection of aflatoxin B2 based on NaCl-induced aggregation of aptamer-modified gold nanoparticles. *Microchimica Acta* 182:995–1001.

Luckham R.E., Brennan J.D. (2010) Bioactive paper dipstick sensors for acetylcholinesterase inhibitors based on sol–gel/enzyme/gold nanoparticle composites. *Analyst* 135:2028–2035.

Luechinger N.A., Loher S., Athanassiou E.K., Grass R.N., Stark W.J. (2007) Highly sensitive optical detection of humidity on polymer/metal nanoparticle hybrid films. *Langmuir* 23:3473–3477. doi:10.1021/la062424y.

Maurer E.I., Comfort K.K., Hussain S.M., Schlager J.J., Mukhopadhyay S.M. (2012) Novel platform development using an assembly of carbon nanotube, nanogold and immobilized RNA capture element towards rapid, selective sensing of bacteria. *Sensors* 12:8135–8144.

Mills A. (2005) Oxygen indicators and intelligent inks for packaging food. *Chemical Society Reviews* 34:1003–1011.

Mills A., Hazafy D. (2009) Nanocrystalline SnO_2-based, UVB-activated, colourimetric oxygen indicator. *Sensors and Actuators B: Chemical* 136:344–349. doi:10.1016/j.snb.2008.12.048.

Mills A., Hazafy D., Lawrie K. (2011) Novel photocatalyst-based colourimetric indicator for oxygen. *Catalysis Today* 161:59–63. doi:10.1016/j.cattod.2010.10.073.

Navrátilová I., Skládal P. (2004) The immunosensors for measurement of 2,4-dichlorophenoxyacetic acid based on electrochemical impedance spectroscopy. *Bioelectrochemistry* 62:11–18. doi:10.1016/j.bioelechem.2003.10.004.

Pal S., Alocilja E.C., Downes F.P. (2007) Nanowire labeled direct-charge transfer biosensor for detecting *Bacillus* species. *Biosensors and Bioelectronics* 22:2329–2336.

Ping H., Zhang M., Li H., Li S., Chen Q., Sun C., Zhang T. (2012) Visual detection of melamine in raw milk by label-free silver nanoparticles. *Food Control* 23:191–197. doi:10.1016/j.foodcont.2011.07.009.

Ramachandraiah K., Han S.G., Chin K.B. (2015) Nanotechnology in meat processing and packaging: potential applications: A review. *Asian-Australasian Journal of Animal Sciences* 28:290–302. doi:10.5713/ajas.14.0607.

Ravindranath S.P., Mauer L.J., Deb-Roy C., Irudayaraj J. (2009) Biofunctionalized magnetic nanoparticle integrated mid-infrared pathogen sensor for food matrixes. *Analytical Chemistry* 81:2840–2846.

Soh J.H., Lin Y., Rana S., Ying J. Y., Stevens M.M. (2015) Colorimetric detection of small molecules in complex matrixes via target-mediated growth of aptamer-functionalized gold nanoparticles. *Analytical Chemistry* 87(15): 7644–7652.

Sporty J.L., Lemire S.W., Jakubowski E.M., Renner J.A., Evans R.A., Williams R.F., Schmidt J.G., Schans M.J.V.D., Noort D., Johnson R.C. (2010) Immunomagnetic separation and quantification of butyrylcholinesterase nerve agent adducts in human serum. *Analytical Chemistry* 82:6593–6600.

Stout II D.M., Bradham K.D., Egeghy P.P., Jones P.A., Croghan C.W., Ashley P.A., Pinzer E., Friedman W., Brinkman M.C., Nishioka M.G. (2009) American healthy homes survey: A national study of residential pesticides measured from floor wipes. *Environmental Science & Technology* 43:4294–4300.

Su X.L., Li Y. (2004) Quantum dot biolabeling coupled with immunomagnetic separation for detection of *Escherichia coli* O157: H7. *Analytical Chemistry* 76:4806–4810.

Szinicz L., Worek F., Thiermann H., Kehe K., Eckert S., Eyer P. (2007) Development of antidotes: Problems and strategies. *Toxicology* 233:23–30.

Urusov A.E., Petrakova A.V., Vozniak M.V., Zherdev A.V., Dzantiev B.B. (2014) Rapid immunoenzyme assay of aflatoxin B1 using magnetic nanoparticles. *Sensors* 14:21843–21857.

Valera E., Ramón-Azcón J., Rodríguez Á., Castañer L.M., Sánchez F.J., Marco M.P. (2007) Impedimetric immunosensor for atrazine detection using interdigitated μ-electrodes (IDμE's). *Sensors and Actuators B: Chemical* 125:526–537. doi:10.1016/j.snb.2007.02.048.

Wang C., Irudayaraj J. (2008) Gold nanorod probes for the detection of multiple pathogens. *Small* 4:2204–2208.

Wang R., Ruan C., Kanayeva D., Lassiter K., Li Y. (2008) TiO_2 nanowire bundle microelectrode based impedance immunosensor for rapid and sensitive detection of *Listeria monocytogenes*. *Nano Letters* 8:2625–2631.

Wang Y.K., Wang Y.C., Wang H.A., Ji W.H., Sun J.H., Yan Y.X. (2014) An immunomagnetic-bead-based enzyme-linked immunosorbent assay for sensitive quantification of fumonisin B1. *Food Control* 40:41–45.

Xiao R., Wang D., Lin Z., Qiu B., Liu M., Guo L., Chen G. (2015) Disassembly of gold nanoparticle dimers for colorimetric detection of ochratoxin A. *Analytical Methods* 7:842–845. doi:10.1039/c4ay02970d.

Yamada K., Kim C.T., Kim J.H., Chung J.H., Lee H.G., Jun S. (2014) Single-walled carbon nanotube-based junction biosensor for detection of *Escherichia coli*. *PLoS One* 9:e105767.

Yamada K., Watanabe A., Takeshita H., Matsumoto K.I. (2016) A method for quantification of serum tenascin-X by nano-LC/MS/MS. *Clinica Chimica Acta* 459:94–100.

Yang H., Qu L., Wimbrow A.N., Jiang X., Sun Y. (2007) Rapid detection of *Listeria monocytogenes* by nanoparticle-based immunomagnetic separation and real-time PCR. *International Journal of Food Microbiology* 118:132–138.

Yang R., Zhou Z., Sun G., Gao Y., Xu J., Strappe P., Blanchard C., Cheng Y., Ding X. (2015) Synthesis of homogeneous protein-stabilized rutin nanodispersions by reversible assembly of soybean (Glycine max) seed ferritin. *RSC Advances* 5(40):31533–31540.

Zhao S.Q., Lu Y., Zhang Y.C., Lu W.G., Liang W.J., Wang E.G. (2014) Piezo-antiferromagnetic effect of sawtooth-like graphene nanoribbons. *Applied Physics Letters* 104:203105.

Zhao X., Hilliard L.R., Mechery S.J., Wang Y., Bagwe R.P., Jin S., Tan W. (2004) A rapid bioassay for single bacterial cell quantitation using bioconjugated nanoparticles. *Proceedings of the National Academy of Sciences of the United States of America* 101:15027–15032.

5

Application in the Beverages Industry

Introduction

Beverages industry is one of the major sectors of the food industry, and the growth in beverages industry was enormous in the 1980s and 1990s. A trend of decline in sale of beverages has been noticed in the last five years, especially the sale of carbonated soft drinks declined about 1%–2% annually over the last decade. The Beverages Marketing Industry (BMI), New York, predicted a 1.4% annual reduction in carbonated soft drinks volume between 2015 and 2020 (Beverage Industry 2017). The consumers' demand for nutrient-rich beverages and organic products is the major reason for this declining trend. The decline in soft drink sales hugely impacts the beverages industry because soft drink sales make nearly 20% of the beverages market share throughout the world. Therefore, the beverages industry is spending billions of dollars in research and development of new healthy beverages with natural ingredients, naturally functionalized drinks, replacement of artificial ingredients and preservatives, and environment-friendly packaging in order to meet consumer's expectations and demands (Evergreen Packaging 2017). Similar to other sectors of the food industry, the development in nanoscience and nanotechnology can help the beverages industry to achieve their goal of developing new products, processes, and packaging techniques. Nanotechnology can be applied in manufacturing processes, quality control, and safety assessment as well as packaging of beverages.

New nutrient-rich and naturally functionalized beverages can be developed using nanotechnology by using nanosize ingredients and new processes based on nanotechnology. Nanofortification, as well as nanoemulsions techniques, can be applied for enriching the nutrient availability of beverages, and the nanoencapsulation technique can help to improve nutrient intake by the targeted delivery of the nutritional compounds in the beverages. The processing techniques such as nanopulverization and nanofiltration help to improve the nutrient content and functionality of the beverages. The size reduction of ingredients into nanoscale level using nanopulverizing techniques can help to improve the nutrient intake by mixing or extracting more nutrients into beverages compared to traditional methods (e.g., increase selenium level in tea made from nanotea). The membrane filtration techniques based on nanotechnology (nanofiltration) can help for retention of macro and micro nutrients, and antioxidants in fruit juices and drinks, which can be easily damaged with the other processing techniques like heat treatment.

Packaging is the major area, where nanotechnology can be applied in the beverages industry. The nanocomposites made with nanoparticles (NPs) of silver (Ag), gold (Au), titanium dioxide (TiO_2), zinc oxide (ZnO), and clay can be used for packaging and bottling of fruit juices, carbonated soft drinks, wine, and other alcoholic beverages. The application of nanocomposite as a packaging film or layer of the package helps to improve the thermal, mechanical, gas barrier, and moisture barrier properties of packaging films and bottles. The application of Ag, Au, or zinc nanoparticles for beverage packaging materials can improve the antimicrobial properties of the packaging material, by which the foodborne pathogens, such as mold, yeast, and bacteria, can be inhibited and shelf life of the product can be increased. The nanocomposites with nanoclay have excellent gas barrier properties (low oxygen and carbon dioxide transmission rate), so this package can retain the carbon dioxide inside the package for a long time as well as would not allow atmospheric oxygen entry into the product. For quality control and safety, nanotechnology-based nanosensors and bionanosensors can be used to detect and control microbial growth in the beverages. The traces of chemical residues in the fruit juices, nutritional compounds of the beverages, and presence of other microorganisms like mold and yeast in the beverages can also be analyzed and detected in a rapid and accurate manner using the nanosensors. Nanotechnology-based processes, packaging techniques, and sensors are discussed in other chapters of this book, and the application of nanotechnology in beverages industry is discussed in detail in this chapter.

Applications During Processing

Selenium is one of the mineral nutrients naturally available in the soil and some food products like Brazil nuts, tuna, cheese, mushrooms, and oysters. Tea plants grown in some parts of China, where soil is selenium rich, has the selenium in it. Selenium has excellent antioxidant properties, and it has several health benefits, such as helping thyroid function, preventing cell damage, and also treating asthma. The recommended intake level of selenium for adults is 55 μg/day (NIH 2016). Tea drinks made from soaking of selenium-rich tea leaves and processed through regular processing methods only capture 10% of selenium, and nearly 90% of selenium is wasted. The Qinhuangdao Taiji Ring Nano-Products Co., Ltd, China, used a three-step processing technique: (1) traditional pulverization of leaves, (2) air current pulverization, and (3) nanopulverization to produce nanosize tea powder with an average particle size of 160 nm, and this nanotea powder increased the selenium intake tenfold from the conventional tea powder. This company produces six different types of nanotea products (Nano-Green Tea, Nano-Dark-Green Tea [selenium-rich tea], Nano-White Tea, Nano-Black Tea, Nano-Yellow Tea, and Nano-Dark Tea) using this three-step nanotea pulverizing technique and commercially markets in Asian countries (Vance et al. 2015).

Nanofiltration

Nanofiltration is the recently developed membrane filtration technique, which uses a membrane with nanosize (1–10 nm) pores. Nanofiltration technology was mainly developed to remove organic and inorganic substances from water and to soften water. But the advantages of the nanofiltration technique led to applications in various sectors of the food industry (especially dairy and beverage industry). In the nanofiltration process, the separation occurs based on the molecular size of the components (Mohammad et al. 2015). The pore size of the membranes in nanofiltration is smaller than in ultrafiltration or microfiltration. The main parameters affecting the nanofiltration are pH, the temperature of the solution, and molecular weight cutoff (MWCO). The membranes used for nanofiltration of food products are fabricated using the methods of incorporation of NPs, interfacial polymerization (IP), and ultraviolet (UV) treatment. Nanoparticles of TiO_2, Ag, silica, and ZnO are commonly incorporated with polymer materials in the fabrication of nanofiltration membranes. In the IP method, the thin-film composite (TFC) membrane with a thickness of 50 nm range is fabricated through the reaction and copolymerization of two or more monomeric polymers (Mohammad et al. 2015). A detailed discussion about the nanofiltration process and membranes used for nanofiltration is available in Chapter 9 as well as in the review manuscript by Mohammad et al. (2015); however, applications of nanofiltration technology related to the beverages industry are discussed in this chapter.

Most of the fruit juices and beverages are processed using thermal processing before packaging to reduce microbial load and to increase the shelf life of the product (Cheng et al. 2016; Sui et al. 2014, 2016). The phenolic components (like anthocyanins) in the fruits are the main components responsible for the antioxidant properties of the fruit. These phenolic compounds are very susceptible to degradation due to heat and reduce the antioxidant properties of fruit juices when they are treated by thermal processes (Patras et al. 2010; Sui et al. 2014). Therefore, the beverages and juice processing industries are applying alternate non-thermal technologies like membrane filtration to remove unnecessary components from the concentrate as well as to retain functional and nutritive compounds. Nanofiltration is one of the membrane separation processes researched by the beverages industry in recent times. Its applications in the beverages industry have increased considerably in last decade due to its advantages of producing phytochemical (anthocyanins and ellagitannins)-enriched fruit juices from the concentrate at less cost and energy consumption during processing when compared with other filtration techniques (microfiltration and reverse osmosis) (Acosta et al. 2017). Strawberry fruit is consumed a lot as raw and due to its short shelf life, it is processed as juice and jelly. Thermal processing of strawberry juice and jelly reduces the antioxidant properties. Arend et al. (2017) tested the phenolic components recovery using a nanofiltration technique. Fresh strawberry fruits were procured from the local grocery store in Florianopolis, Brazil, and processed using a commercial juicer (Model: Walita RI1855, Philips, Amsterdam, Netherlands). Then the juice was microfiltered using a polyamide membrane (pore diameter of 0.4 mm, a filtration area of 0.7 m^2; PAM Selective Membranes, Rio de Janeiro, Brazil) at a temperature and pressure of 20°C ± 2°C and 300 kPa, respectively. The microfiltered juice and raw juice were then nanofiltered using polyvinylidene difluoride membrane (filtration area 1.2 m^2, MWCO 150 to 300 Da, GE Osmonics, Philadelphia, PA, USA) at an operating temperature and pressure of 20°C ± 2°C and 600 kPa, respectively. The nanofiltered strawberry juice was analyzed for total phenolic content (TPC), anthocyanins content (AC), the color of the juice and antioxidant activity. The permeate fluxes of nanofiltered natural juice (NFR) and microfiltered juice (NFMF) were 4.0 L h^{-1} m^{-2}, and 3.0 L $h^{-1}m^{-2}$, respectively. The color measurement using a colorimeter (Model: MiniScan EZ, Hunterlab, Reston, VA) with $L^*a^*b^*$ color values showed that, lightness (L^*) values were high in permeate from the both the juices (NFR, NFMF), and there were no significant differences in $L^*a^*b^*$ values between NFR and NFMF permeates. But there were significant differences for the concentrates in the $L^*a^*b^*$ values between NFR and NFMF. Arend et al. (2017) attributed the removal of suspended solids and color compounds during microfiltration as a possible reason for this difference in $L^*a^*b^*$ values. The TPC content of the concentrate of NFR and NFMF juices were 7.68 ± 0.53, and 16.6 ± 0.3 mgGAE mL^{-1}, respectively. The AC contents of the concentrate of NFR and NFMF juices were 29.5 and 16.6 mg 100 mL^{-1}, respectively. Application of nanofiltration retained more than 97.0%, and 97.8% of phenolic and anthocyanins compounds in the strawberry juice (Arend et al. 2017). The antioxidant activity measured using 2, 2′-azino-bis-3-ethyl benzothiazoline-6-sulphonic acid assay showed that nanofiltration increased the antioxidant activity of NFR and NFMF juices by 99% and 51%, respectively.

Blackberry (*Rubus Adenotrichos* Schltdl.) juice is another major juice consumed for its nutritional benefits, and blackberry is rich in phytochemicals (anthocyanins and ellagitannins), which are the main reason for its antioxidant properties. Similar to strawberry juice, application of thermal processes can reduce the antioxidant properties of blackberry juice, and the increased demand for antioxidant-rich products drives the food industry to apply alternate techniques, such as membrane separation processes, to purify and sterilize the fruit juices without affecting the polyphenolic compounds (anthocyanins and ellagitannins) concentration. Acosta et al. (2017) tested nanofiltration technique for clarification of blackberry juice and measured the retention of anthocyanins and ellagitannins compounds. The frozen ribbed blackberries purchased from the commercial store were pressed using a hydraulic press (model: OTC 25- Ton, SPX Corporation, Owatonna, MN) for extraction of juice and then preclarified using microfiltration with a cross-flow ceramic membrane (average pore size 0.2 μm). Then the concentrate was nanofiltered using seven different (NF270, Dow FILMTEC; UTC60, Toray; MPF36, Koch Membrane Systems; DL, GE Osmonics; DK, GE Osmonics; NP010, Microdyn-Nadir; and NP030, Microdyn-Nadir) commercially available nanofiltration membranes, and the efficiency of each membrane was calculated based on the retention of anthocyanins and ellagitannins at different transmembrane pressures (0.5, 1.0, 1.5, 2.0, 2.5, and 3.0 MPa). There were no fouling issues, and the permeate flux variation rate against time was below 13% in all the transmembrane pressures and all the nanofiltration membranes. Acosta et al. (2017) interpolated the preclarification of the juice with the microfiltration to remove all major compounds causing fouling to the no fouling in the nanofiltration process. The increase in transmembrane pressure from 2 to 3 MPa significantly increased the permeate flux. All seven nanofiltration membranes retained 100% of anthocyanins and ellagitannins at a transmembrane pressure of 3 MPa (Figure 5.1), and the NF270 membrane (Dow FILMTEC, Midland, MI) retained more than 90% of total solid sugars (TSS), but the TSS retention rate was lower for other membrane types. The NF270 membrane also had the maximum permeate flux of 30.4 to 60.7 kg $h^{-1}m^{-2}$. The solution-diffusion model developed from these results for selecting most suitable membrane for nanofiltration also predicted NF270 membrane for clarification of blackberry juice.

Wine is one of the main alcoholic beverages consumed around the world, and sensible consumption of red wine has health benefits like prevention of cardiovascular diseases (Szmitko and Verma 2005). Even though red wine has increased antioxidant properties due to the presence of polyphenolic compounds (flavonoids), consumption of wine is banned because of its alcohol content in some part of the world due to cultural and religious reasons (Labanda et al. 2009; Takács et al. 2007). Researches carried out in last decade proved that reduction and removal of the alcohol level (ethanol) in red and white wines

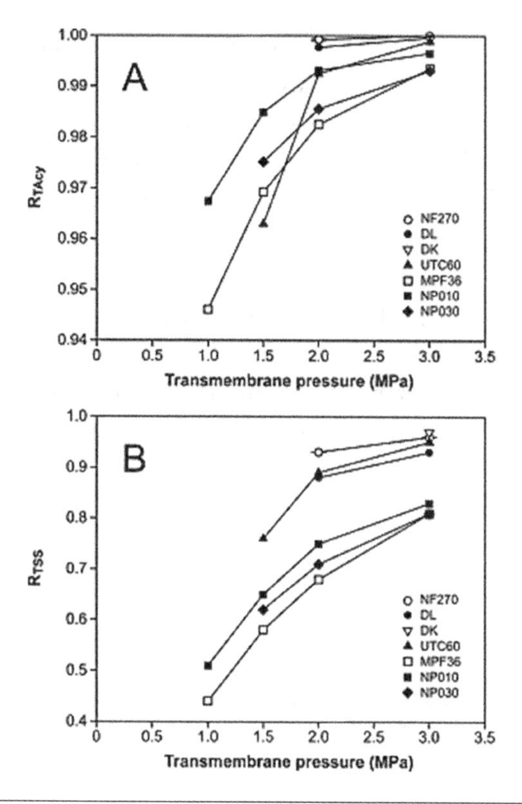

Figure 5.1 Retention rates of total anthocyanins (RTacy) (A) and total soluble solids (RTss) (B) for different nanofiltration membranes at different transmembrane pressures. (Reproduced from Acosta, O. et al., *J. Food Process Eng.*, 40, e12343, 2017.)

had no effect on antioxidant as well as cardiovascular disease prevention properties of wine (Labanda et al. 2009; Lecour et al. 2006). An increase of alcohol level in the wine due to the new grape cultivars, global warming, and the advanced wine production processes was noticed in the last decade, and the wine industry wanted to bring down the alcohol level in the wine below 14% because of the levy imposed on wine products with high alcohol content (Massot et al. 2008). Use of unripe grape fruits or enzymes to reduce the sugar content of the juice are some techniques for the production of low alcohol content wine. Use of membrane separation for removing ethanol from the fermented juice is now becoming popular since this technique needs only one filtering process, and there is no need to modify the conventional wine production processes resulting in complete retention of polyphenolic compounds (Labanda et al. 2009; Lisanti et al. 2013). Catarino and Mendes (2011) tested five different types (four nanofiltration membranes, one reverse osmosis membrane) of membranes (Model: CA995PE, Alfa Laval, Lund, Sweden; NF99 HF, Alfa Laval, Lund, Sweden; NF97, Alfa Laval, Lund, Sweden; NF99, Alfa Laval, Lund, Sweden; and YMHLSP1905, GE Osmonics, Minnetonka, MN) for dealcoholization of wine. The wine with the alcohol content of 12% (vol) was fed from the feed tank and filtered using the membrane cell of the lab scale dealcoholization unit (Figure 5.2) at a feed rate, feed pressure, and feed temperature of 2 L min^{-1}, 16 bar (1600 kPa), and 30°C, respectively for 150 min. The performance of the membrane was analyzed using the permeate flux, ethanol rejection, aroma

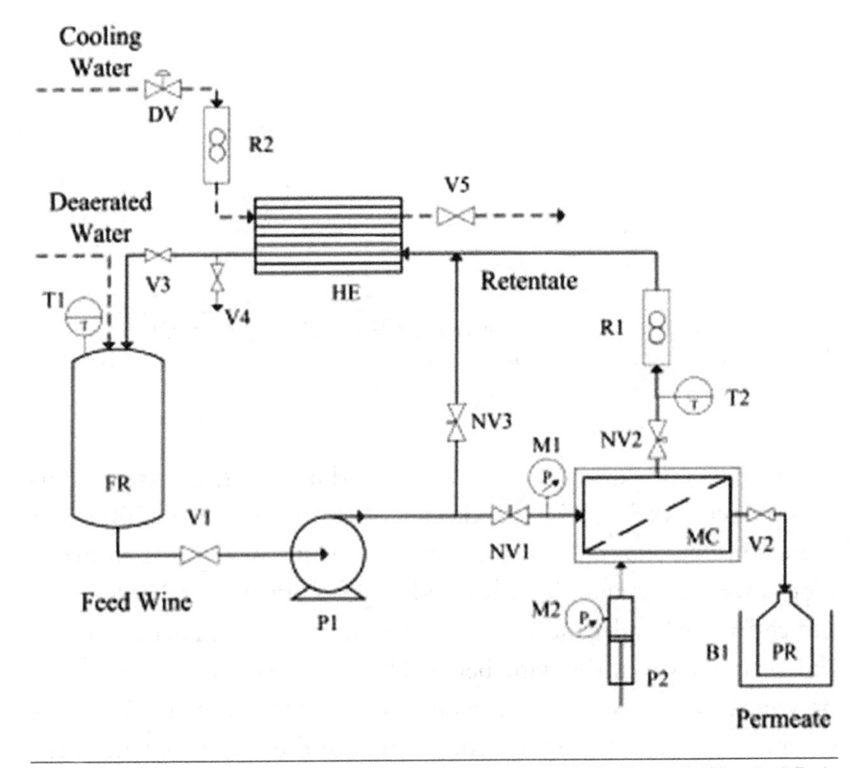

Figure 5.2 Schematic diagram of the lab scale wine dealcoholization unit (FR—feed Tank; P1—centrifugal pump; MC—membrane cell; HE—heat exchanger for control flow temperature; PR—permeate tank; B1—ice bath for preventing volatilization; R1 and R2—rotameters; NV1, NV2 and NV3—needle valves for controlling flow; DV—diaphragm valve; V1 to V5—on/off valves; T1—PT100 sensor for feed temperature measurement; T2—thermocouple for membrane cell temperature measurement; M1 and M2—manometers for pressure measurement; P2—pneumatic sealing pump). (Reproduced from Catarino, M., and Mendes, A., *Innov. Food Sci. Emerg. Technol.*, 12, 330–337, 2011.)

rejection, and sensory evaluation. The nanofiltration membranes NF99, NF99 HF, and the reverse osmosis membrane YMHLSP1905 yielded better permeate flux and ethanol rejection (Figure 5.3). The nanofiltration membrane NF97 showed the highest aroma rejection than the other membranes. The aroma rejection results showed that all the membranes showed a tendency of higher aroma rejection than the ethanol rejection, which helps to produce low alcoholic wines. But wines filtered by all the membranes failed in the sensory analysis due to higher astringency or unbalanced aroma and taste. The NF99 HF produced low alcohol wine with good aromatic properties like the original wine, but the astringency caused the failure score in sensory tests. The NF99 and YMHLSP1905 also produced low alcohol wine with aromatic properties closer to the original wine. Catarino and Mendes (2011) suggested collecting the aromatic compounds during the pervaporation process, and adding these back to the dealcoholized wine after nanofiltration process might help to produce low alcohol wines with sensory attributes closer original wine.

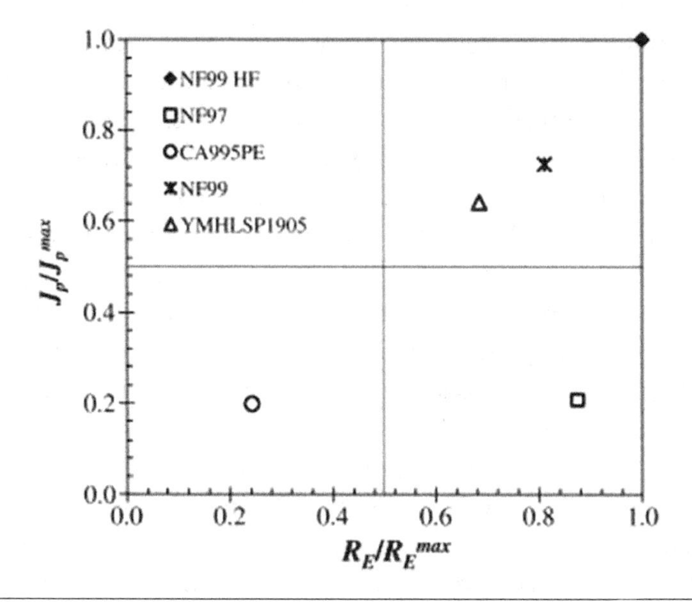

Figure 5.3 Normalized ethanol rejection vs. normalized permeate flux under nanofiltration with different membranes. (Reproduced from Catarino, M., and Mendes, A., *Innov. Food Sci. Emerg. Technol.*, 12, 330–337, 2011.)

Enzyme Immobilization

Enzyme immobilization is used in the brewing processes by the beverages industry to make the continuous production of alcoholic beverages easy as well as additional advantages of easy separation of yeast from the fermented product, less amount of yeast used, maximum conversion of substrate to product, and lower fermentation time (Duarte et al. 2013). Nanotechnology was used to immobilize yeast during wine and beer making using nanotubular cellulose (Koutinas et al. 2012) and Fe_2O_3 magnetic nanoparticles (MNPs) (Berovic et al. 2014). The *Saccharomyces cerevisiae* is the major yeast used in wine and sparkling wine production and Berovic et al. (2014) tested the effect of immobilization of *S. cerevisiae* yeast using γ-Fe_2O_3 MNPs on metabolism activity of the yeast and the separation of yeast from the wine. The γ-Fe_2O_3 MNPs were prepared by precipitating the solution containing Fe(II) sulphate ($FeSO_4$), Fe(III) sulphate ($Fe_2[SO_4]_3$), and aqueous ammonia. The average size of the synthesized γ-Fe_2O_3 MNPs was 13.7 ± 2.9 nm. Then the MNPs were coated with a thin layer (about 1 nm) of silica after dispersing MNPs into the water with citric acid solution; and the transmission electron microscopy images acquired from these silica-coated MNPs showed the average diameter of these MNPs was 16 ± 5 nm. Silica-coated MNPs were functionalized with 3-(2- aminoethylamino) propylmethyldimethoxysilane and then washed with distilled water. The *S. cerevisiae* purchased from Daystar Ferment AG, Switzerland, was mixed in water at a concentration of 10^7 yeast cells/mL, and prepared γ-Fe_2O_3 MNPs were mixed at a ratio of 1:10 under constant stirring. After 10 minutes of stirring, the yeast-loaded MNPs (magnetized yeast) were separated from the solution using a permanent magnet and washed with distilled water. For sparkling wine production, magnetized yeast, and control yeast cells at a concentration of 2×10^7 cells/mL were added to 10 L of grape must and inoculated at 22°C for primary fermentation. For secondary fermentation, yeast cells (magnetized and non-magnetized) at a concentration of 0.3 g/L in each bottle were added and inoculated at 17°C for 24 days. The metabolism of the yeast was measured by biomass production, redox potential, and pH. During primary fermentation, the grape must, fermented with magnetized yeast, produced 4.22 g/L of biomass in 88 h, and non-magnetized yeast produced 4.14 g/L biomass in 120 h. The redox potential of grape must before primary fermentation was +212 mV, and the anaerobic ethanol production phase started after 2 and 16 h in magnetized yeast and non-magnetized yeast, respectively. This more intensive fermentation phase (anaerobic ethanol production) lasted for 190 and 160 h in grape must fermented with magnetized yeast and non-magnetized yeast, respectively. The initial pH of grape must with non-magnetized yeast was 3.55, and it reduced to 3.37 after 50 h due to the lactic acid formation, and then increased to 3.72 after 160 h. In grape must with magnetized yeast, initial pH was 3.94 and reduced to 3.40 after 100 h, and the final pH after 162 h was 3.62 (Figure 5.4). The consumption of sugar components (glucose and fructose) was also quicker (about 50 h less) in magnetized yeast fermentation when compared with the fermentation with non-magnetized yeast. There was no significant difference in ethanol accumulation between both the fermentation processes (110 g/L), as well as no significant difference in the composition of wine produced. After secondary fermentation in bottles for 24 days, the carbon dioxide concentration, measured as pressure increase in the bottles, was 525 ± 10 kPa and 530 ± 10 kPa in non-magnetized and magnetized yeast samples, respectively. In the magnetized yeast bottles, the sediment with the yeast was easily removed using a permanent magnet (turning the bottles upside down and placing a magnet on the outside of bottle neck) within 20 min (Figure 5.5), whereas sediment separation was difficult in the non-magnetized yeast (control) samples. Berovic et al. (2014) also measured the iron residual in the sparkling wine produced using γ-Fe_2O_3 MNPs, and the iron concentration was 8.30 ± 0.16 mg/L, which was below the European council's maximum allowable limit (10.00 mg/L).

Li et al. (2010) developed a technique to immobilize the yeast alcohol dehydrogenase (YADH) derived from *S. cerevisiae* using chitosan-coated MNPs. The YADH accelerates the oxidation of alcohol and reduction of carbonyl compounds like aldehydes and ketones (Chen and Liao 2002). The iron oxide nanoparticles (Fe_3O_4NPs) were prepared from a $FeSO_4$ and H_2O_2 solution using the hydrothermal process, and then 0.2 g of Fe_3O_4NPs were dispersed into the solution of paraffin (30 mL) and span-80 (0.5 mL). Then the alpha-ketoglutaric acid chitosan (KCTS) (15 mL) dispersed in acetic acid

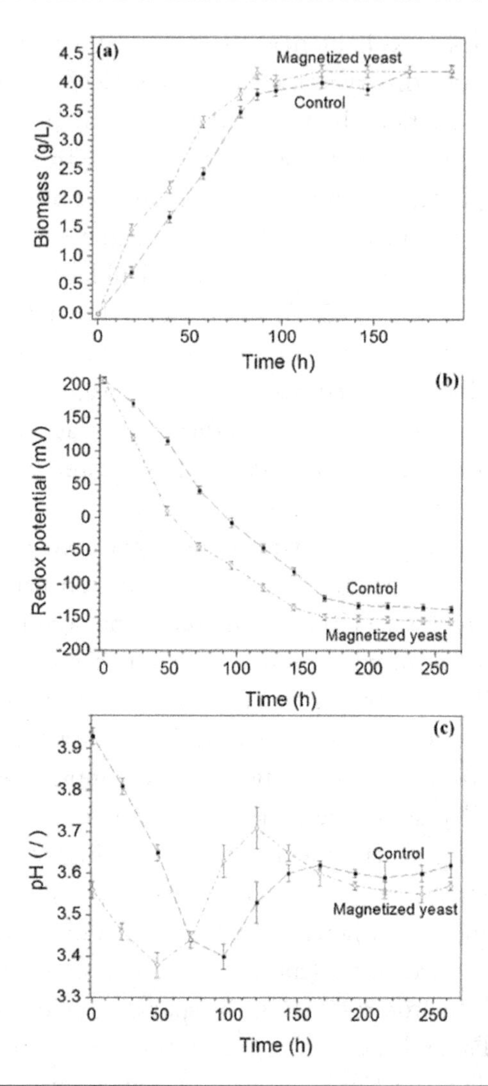

Figure 5.4 Assessment of metabolic reaction during primary fermentation by magnetized yeast (•) and non-magnetized (control) yeast (○) using (a) biomass production, (b) redox potential, and (c) pH. (Reproduced from Berovic, M. et al., *Biochem. Eng. J.*, 88, 77–84, 2014.)

(2% concentration) was mixed with the solution and ultrasonicated for 30 min. Finally, 1 mL of carbodiimide solution was added to the solution and stirred for 4 h. After the reaction, chitosan-coated nanomagnetic particles (Fe_3O_4/KCTS nanoparticles) were separated from the solution using a permanent magnet (6000 G surface magnetization capacity) and washed with water and ethanol. Then the Fe_3O_4/KCTS nanoparticles (average size of 26 nm) were dried using a vacuum dryer

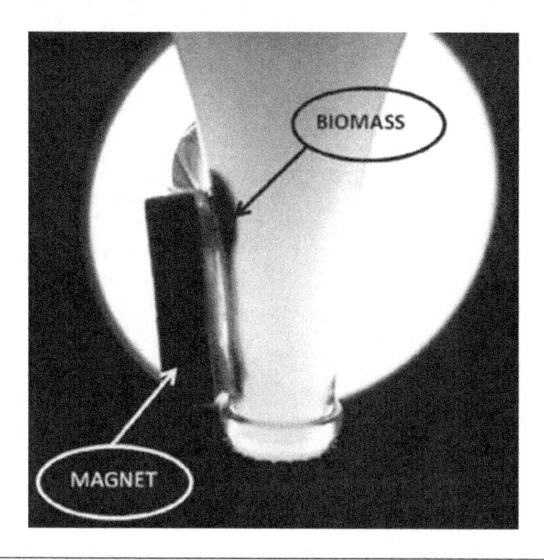

Figure 5.5 **(See color insert.)** Separation of the sediment using a permanent magnet from the wine bottle when wine was produced using fermentation by magnetized yeast. (Reproduced from Berovic, M. et al., *Biochem. Eng. J.*, 88, 77–84, 2014.)

at 50°C. A 50 mg of above prepared Fe_3O_4/KCTS MNPs were mixed with 30 mL of 5% glutaraldehyde phosphate buffer solution (pH 6.8), stored for 10 h at room temperature (24°C), magnetically separated, and then washed with deionized water. Then the Fe_3O_4/KCTS MNPs were mixed with 30 mL YADH diluent solution (YADH solution in phosphate buffer at a concentration of 0.23 mg/mL) at 25°C and shaken for 2 h at a speed of 150 rpm. Finally, the immobilized YADH on the Fe_3O_4/KCTS MNPs were collected using magnetic separation and analyzed. The results showed that 65% of enzyme activity was recovered after immobilizing the YADH with Fe_3O_4/KCTS MNPs. The maximum activity of the free YADH was at 6.8 pH, and then its activity reduced with an increase in pH. But when the YADH was immobilized with chitosan-coated MNPs, the enzyme activity was high even at high pH range. The YADH's activity under different temperatures (4 to 50°C) was tested, and the YADH immobilized with Fe_3O_4/KCTS MNPs showed higher activity (68%) at the high temperature end than the free YADH (47%) (Figure 5.6). So immobilizing YADH with chitosan-coated MNPs could help to use the YADH under a large range of temperatures and as well as solutions with high pH values.

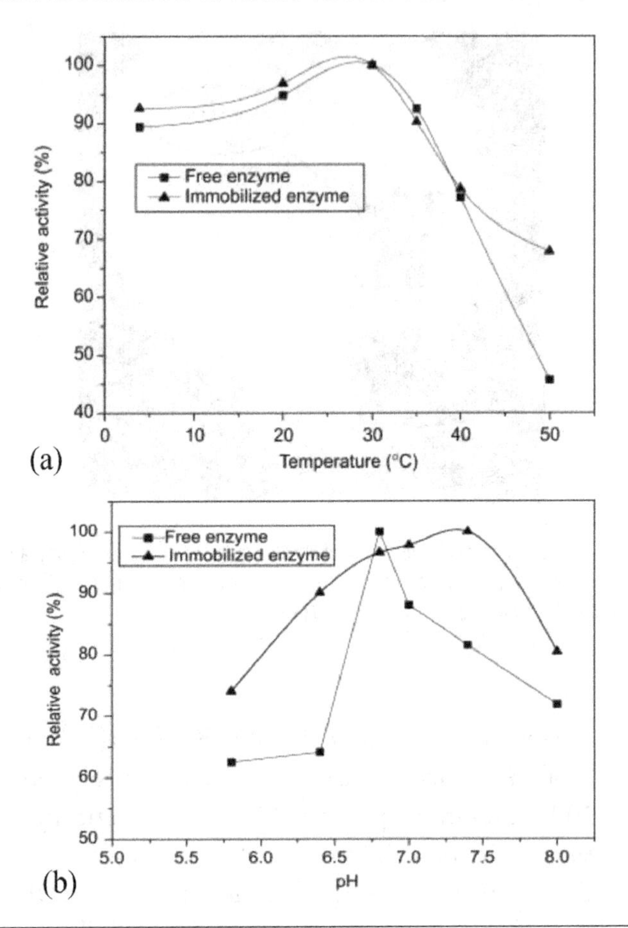

Figure 5.6 (a,b) Relative activity of YADH enzyme (free and immobilized with chitosan-coated MNPs). (Reproduced from Li, G.Y. et al., *J. Magn. Magn. Mater.*, 322, 3862–3868, 2010.)

Applications During Quality Control and Safety

Astringency is one of the sensory parameters used in wine quality testing, which depends mainly on polyphenols in the red wine. The interaction between polyphenols and the proteins in saliva is the reason for this sensory property. Guerreiro et al. (2014) developed a bionanosensor using Au nanoparticles based on localized surface plasmon resonance (LSPR) technique. The glass disks were cleaned by acetone using plasma cleansing technique, and using the mask colloidal lithography, a triple layer of polyelectrolytes [2% poly(diallyldimethylammonium chloride) (PDDA), 2% poly(sodium-4-styrenesulfonate) (PSS), and 5% poly(ammonium chloride) (PAX-XL60)] were deposited on the disk. Then the disk was washed with Milli-Q water and dried with nitrogen. Then a layer of Ti (20 nm size) was coated on the disk and then etched for 10 min to remove polymethylmethacrylate (PMMA A4) from the glass disk. Again, this disk was coated with Ti (2 nm size), Au (20 nm size), and then rinsed in a sonicator using ethanol, acetone, and Milli-Q water (cleaned for 3 min each in each solution). Cleansed disks were dried using nitrogen gas and then treated with UV/Ozone and Milli-Q water (1 h each) to prevent oxidation of Au. The α-amylase (AMY) was used to mimic the actions of saliva and immobilized on the Au-coated nanodisks using amine coupling N-Hydroxysuccinimide (NHS) and 1-Ethyl-3-(3-dimethylaminopropyl) carbodiimide (EDAC) activation. The schematic view of the sensor is shown in Figure 5.7. The AMY-immobilized nanobiosensors were placed in buffer samples and the spectra were obtained using spectropolarimeter (model: J-81, Jasco corporation, Tokyo, Japan). The spectra from the buffer and red wine samples showed the plasmic peak due to the polyphenols at 760 nm, and the red wine showed a different peak at a wavelength of 540 nm (Figure 5.8), which related to its color. Then two white wine samples with astringency levels of 1 and 2 (analyzed by the sensory panel), and two red wine samples (astringency levels 3 and 4) were analyzed using the LSPR bionanosensor with Au and the peak shift at 760 nm was used to predict the pentagalloyl glucose (PGG) polyphenol concentration, which was then used for assigning astringency levels of wine. The results showed that the white wine samples (astringency levels 1 and 2) had lower PGG concentrations, and red wine samples (astringency levels 3 and 4) had higher PGG concentrations, and the error in

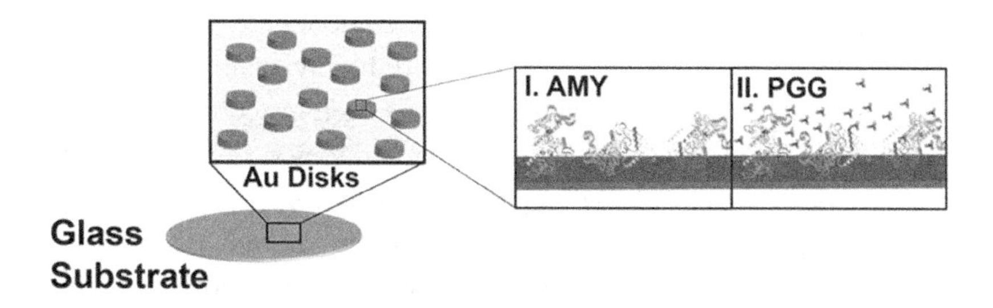

Figure 5.7 (See color insert.) Schematic view of LSPR-based bio nanosensors with Au nanodisks and its reaction with α-amylase (AMY) and polyphenol pentagalloyl glucose (PGG). (Reproduced from Guerreiro, J.R.L. et al., *Acs Nano*, 8, 7958–7967, 2014.)

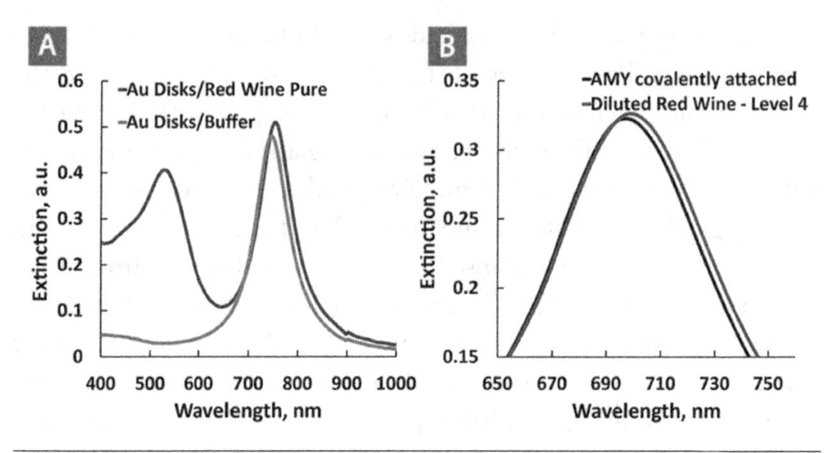

Figure 5.8 **(See color insert.)** Spectra of wine samples analysis by the Gold nanodisk-localized surface plasmon resonance (LSPR) sensor: (A) spectra of red wine and buffer solution; (B) red shift of red wine analysis. (Reproduced from Guerreiro, J.R.L. et al., *Acs Nano*, 8, 7958–7967, 2014.)

prediction was 15%. Even though the error level was high, Guerreiro et al. (2014) suggested the significant differences in PGG concentration between different astringency level wines showed the potential of this bionanosensor based on LSPR technology to apply for non-subjective sensory evaluation of wines.

Red and white wines are rich in phenolic components like tannin and measurement of total phenolic compounds helps to measure the antioxidant properties of the wine. Red wine has higher antioxidant property because of the higher phenolic compounds concentration (2.0–5.0 g/L) than the white wines (0.1 g/L). The regular quantification techniques used to measure phenolic compounds in wine are gas chromatography coupled with mass spectroscopy (GC-MS), spectrophotometer and high-performance liquid chromatography (HPLC). As we discussed in other chapters, these techniques are expensive, and also need more time for analysis, along with highly qualified personnel for analysis. Amperometric biosensors have been tested for analysis of major components of food and beverages, and they are simple in operation, less expensive, and analysis time is shorter than the conventional analytical methods. Advancement in nanoscience and nanotechnology helped researchers to develop amperometric biosensors with silver and zinc oxide nanoparticles (AgNPs and ZnONPs) for quantification of components of food and beverages due to their abilities of high chemical stability, stronger biomolecules absorption, and higher catalytic activity to reduce hydrogen peroxide (H_2O_2). Chawla et al. (2012) developed an amperometric biosensor with Ag and ZnONPs for measuring the total phenolic compound content in the wine. The sodium hydroxide (NaOH) solution (0.9 M) in double-distilled water was heated to 50°C, and the zinc nitrate $Zn(NO_3)_2 \cdot 4H_2O$ solution (0.45 M with double-distilled water) was added into NaOH solution slowly (drop by drop) under constant stirring for 45 min. The container was sealed and kept for 2 h, and then the ZnONPs were collected by washing the particles in the container with ethanol and air drying at 60°C. The AgNPs were prepared by mixing silver nitrate (1.0 mM, 10 mL) and sodium borohydride (2.0 mM, 30 mL) under high speed stirring for 3 min. An Au electrode (1 mm diameter) of the potentiostat/galvanostat (PGSTAT) (Model: AUT83785, Eco Chemie, Metrohm Autolab B.V., Netherlands) was first cleaned with H_2SO_4, H_2O_2 (3:1 ratio) solution for 20 min, followed by rinsing and polishing with double-distilled water and alumina slurry, respectively. Then the electrode was placed inside the solution containing $K_3Fe(CN)_6/K_4Fe(CN)_6$(1:1 ratio) (23 mL of 2.5 mM) and ZnONPs (2 mL). The ZnONP-coated sensor was rinsed with double-distilled water and then dried at room temperature. The AgNP-coated electrodes were also prepared in similar manner (immersing inside 23 mL of 2.5 mM $K_3Fe(CN)_6/K_4Fe(CN)_6$(1:1 ratio), and 2 mL of AgNPs). The ZnONPs or AgNP-coated electrodes were immersed in a cysteine ethanol solution (1.0 mM) at room temperature for 10 h. The ZnONPs/AgNP-coated gold electrode was activated with 5% glutaraldehyde (GA) solution, and then an electrode was dipped to acetate buffer (2.5 mL, 0.1 M, pH 5.0) solution with protein enzyme (50 μL, concentration 40 mg/mL) for 24 h, and then rinsed with the phosphate buffer (0.1 M, pH 7.0) to remove unbound enzyme particles. This enzyme immobilized, the ZnONPs/AgNP-coated gold electrode was used to measure the total phenolic content using the measured current. Totally five commercial wine samples were analyzed by placing the ZnONPs/AgNPs gold electrode as a working sensor of three electrode PGSTAT units. The Pt wire and Ag/AgCl electrodes acted as auxiliary and reference electrodes, respectively, of the PGSTAT unit. The total phenolic content measured with this newly developed sensor and the regular spectrophotometric methods had a correlation coefficient of 0.98, and the response time for analysis of the total phenolic content of wine samples was 8 s. This result proved this sensor can be used for the determination of total phenolic content of wines in a short time with good accuracy, but the activity of the sensor was reduced to 75% after 5 months of storage and after its use for 200 times (Chawla et al. 2012).

Acetaldehyde (CH_3CHO) is one of the main flavor compounds found in alcoholic beverages, which is formed by the reaction between acetic acid bacteria and yeast during fermentation by oxidizing the ethanol. Acetaldehyde is the major carbonyl compound (aldehydes), which creates the aroma of wine, and acetaldehyde shares more than 90% of all the carbonyl compounds in the wine (Liu and Pilone 2000; van Jaarsveld and October 2015). The acetaldehyde concentration in white and red wines are 80 and 30 mg/L, respectively. Controlling acetaldehyde level in wine is important during wine production because a higher amount of acetaldehyde can cause simple behavioral problems like nausea, restlessness, and headaches to severe health hazards to consumers like liver fibrosis and pregnancy fetal injury (Mello et al. 2008; Quertemont et al. 2005; Salaspuro 2011). Besides health concerns, an elevated acetaldehyde level also affects the quality of alcoholic beverages. Sulfur dioxide (SO_2) is added to wine and other alcoholic beverages during processing as a preservative due to its excellent anti-enzymatic and antimicrobial properties. Acetaldehyde binds with SO_2 by forming a complex compound hydroxy-sulfonate, which changes the anti-enzymatic and antimicrobial properties of the SO_2 and spoils the wine in short time. Usage of excess SO_2 to counterattack this ill effect of acetaldehyde and SO_2 binding may also pose health risks. Therefore, the beverages and wine production industry are implementing several acetaldehyde reduction techniques like using yeast strains like *S. cerevisiae* for the fermentation process and restricting oxidation during processing (Jackowetz et al. 2011). Nanotechnology-based techniques like the application of AuNPs for degradation of acetaldehyde in wine have been tested to control the acetaldehyde level in beverages. Yu et al. (2010) prepared a nanocomposite with AuNPs and chitosan (Ch) to control the acetaldehyde content in wine. The AuNP-absorbed Ch were prepared by an electrochemical reaction using a potentiostat (model: PGSTAT30, Eco Chemie, Metrohm Autolab B.V., Netherlands). A sheet of gold, a sheet of platinum, and the silver–silver chloride (Ag/AgCl) rod were connected as working, auxiliary, and reference electrodes, respectively. The Au electrode was first polished with an aluminum slurry, and then the Au sheet was placed in a deoxygenated aqueous solution of 40 mL NaCl (0.1N) and Ch (1 g/L) and cycled for 200 scans from −0.28 to +1.22 V against Ag/AgCl at 500 mV/s under slow stirring. The size controllable AuNPs on the Ch biopolymer were synthesized by irradiating this solution at 254, 310, and 365 nm using a UV light for 1 h. The results showed that the sizes of AuNPs on the Ch biopolymer prepared at 254, 310, and 365 nm wavelength UV light were 50, 30, and 10 nm, respectively. The above prepared Ch biopolymer solution with AuNPs was added into commercial white wine samples with a total acetaldehyde content of 95 ppm and kept at room temperature for seven days. Then the final acetaldehyde content was analyzed using a GC-MS system (model: Micromass TRIO-2000, Micromass, Manchester, UK). The results showed the total acetaldehyde content was reduced to 68, 39, and 24 ppm when using AuNPs absorbed Ch solution prepared using 254, 310, and 365 nm wavelength UV lights, respectively. Yu et al. (2010) also used the regular AuNPs not capped with Ch with a size of 10 nm to reduce the total acetaldehyde content of commercial wine in a similar manner and found the acetaldehyde content reduced from 95 to 70 ppm after 7 days of storage at room temperature. These results showed the potential of AuNPs for reducing total acetaldehyde content in wine in a simple way, and also the Ch capping of AuNPs helped to control the size of AuNPs as well as increased the acetaldehyde decomposition rate.

Beverages production units produce hundreds of liters of wastewater every day from the processing operations like the washing of fruits and raw materials before and during crushing or pressing, cleaning storage and fermentation tanks, bottling and packaging of final products. Treatment of wastewater produced from beverages industry is one of the toughest processing steps since the wastewater contains large amount phenols and other toxic substances, which cause environmental and health hazards. Biological treatments (aerobic and anaerobic) using activated sludge process or fluidized bed process with the use of biological agents like bacteria are the common way of treating wastewater from the food processing industry. But the high phenolic content in the wastewater from the winery and other beverages industry restricts the use of biological treatments for the wastewater produced from winery and beverages industry due to their higher phytotoxicity and bacterial toxicity. Pretreatment of wastewater using nanoclay, TiO_2NPs, AgNPs, and AuNPs degraded toxic organic pollutants in the wastewater and helped improve the efficacy of the wastewater treatment units in the beverages industry by neutralizing, flocculating, and precipitating the colloids in the industry wastewater (Rytwo 2012). The TiO_2NPs is one of the ideal photocatalyst materials, which has high chemical stability and economically cheap to use for wastewater treatment. The TiO_2NPs absorb the UV light and create photogenerated holes, which improve the reaction with the wastewater molecules to create highly reactive hydroxyl radicals. These reactions enhance the oxidation of organic pollutants in the wastewater on the surface of the TiO_2 NPs (Monge and Moreno-Arribas 2016). Rytwo et al. (2013) developed a wastewater treatment process using nanocomposites with polymer and nanoclay to clarify the wastewater from winery and olive oil production units. The nanocomposite was prepared using poly-DADMAC polymer and sepiolite (soft white clay mineral) using the following procedure. First, a solution of poly-DADMAC in warm water (concentration 1 g/100 mL water) was made and ultrasonicated to a homogenous solution. Then a 50 g of sepiolite was placed inside a container and a 50 mL of the polymer solution was poured into it and then mixed for 2 h to obtain the polymer-clay nanocomposite. A 20 mL of above-prepared polymer-clay nanocomposite (5% concentration [50 g clay/1000 g polymer]) was mixed with 1000 mL of raw effluent from the winery and stirred for 30 s. Then the mix was centrifuged for 5 min at a speed of 30 *g*, which produced clear sludge and effluent. The effluent at the top (about 900 mL) was removed and then another 900 mL of raw effluent was added to the remaining sludge. A booster dose of 4 mL of polymer-clay nanocomposite was added and again stirred for 30 s, followed by centrifugation at 30 *g* for 5 min. The procedure was repeated until the relative turbidity (ratio of turbidity between raw and treated wastewater) measured using turbidity meter (model: LaMotte 2020i, LaMotte Company, Chestertown, Maryland) reached 10% level. The treatment procedure is schematically shown in Figure 5.9. The light intensity analysis was carried out to measure the clarification level of wastewater using a dispersion analyzer (model: LUMiSizer 6110, LUM GmbH, Berlin, Germany). The light intensities through distilled

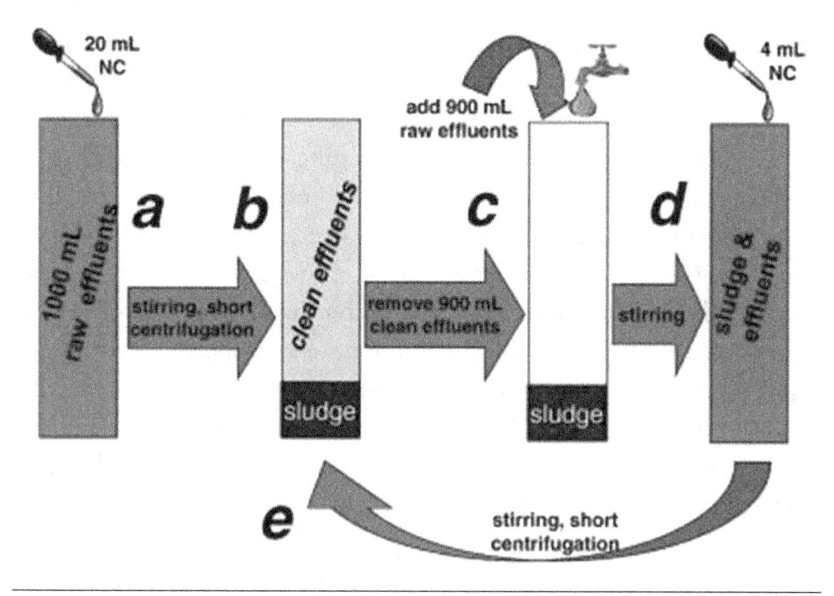

Figure 5.9 Schematic representation of winery wastewater pretreatment using polymer-clay nanocomposites (NC). (Reproduced from Rytwo, G. et al., *Sci. Total Environ.*, 442, 134–142, 2013.)

water, raw wastewater from the winery, and treated wastewater using polymer-clay nanocomposite were 94%, 22%, and 80%, respectively. Rytwo et al. (2013) also found that the pretreatment (clarification) of winery wastewater using the polymer-clay nanocomposite reduced the wastewater treatment time to 15 min, whereas the regular sedimentation treatments took 2 h of the clarification process.

Applications During Packaging

The advances in nanoscience and nanotechnology have helped the packaging industry to develop nanocomposite polymers incorporating nanoparticles of TiO_2, Ag, ZnO, or nanoclay with improved thermal, mechanical, moisture transmission, and gas barrier capacities for food and beverages packaging. Beverages like soft drinks and beer need to maintain carbon dioxide inside the package at a certain level in order to maintain the quality of the product. The nanoparticles of oxygen-scavenging materials like TiO_2 can be used along with regular polymer to produce containers and packaging materials for these types of beverages. The oxygen level inside the beverages also needs to keep at a certain level in order to maintain the quality, flavor, and taste of the product and at the same time inhibit the microbial growth in the product. Nanoclay is an excellent nanocomposite material for beverages packaging due to its low transmission rate for carbon dioxide and oxygen, as well as moisture. The packaging industry is using these advantages and commercially producing polyethylene terephthalate (PET) bottles, plastic films for packing beer, carbonated soft drinks, and fruit juices (Chaudhry et al. 2008). The Nanocor Inc., Hoffman Estates, IL, developed multilayer PET bottles with Imperm® resins, which have excellent gas and moisture barrier properties for packaging of beer and fruit juices. The Imperm® resins are made with the polymer Nylon MXD6 and nanoclay particles. The Nylon MXD6 polymer was used as a gas barrier material in the packaging films and containers for a long time, and the regular monolayer PET bottles, and PET bottles produced with Nylon MXD6, had oxygen transmission rates (OTR) of 0.025 and 0.013 cc/bottle day 0.21 atm (at room temperature), respectively. A significant reduction in OTR was observed when the Imperm® resin (nanocomposite of Nylon MXD6 and nanoclay) was used for the production of multilayer PET bottles (0.004 cc/bottle day 0.21 atm). The shelf life of CO_2 (10% loss of initial level) in the regular PET bottle, PET bottle with Nylon MXD6, and multilayer PET bottle with Imperm® resin were 7, 14, and 23 days, respectively (Nanocor 2017). The Bayor AG also developed plastic polymer films with nanoclay coating with improved mechanical and gas barrier properties with a commercial name of "Durethan® KU2-2601" to use as the inside layer of paperboard juice containers. Honeywell, Morris Plains, NJ, developed a nanocomposite material called Aegis® OXCE with nylon and nanoclay to use as oxygen and carbon dioxide barrier material for making PET bottles for beer and soft drinks using a co-injection process. The nylon is used as an oxygen scavenger, and the nanoclay acted as a passive barrier to maintain the CO_2 inside the bottle and restrict the O_2 entry into the PET beer bottles (Honeywell 2016).

Orange juice is one of the most consumed fruit juices in the world, and it is liked by consumers around the world for its excellent nutritional characteristics, unique taste, and flavor. Freshly squeezed orange juice is packed in pouches, tetra packs, and bottles using various packaging materials and distributed throughout the world. The acidity of the citric juices like orange juice is a natural antimicrobial agent to control microorganism growth in these kinds of juices, but some microorganisms (Acidophilic in nature) like yeast and lactic acid bacteria have the capability to grow in the highly acidic environment. Among the lactic acid bacterium, *Lactobacillus plantarum* has the ability to grow in freshly squeezed and minimally processed orange juice and produce an unpleasant flavor in a short time. Nanotechnology-based antimicrobial active packaging can be used to prevent the growth of harmful lactic acid bacteria like *L. plantarum* in orange juice. Emamifar et al. (2011) developed low-density polyethylene (LDPE) packaging films incorporating AgNPs and ZnONPs to inhibit the *L. plantarum* development in minimally processed orange juice. Commercially available ZnONPs powder (average particle size of 70 nm) and P105 powder (contains 5% AgNPs and 95% TiO_2 [average particle size of 10 nm]) were purchased from Pars Nanonasb, Tehran, Iran, and mixed with LDPE resin pellets (film grade), and packaging film was extruded using a twin-screw extruder (Cincinnati Milacron, Batavia, OH) at a temperature of 190°C. Fresh oranges were purchased from local market and squeezed using a pilot scale juice extractor (Model: M2000A-1, CMEC Food Machinery, Suzhou, China) and filtered with 1 mm mesh filter, then sterilized for 15 min at a temperature of 121°C. The sterilized juice was kept in refrigerator (at 4°C) for all the time before packaging. The active *L. plantarum* culture was produced from the pure *L. plantarum* spores obtained from the Food Science Laboratory, Isfahan University of Technology, Tehran, Iran, by inoculating at 37°C for 24 h under 5% CO_2 atmosphere and then mixed with the above prepared orange juice to a microbial load level of 8.5 log CFU/mL. Then the juice was packed using pure LDPE, LDPE+0.25% ZnONPs, LDPE+1.0% ZnONPs, LDPE+1.5% P105 powder, and LDPE+ 5.0% P105 powder-packing films and kept under refrigeration temperature (4°C). Samples were collected on days 0, 7, 28, 56, 84, and 112 days of storage, and the microbial load was measured using a plate-culturing technique (plated on MRS at an incubation temperature 37°C for 48 h under 5% CO_2 atmosphere). The microbial count results showed that the LDPE packing film with nanoparticles (Ag and ZnO) controlled the growth of *L. plantarum* in orange juice except for the LDPE film with 1% ZnONPs better than the pure LDPE film. Increase in AgNPs concentration in packaging film reduced the growth of *L. plantarum*, but an increase

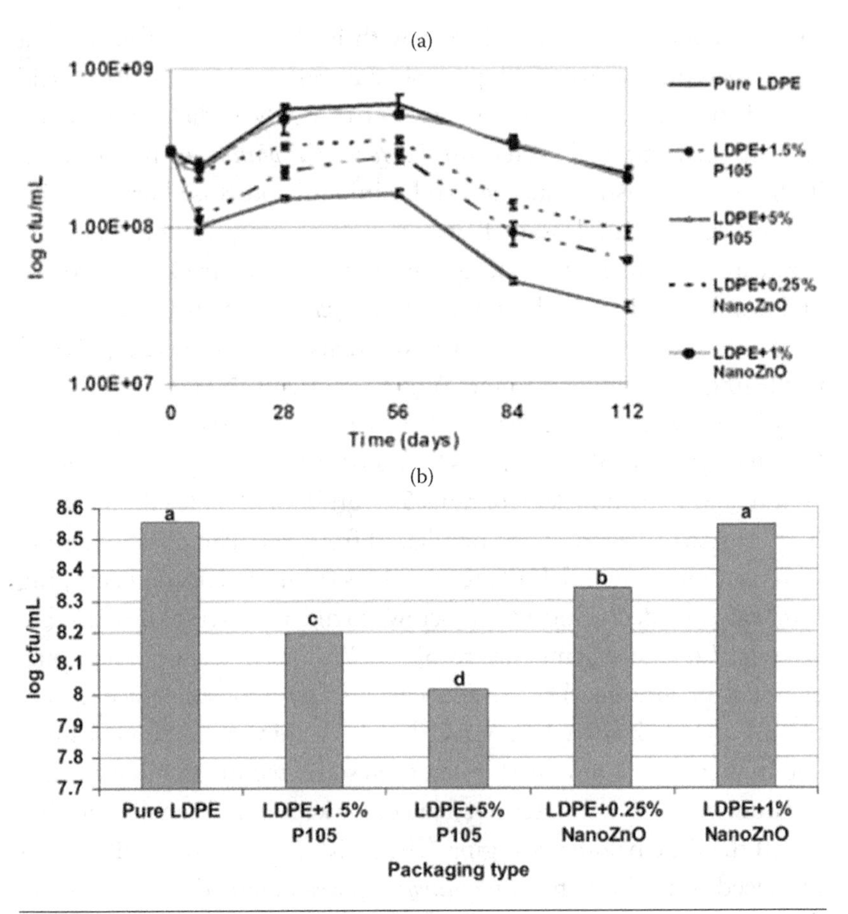

Figure 5.10 (a) *Lactobacillus plantarum* growth in orange juice packaging with Ag and ZnO nanoparticles stored at 4°C for 112 days; (b) microbial load of *L. plantarum* in orange juice packed in different types of packaging. (Reproduced from Emamifar, A. et al., *Food Control*, 22, 408–413, 2011.)

of ZnONPs concentration from 0.25% to 1% increased the microbial growth (Figure 5.10a). Emamifar et al. (2011) attributed this increase in microbial growth to the increase in ZnONP concentration resulted in the higher particle density in the film, which might reduce the generation of H_2O_2 from ZnONPs. The LDPE film with 5% AgNPs (P105 powder) showed the highest inhibition rate among all the treatments. The initial microbial load in the orange juice was 8.50 log CFU/mL, and the microbial load after 56 days storage at 4°C packed with pure LDPE film and LDPE+5% AgNPs was 8.82 and 8.23 log CFU/mL, respectively (Figure 5.10b). Results also showed that the AgNP-incorporated LDPE packaging films had higher inhibition rate than the ZnONP-incorporated LDPE films throughout the storage time at refrigeration temperature.

Emamifar et al. (2010) tested the shelf life of orange juice packed in LDPE polymers incorporated with AgNPs, and ZnONPs using the color, browning index, sensory, ascorbic acid (AA) degradation, and microbial count measurements while stored at refrigeration temperature (4°C). Packaging films (pure LDPE, LDPE+1.50% P105 powder [AgNPs], LDPE+5.00% P105 powder, LDPE+0.25% ZnONPs, LDPE+1.00% ZnONPs) were manufactured using the procedure described above (Emamifar et al. 2011), and 175 mL of freshly squeezed and minimally processed (filtered using 1 mm mesh filter and sterilized at 90°C) orange juice was packed in these packages. Then the packages were stored at 4°C, and samples were collected and analyzed for quality parameters on days 0, 7, 28, and 56 of storage. The initial microbial loads were 4.93 log CFU/mL and 4.84 log CFU/mL for mold (yeast+mold) and aerobic bacteria (*L. plantarum*), respectively. This microbial load increased above 6 log CFU/mL (the cutoff limit for defining shelf storage time for orange juice) after 28 days of storage in pure LDPE packages; the microbial load remained below the cutoff value in all nanoengineered packages except LDPE+1.00% ZnONPs. LDPE film with AgNPs showed a better inhibition rate than the film with ZnONPs against mold, yeast, and aerobic bacteria. The browning index test showed that the development of brown pigments in orange juice increased with increase in storage time in all the packages, and the orange juice packed in LDPE+0.25% ZnONPs packaging film had lower browning index than all other packages. The initial AA value was above 85 mg/100 g in all packages, and AA content decreased with increase in storage time, and the AA loss was greater in the orange juice packed in LDPE+5.00% P105 powder when compared with the other packages. The AA degradation was also lower in the LDPE+0.25% ZnONPs packages than other packages. The color test of orange juices also showed that the packages with ZnONPs had lower color change than the packaging film with AgNPs. The sensory test performed with experienced sensory panel gave a high score for orange juice packed in LDPE+0.25% ZnONPs in all four categories (odor, taste, color, overall) (Figure 5.11).

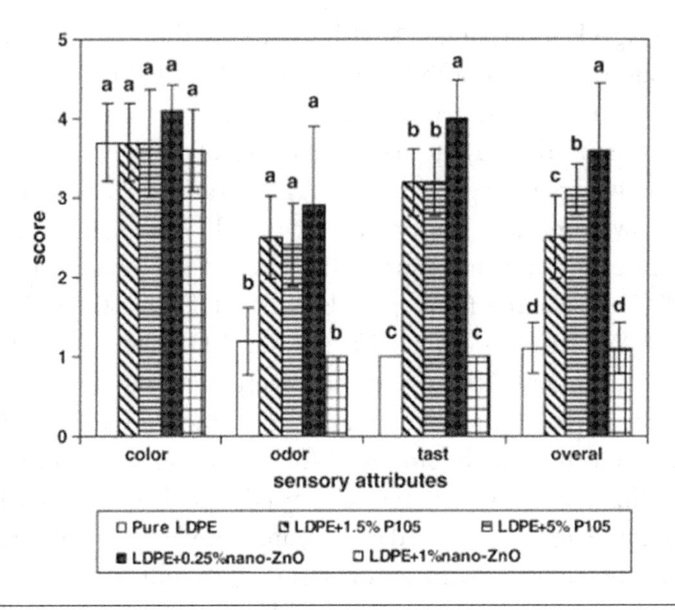

Figure 5.11 Sensory parameters of orange juice packed in different types of packaging films stored at 4°C for 28 days. (Reproduced from Emamifar, A. et al., *Innov. Food Sci. Emerg. Technol.*, 11, 742–748, 2010.)

Even though packaging films incorporated with AgNPs had better microbial load control than the packaging film with ZnONPs, the other quality parameters (color change, browning index, AA degradation) and sensory tests results showed packaging films incorporated with 0.25% ZnONPs was better suited for packaging of orange juice. Emamifar et al. (2010) suggested a mild heat treatment before packaging in the LDPE+0.25% ZnONPs film could significantly increase the shelf life of orange juice.

Antimicrobial properties of AgNPs are well known and used to control most of the foodborne pathogens in all sectors of the food industry. Nanocomposite materials developed with AgNPs have been tested for improving the shelf life of apple juice, watermelon juice, and kiwi juice as well as controlling foodborne pathogens (Del Nobile et al. 2004; Lloret et al. 2012). *Alicyclobacillus acidoterrestris* is one of the major gram-positive bacteria, which grows in acidic food products like fruit juices and tomato juice. Polyethylene packaging film was coated with AgNPs using parallel plasma reactor using an Ag electrode as a cathode. The pure *A. acidoterrestris* was cultured in 10 mL of malt extract agar and incubated for 2 days at 44°C. Then the solution was diluted to get a microbial load of 10^5 CFU/mL, and a 1 mL of this solution was mixed with the 10 mL of pasteurized commercial apple juice and packed with AgNP-coated nanopolymer film, regular polyethylene film without AgNPs, and no package. All the samples were kept at 44°C, and the microbial load was analyzed at regular intervals. The results showed that the apple juice in AgNP-coated polyethylene film had lower microbial (*A. acidoterrestris*) count (almost 2 log cycle lower load) than the control (no packaging film) and polyethylene film without AgNPs (Del Nobile et al. 2004). The cellulose absorbent pads used in the food industry was synthesized with AgNPs and tested for their antimicrobial effect against the mold, yeast, and lactic acid bacteria growth in kiwi and watermelon juices (Lloret et al. 2012). A 2 g of fluff pulp cellulose fibers were collected from the absorbent pads and mixed along with 200 mL of silver nitrate solution and stirred for 1 h using a magnetic stirrer and then dried for 48 h at 37°C. The dried cellulose-Ag solution was then heated for 10 min at 155°C and UV treated for 20 min followed by grinding for 2 min. The cellulose-AgNPs nanocomposite film was prepared from these homogenized particles, and the transmission electron microscopy images obtained from this film showed that this film had average particle size of 90 nm. Minimally processed kiwi and watermelon juices were stored in the polyethylene packages with and without the above-prepared cellulose-AgNPs nanocomposite for 11 days at 4°C. Samples were collected on day 1, 3, 7, and 10 and analyzed for yeast, mold, and lactic acid bacteria counts. The results showed that the kiwi and watermelon juices stored with cellulose-AgNPs composite had 99% lower yeast and mold growth and 90% lower lactic acid bacteria than the regular package after 10 days of storage at refrigeration temperature (Lloret et al. 2012).

Conclusion

Applications of nanotechnology are increasing dramatically for last decade in the beverages industry. The recent trend of consumer's preference towards beverages with natural and organic ingredients and a decline in carbonated drinks sale pressures the beverages industry to use advanced techniques like nanotechnology to produce beverages with natural products and enriched nutrient content.

The nanotechnology-based processes like nanofortification, nanofiltering, and nanoencapsulation have been used to enrich the nutrient value, the bioavailability of the nutrients, as well as nutrient intake. The beverages industry is applying nanotechnology-based products and processes in quality control and food safety of the beverages by controlling the growth of mold, yeast, and bacteria (especially lactic acid bacteria), which can increase the shelf life of the beverages. The nanotechnology-based sensors can be used to detect foodborne pathogens in beverages quickly with high accuracy, which can help the industry to take control measures to avoid quantity and quality losses. The major application of nanotechnology in the beverages industry is in packaging. Nanotechnology-based biopolymers, nanocomposites with the combination of regular packaging polymer films, and nanoparticles are helping to manufacture packaging films, layers of packaging materials, and containers for keeping the produce fresh and safe for a longer time. A recent survey by EcoFocus Worldwide showed that consumers want to avoid beverages with preservatives and artificial ingredients, and most prefer beverages with nutrient-rich natural ingredients (Evergreen Packaging 2017). Which is advantageous to a nanotechnology-based process where the naturally available micro and macro nutrients, and antioxidants in the beverages (like fruit juices), can be kept in the final product without any damage when the technologies like nanofiltration are used, which is not the case in traditional processing techniques like thermal treatment.

Even though the application of nanotechnology-based packaging methods like active packaging and smart packaging is increasing day by day in the beverages industry, the consumer's acceptance of newer technologies is less at present. The survey also showed a majority of the consumers lean towards eco-friendly packaging materials (nearly 73% of the consumers preferred recyclable packaging, and 59% preferred the packaging films made with renewable materials). Nearly 86% of the consumers believe the packaging material leaves some undesirable chemicals in beverages (Evergreen Packaging 2017), and consumers are still hesitant to accept the packaging films with nanoparticles (like Ag, Au, TiO_2) when these have a direct contact with the product inside the package. Many studies proved that these nanoparticles showed excellent antimicrobial properties towards major foodborne pathogens, but there is a risk of leaching of Ag or other metal oxide nanoparticles in the packaging film into the product, and this may lead to health issues. Therefore, the beverages packaging industry has invested considerably to develop beverage containers with nanoparticles as non-food contact materials to improve gas-barrier properties. Some of the commercially available fruit juice and beverages packaging multilayer PET containers use nanoclay as a gas-barrier material and in a layer where it would not have a direct contact with the product inside the package. Similar to other sectors of the food industry, winning consumers' hearts by proving the advantages of nanotechnology-based products and processes in beverages and eliminating the consumers' doubts about the health risks of nanotechnology by well-defined scientific research and technology transfer is the key to the successful adaptation of nanoscience and nanotechnology in the beverages industry.

References

Acosta O., Vaillant F., Pérez A.M., Dornier M. (2017) Concentration of polyphenolic compounds in blackberry (*Rubus adenotrichos* Schltdl.) juice by nanofiltration. *Journal of Food Process Engineering* 40:e12343.

Arend G.D., Adorno W.T., Rezzadori K., Di Luccio M., Chaves V.C., Reginatto F.H., Petrus J.C.C. (2017) Concentration of phenolic compounds from strawberry (*Fragaria X ananassa* Duch) juice by nanofiltration membrane. *Journal of Food Engineering* 201:36–41.

Berovic M., Berlot M., Kralj S., Makovec D. (2014) A new method for the rapid separation of magnetized yeast in sparkling wine. *Biochemical Engineering Journal* 88:77–84.

Beverage Industry. (2017) 2017 State of the industry, Beverage Industry Magazine, BNP Media, Troy, MI. Available from http://www.bevindustry.com/articles/90361-state-of-the-industry.

Catarino M., Mendes A. (2011) Dealcoholizing wine by membrane separation processes. *Innovative Food Science & Emerging Technologies* 12:330–337.

Chaudhry Q., Scotter M., Blackburn J., Ross B., Boxall A., Castle L., Aitken R., Watkins R. (2008) Applications and implications of nanotechnologies for the food sector. *Food Additives & Contaminants: Part A* 25:241–258. doi:10.1080/02652030701744538.

Chawla S., Rawal R., Kumar D., Pundir C.S. (2012) Amperometric determination of total phenolic content in wine by laccase immobilized onto silver nanoparticles/zinc oxide nanoparticles modified gold electrode. *Analytical Biochemistry* 430:16–23. doi:10.1016/j.ab.2012.07.025.

Chen D.H., Liao M.H. (2002) Preparation and characterization of YADH-bound magnetic nanoparticles. *Journal of Molecular Catalysis B: Enzymatic* 16:283–291.

Cheng H., Chen J., Chen S., Xia Q., Liu D., Ye X. (2016) Sensory evaluation, physicochemical properties and aroma-active profiles in a diverse collection of Chinese bayberry (*Myricarubra*) cultivars. *Food Chemistry* 212:374–385.

Del Nobile M., Cannarsi M., Altieri C., Sinigaglia M., Favia P., Iacoviello G., D'agostino R. (2004) Effect of Ag-containing nano-composite active packaging system on survival of *Alicyclobacillus acidoterrestris*. *Journal of Food Science* 69(8):E379–E383.

Duarte J.C., Rodrigues J.A.R., Moran P.J.S., Valença G.P., Nunhez J.R. (2013) Effect of immobilized cells in calcium alginate beads in alcoholic fermentation. *AMB Express* 3:31–31. doi:10.1186/2191-0855-3-31.

Emamifar A., Kadivar M., Shahedi M., Soleimanian-Zad S. (2010) Evaluation of nanocomposite packaging containing Ag and ZnO on shelf life of fresh orange juice. *Innovative Food Science & Emerging Technologies* 11:742–748.

Emamifar A., Kadivar M., Shahedi M., Soleimanian-Zad S. (2011) Effect of nanocomposite packaging containing Ag and ZnO on inactivation of *Lactobacillus plantarum* in orange juice. *Food Control* 22:408–413.

Evergreen Packaging. (2017) Four key beverage industry trends for 2017 are changing what consumers buy, *PR Newswire*, Cision, New York. Available from http://www.prnewswire.com/news-releases/four-key-beverage-industry-trends-for-2017-are-changing-what-consumers-buy-300384417.html.

Guerreiro J.R.L., Frederiksen M., Bochenkov V.E., De Freitas V., Ferreira Sales M.G., Sutherland D.S. (2014) Multifunctional biosensor based on localized surface plasmon resonance for monitoring small molecule–protein interaction. *Acs Nano* 8:7958–7967. doi:10.1021/nn501962y.

Honeywell. (2016) Aegis® OXCE Barrier Nylon Resin, in: Honeywell (Ed.), http://www51.honeywell.com/sm/aegis/products-n2/aegis-ox.html, Morris Plains, NJ.

Jackowetz J., Dierschke S., de Orduña R.M. (2011) Multifactorial analysis of acetaldehyde kinetics during alcoholic fermentation by *Saccharomyces cerevisiae*. *Food Research International* 44:310–316.

Koutinas A.A., Sypsas V., Kandylis P., Michelis A., Bekatorou A., Kourkoutas Y., Kordulis C., Lycourghiotis A., Banat I.M., Nigam P. (2012) Nano-tubular cellulose for bioprocess technology development. *PLoS One* 7:e34350.

Labanda J., Vichi S., Llorens J., López-Tamames E. (2009) Membrane separation technology for the reduction of alcoholic degree of a white model wine. *LWT-Food Science and Technology* 42:1390–1395.

Lecour S., Blackhurst D., Marais D., Opie L. (2006) Lowering the degree of alcohol in red wine does not alter its cardioprotective effect. *Journal of Molecular and Cellular Cardiology* 40:997–998.

Li G.Y., Li Y.J., Huang K.L., Zhong M. (2010) Surface functionalization of chitosan-coated magnetic nanoparticles for covalent immobilization of yeast alcohol dehydrogenase from *Saccharomyces cerevisiae*. *Journal of Magnetism and Magnetic Materials* 322:3862–3868.

Lisanti M.T., Gambuti A., Genovese A., Piombino P., Moio L. (2013) Partial dealcoholization of red wines by membrane contactor technique: Effect on sensory characteristics and volatile composition. *Food and Bioprocess Technology* 6:2289–2305.

Liu S.Q., Pilone G.J. (2000) An overview of formation and roles of acetaldehyde in winemaking with emphasis on microbiological implications. *International Journal of Food Science & Technology* 35:49–61.

Lloret E., Picouet P., Fernández A. (2012) Matrix effects on the antimicrobial capacity of silver-based nanocomposite absorbing materials. *LWT-Food Science and Technology* 49:333–338.

Massot A., Mietton-Peuchot M., Peuchot C., Milisic V. (2008) Nanofiltration and reverse osmosis in winemaking. *Desalination* 231:283–289.

Mello T., Ceni E., Surrenti C., Galli A. (2008) Alcohol-induced hepatic fibrosis: Role of acetaldehyde. *Molecular Aspects of Medicine* 29:17–21.

Mohammad A.W., Teow Y.H., Ang W.L., Chung Y.T., Oatley-Radcliffe D.L., Hilal N. (2015) Nanofiltration membranes review: Recent advances and future prospects. *Desalination* 356:226–254. doi:10.1016/j.desal.2014.10.043.

Monge M., Moreno-Arribas M.V. (2016) Applications of nanotechnology in wine production and quality and safety control. In: Moreno-Arribas M.V. and Bartolomé Suáldea B. (Eds.) *Wine Safety, Consumer Preference, and Human Health*. Springer International Publishing, Cham, Switzerland. pp. 51–69.

Nanocor. (2017) Imperm® Grade 103 Superior Gas barrier resin, Nanoror, Hoffman Estates, IL. Available from http://www.nanocor.com/tech_sheets/i103.pdf

NIH. (2016) Selenium: Dietary supplement fact sheet, National Institutes of Health, Bethesda, MD.

Patras A., Brunton N.P., O'Donnell C., Tiwari B. (2010) Effect of thermal processing on anthocyanin stability in foods: Mechanisms and kinetics of degradation. *Trends in Food Science & Technology* 21:3–11.

Quertemont E., Tambour S., Tirelli E. (2005) The role of acetaldehyde in the neurobehavioral effects of ethanol: A comprehensive review of animal studies. *Progress in Neurobiology* 75:247–274.

Rytwo G. (2012) The use of clay-polymer nanocomposites in wastewater pretreatment. *The Scientific World Journal* (Article Id 498503):1–7.

Rytwo G., Lavi R., Rytwo Y., Monchase H., Dultz S., König T.N. (2013) Clarification of olive mill and winery wastewater by means of clay–polymer nanocomposites. *Science of the Total Environment* 442:134–142.

Salaspuro M. (2011) Acetaldehyde and gastric cancer. *Journal of Digestive Diseases* 12:51–59.

Sui X., Bary S., Zhou W. (2016) Changes in the color, chemical stability and antioxidant capacity of thermally treated anthocyanin aqueous solution over storage. *Food Chemistry* 192:516–524. doi:10.1016/j.foodchem.2015.07.021.

Sui X., Dong X., Zhou W. (2014) Combined effect of pH and high temperature on the stability and antioxidant capacity of two anthocyanins in aqueous solution. *Food Chemistry* 163:163–170.

Szmitko P.E., Verma S. (2005) Red wine and your heart. *Circulation* 111(2):e10–e11.

Takács L., Vatai G., Korány K. (2007) Production of alcohol free wine by pervaporation. *Journal of Food Engineering* 78:118–125.

van Jaarsveld F., October F. (2015) Acetaldehyde in wine, Oenology Research, WineLand Media, Suider Paarl, South Africa.

Vance M.E., Kuiken T., Vejerano E.P., McGinnis S.P., Hochella Jr M.F., Rejeski D., Hull M.S. (2015) Nanotechnology in the real world: Redeveloping the nanomaterial consumer products inventory. *Beilstein Journal of Nanotechnology* 6:1769–1780.

Yu C.C., Yang K.H., Liu Y.C., Chen B.C. (2010) Photochemical fabrication of size-controllable gold nanoparticles on chitosan and their application on catalytic decomposition of acetaldehyde. *Materials Research Bulletin* 45:838–843.

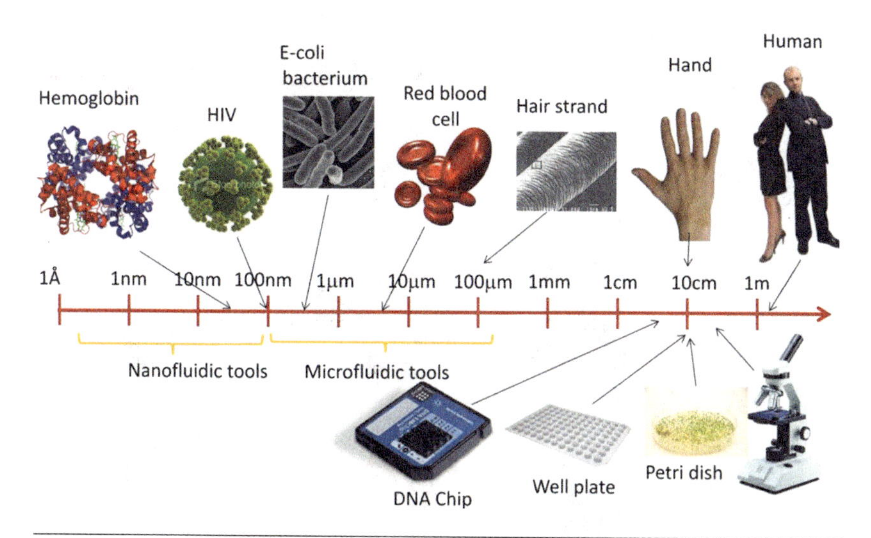

Figure 1.2 Scale of different materials. (Reproduced from Nguyen et al. *Adv. Drug Del. Rev.*, 65, 1403–1419, 2013.)

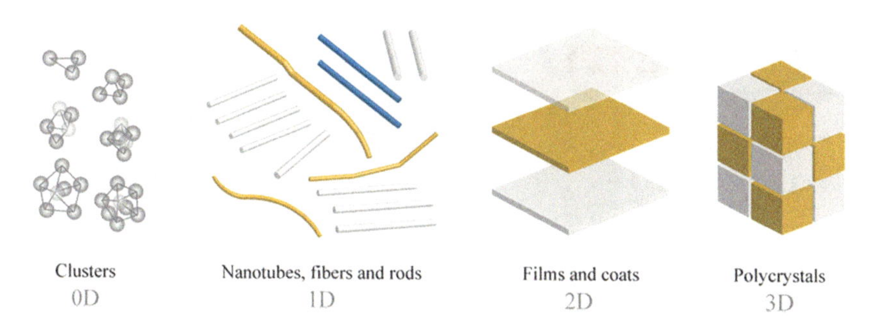

Figure 1.3 Types of nanomaterials based on dimensions. (Reproduced from Gusev, A.I., and Rempel, A.A., *Nanocrystalline Materials*, Cambridge International Science Publishing, Cambridge, UK, pp. 351, 2004.)

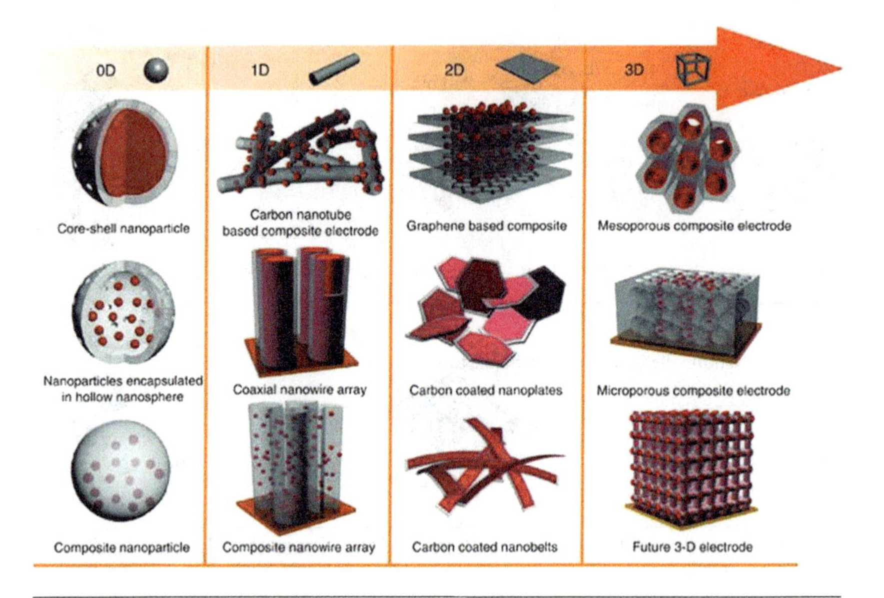

Figure 1.4 Examples of different types of nanomaterials based on the dimension. (Reproduced from Lukatskaya, M.R. et al., *Nat. Commun.*, 7, 1–13, 2016.)

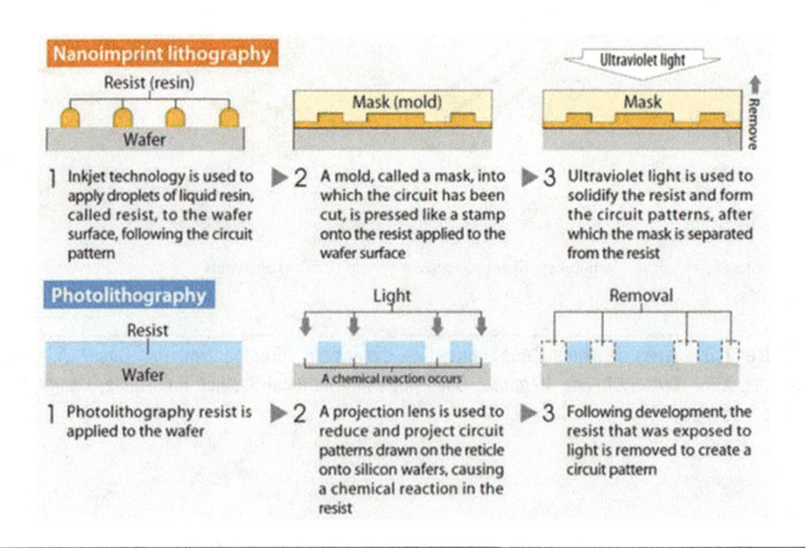

Figure 1.8 Schematic of nanoimprint lithography technique. (Reproduced from Clarke, R., Report: Toshiba adopts imprint litho for NAND production, ee News Analog, Available at http://www.eenewsanalog.com/news/report-toshiba-adopts-imprint-litho-nand-production, 2016.)

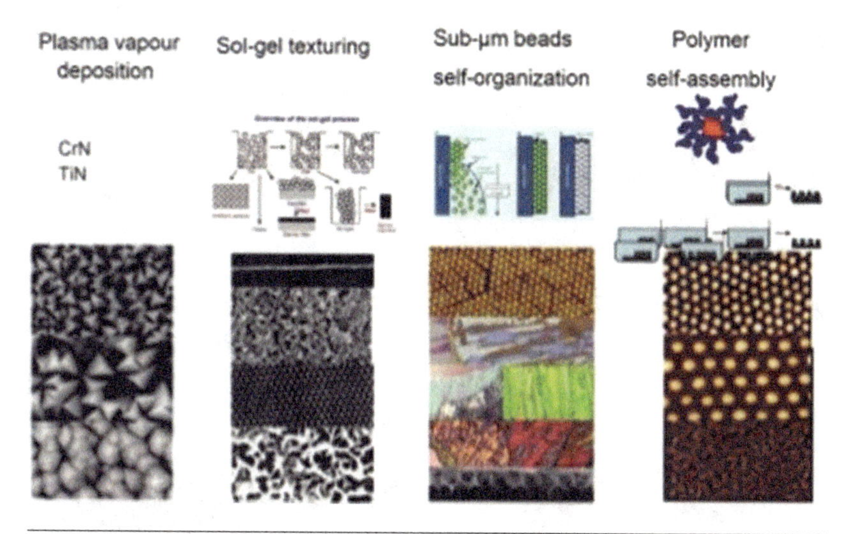

Figure 1.10 Bottom-up fabrication techniques. (Reproduced from CSME, Nanostructure engineering, Accessed on November 21, Available at https://www.csem.ch/Publications, 2017.)

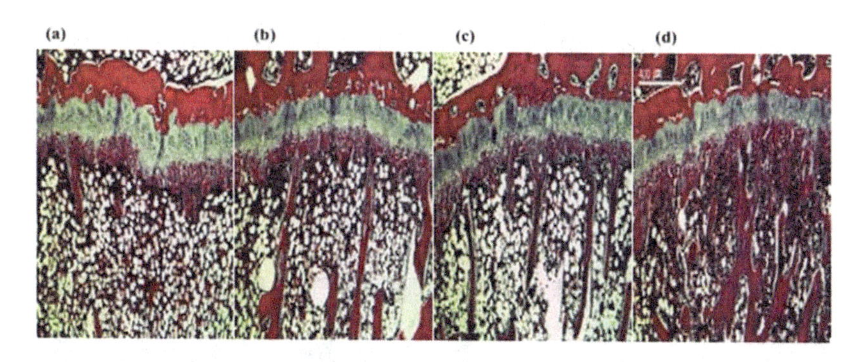

Figure 2.2 Microscopic images of trabecular bone in tibia part of rats fed with milk containing (a) no Ca supplementation (regular milk), (b) carbonated Ca, (c) ionized Ca, and (d) nano Ca. (Reproduced from Park, H.S. et al., *Asian-Aust. J. Anim. Sci.*, 20, 1266–1271, 2007.)

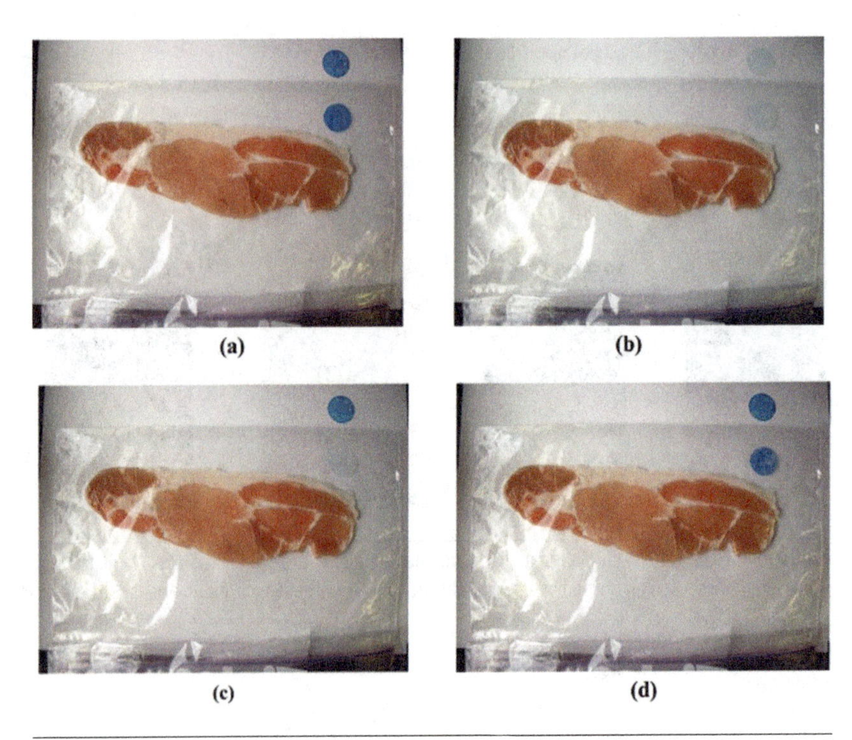

Figure 4.1 Oxygen nanosensors for detection of O_2 levels inside a packed meat (one sensor at each side [inside and outside] of the packaging film) (a) right after sealing, (b) immediately after activation of sensors with UVA light, (c) few minutes after activation of sensors, (d) after opening the package. (Reproduced from Mills, A., *Chem. Soc. Rev.*, 34, 1003–1011, 2005.)

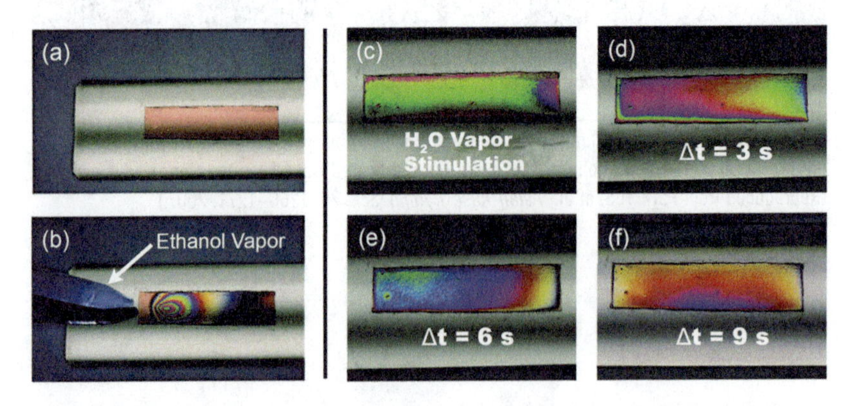

Figure 4.2 Nanosensor for moisture measurement (a) moisture sensor with carbon-coated copper nanoparticles at the start; (b) when exposed to ethanol vapor; (c) when exposed to water vapor; (d–f) moisture dry out by time (3, 6, and 9 s). (Reproduced from Luechinger, N.A. et al., *Langmuir*, 23, 3473–3477, 2007.)

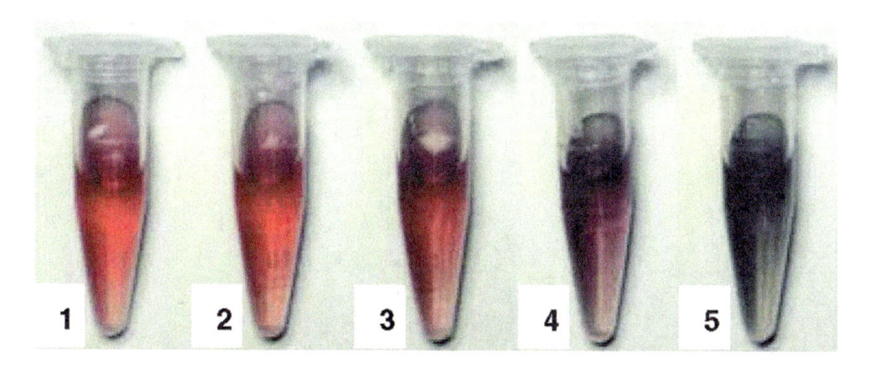

Figure 4.3 Visual color change of the optimized sensor: (1) without any addition of melamine; (2) with the addition of the extract from blank raw milk; (3) with the addition of the extract containing 1 ppm (final concentration: 8 ppb); (4) with the addition of the extract containing 2.5 ppm (final concentration: 20 ppb); (5) with the addition of the extract containing 5 ppm (final concentration: 40 ppb). (Reproduced from Ai, K. et al., *J. Am. Chem. Soc.*, 131, 9496–9497, 2009.)

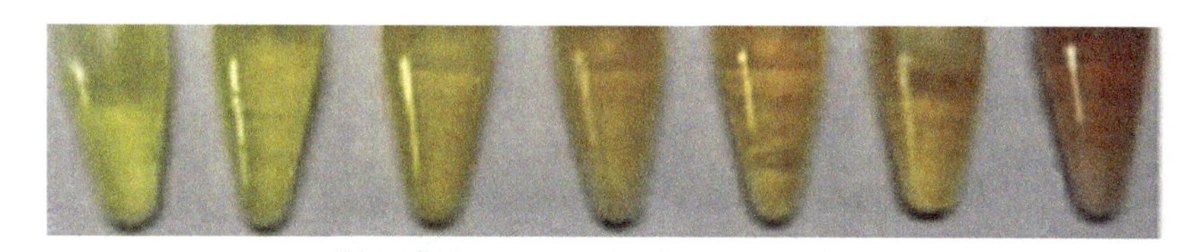

Figure 4.4 Test tubes contain 800 mL Ag NPs solution mixed with 600 mL melamine solution. The concentrations of spiked melamine (from left to right) are 0, 0.002, 0.004, 0.008, 0.017, 0.085, 0.17 mM, respectively. (Reproduced from Ping, H. et al., *Food Control*, 23, 191–197, 2012.)

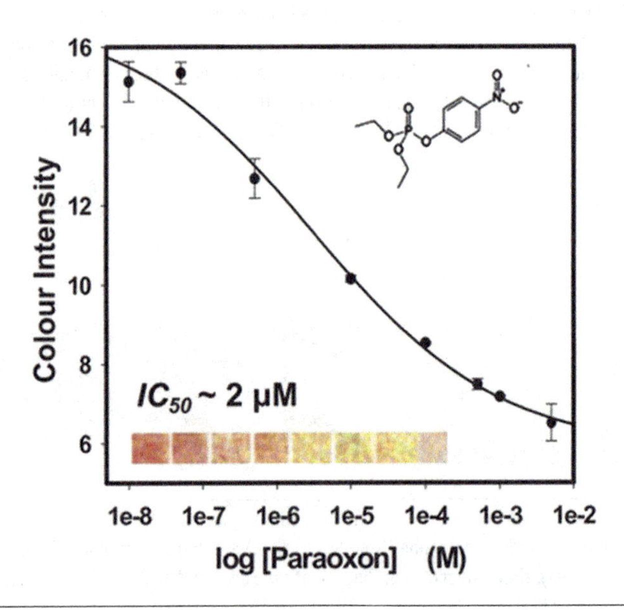

Figure 4.10 Semi-log color intensity curve of dipstick paper sensor with gold nanoparticles tested against paraoxon. (Reproduced from Luckham, R.E., and Brennan, J.D., *Analyst*, 135, 2028–2035, 2010.)

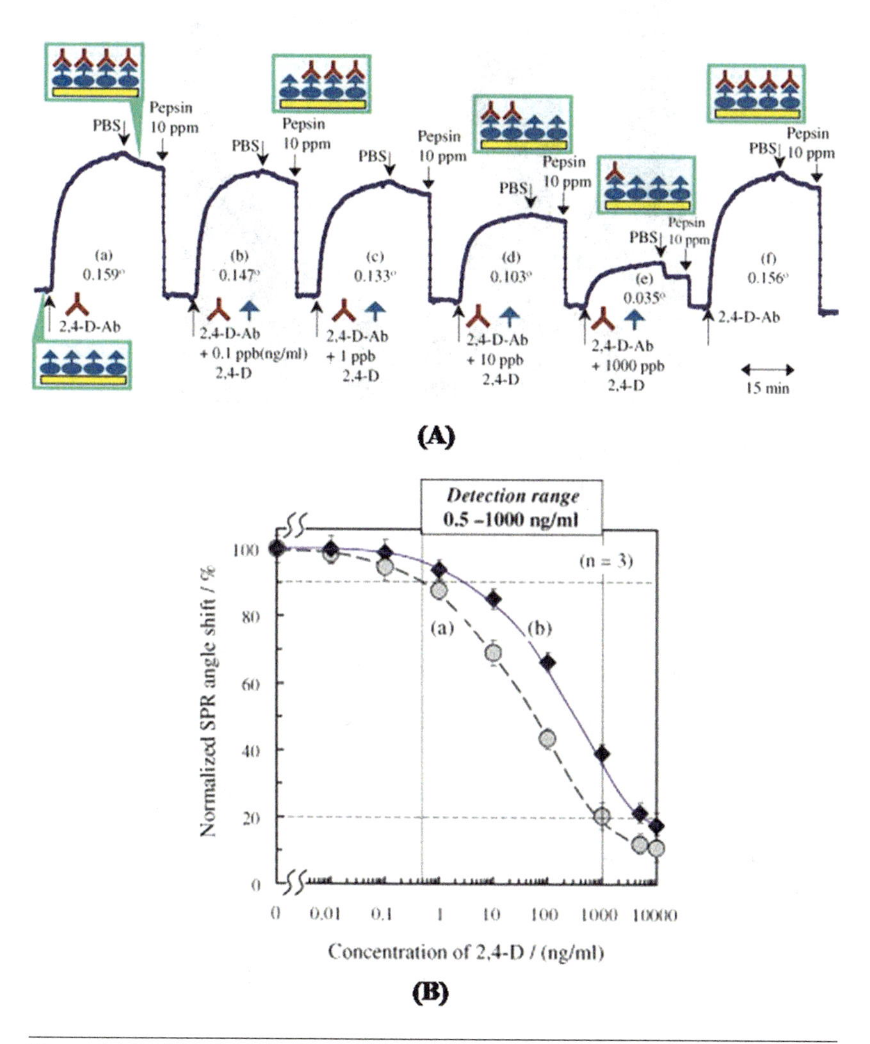

Figure 4.12 (A) Surface plasmon resonance (SPR) sensorgram of flow 0.1, 1, 10 and 1000 ng/mL 2,4-D and fixed concentration (20 ng/mL) and 2,4-D-antibody solutions over the immunosensor chip immobilized with of 2,4-D-BSA; (B) normalized SPR angle shift observed through the immunosensor with various concentrations of 2,4-D. (Reproduced from Gobi K.V. et al., *Sensor. Actuat. B-Chem.*, 111–112:562–571, 2005.)

Figure 5.5 Separation of the sediment using a permanent magnet from the wine bottle when wine was produced using fermentation by magnetized yeast. (Reproduced from Berovic, M. et al., *Biochem. Eng. J.*, 88, 77–84, 2014.)

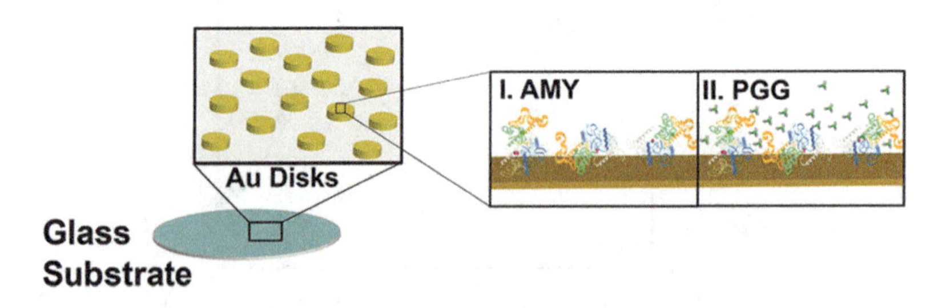

Figure 5.7 Schematic view of LSPR-based bio nanosensors with Au nanodisks and its reaction with α-amylase (AMY) and polyphenol pentagalloyl glucose (PGG). (Reproduced from Guerreiro, J.R.L. et al., *Acs Nano*, 8, 7958–7967, 2014.)

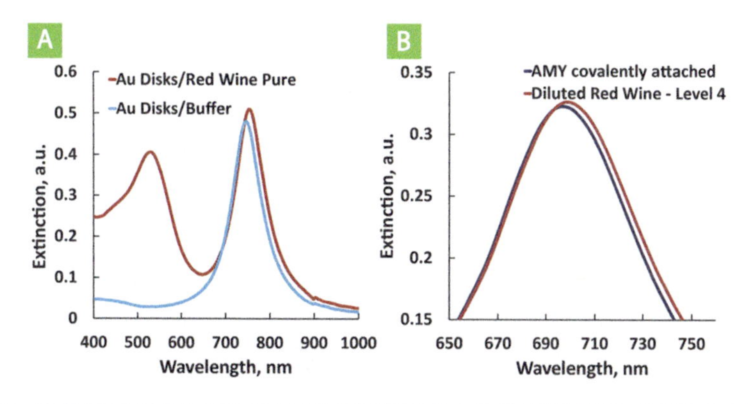

Figure 5.8 Spectra of wine samples analysis by the Gold nanodisk-localized surface plasmon resonance (LSPR) sensor: (A) spectra of red wine and buffer solution; (B) red shift of red wine analysis. (Reproduced from Guerreiro, J.R.L. et al., *Acs Nano*, 8, 7958–7967, 2014.)

Figure 6.1 (a) Crumb, (b) Crust, and (c) Top views of breads incorporated with different amounts of nanoencapsulated high-amylose corn starch (HACS) and Omega-3 particles. (Reproduced from Mogol, B.A. et al., *Agro Food Ind. Hi Tech.*, 24, 62–65, 2013.)

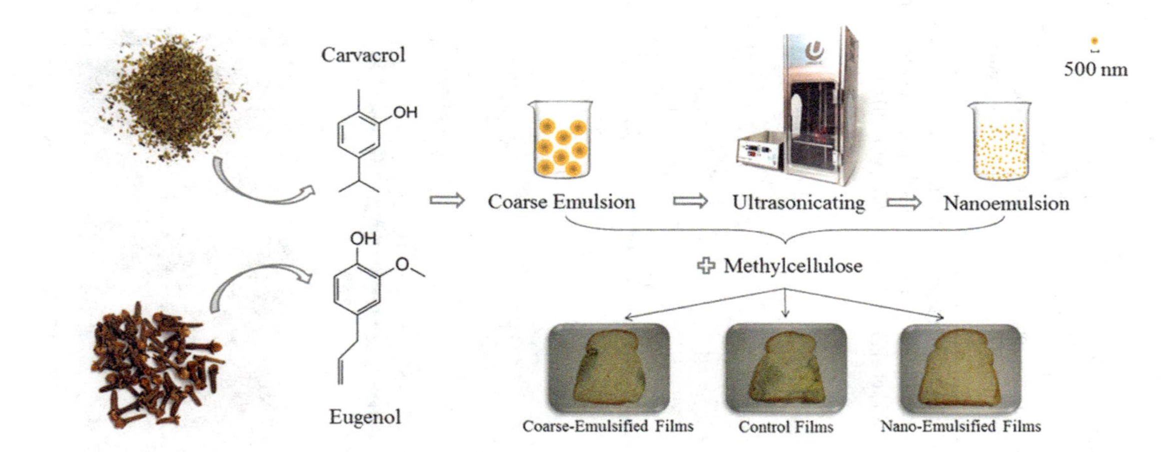

Figure 6.5 Schematic representation of shelf life extension of sliced bread packed using methylcellulose edible film incorporated with clove bud and oregano essential oil nanoemulsions. (Reproduced from Otoni, C.G. et al., *J. Agric. Food Chem.*, 62, 5214–5219, 2014.)

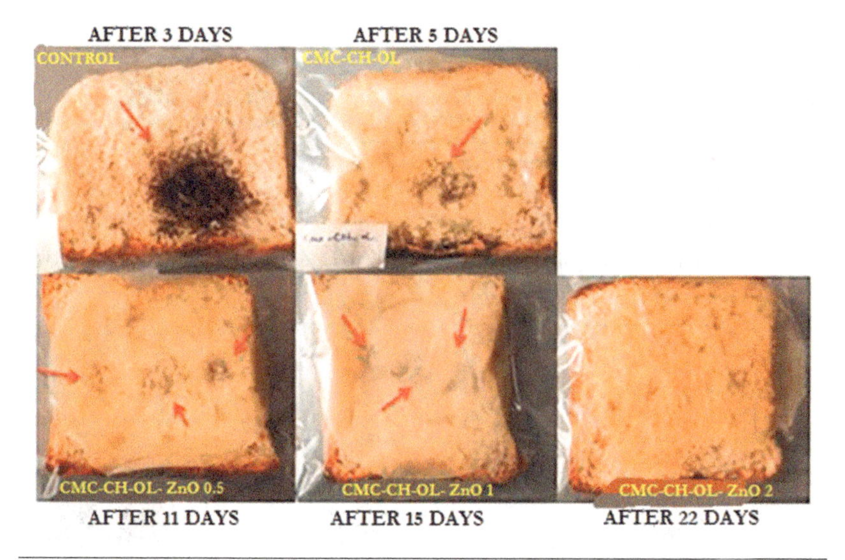

Figure 6.7 Visual appearance of bread stored at 25°C inoculated with *Aspergillus niger* in control and active packaging film (coated with chitosan-carboxymethyl cellulose-oleic acid (CMC-CH-OL) incorporated with 0.5%, 1%, and 2% of zinc oxide nanoparticles). (Reproduced from Noshirvani, N. et al., *Food Packaging and Shelf Life*, 11, 106–114, 2017a.)

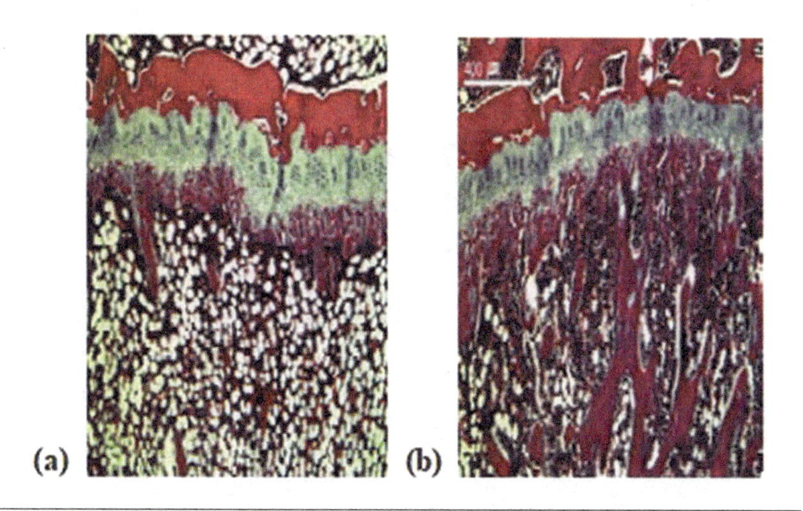

Figure 7.1 Micrographs of trabecular bones of ovariectomized rats after 18 weeks feeding with (a) control and (b) nano calcium supplemented milk. (Reproduced from Park, H.S. et al., *Asian Aust. J. Anim. Sci.*, 20, 1266–1271, 2007.)

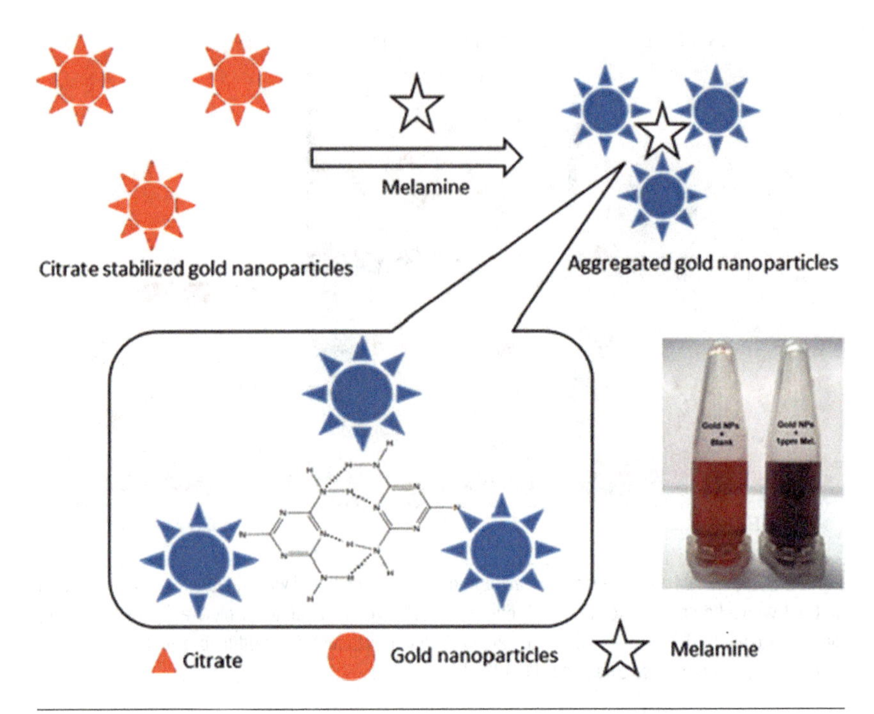

Figure 7.6 AuNPs nanosensors for melamine detection in milk (inset: AuNPs sensing solution in absence and presence of 1 mg/L melamine). (Reproduced from Kumar, N. et al., *Anal Biochem.*, 456, 43–49, 2014.)

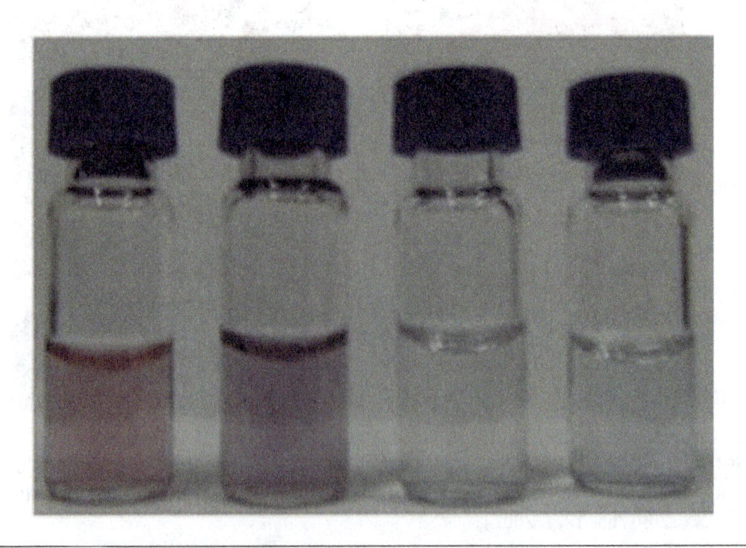

Figure 7.7 MA-functionalized AuNPs nanosensors reaction with addition of melamine at different concentrations (from left to right: 0, 10, 20, and 30 μg/mL). (Reproduced from Cai, H.H. et al., *J. Food Eng.*, 142, 163–169, 2014.)

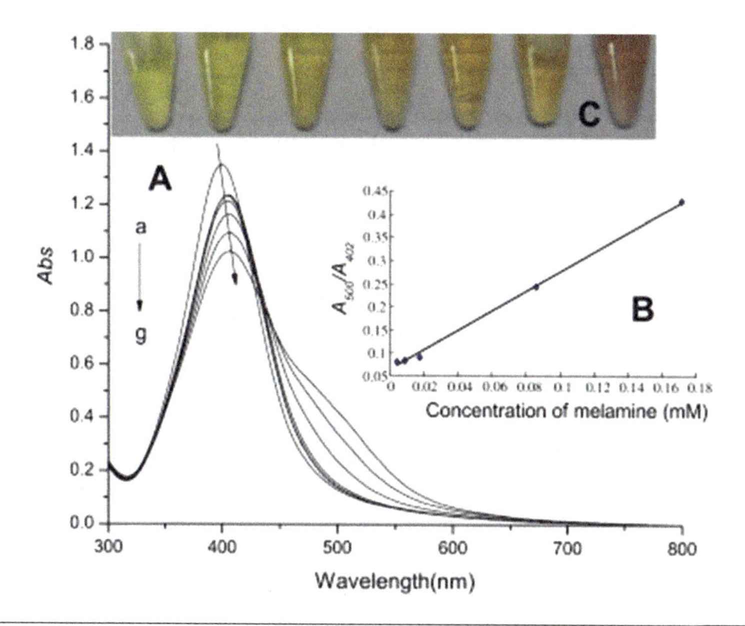

Figure 7.8 (A) Absorption spectra, (B) absorption ratio (A_{500}/A_{402}) vs melamine concentration, and (C) using AgNPs mixed in raw milk and various concentrations of melamine (from left to right 0, 0.002, 0.004, 0.008, 0.017, 0.085, 0.1700 mM, respectively). (Reproduced from Ping, H. et al., *Food Control*, 23, 191–197, 2012.)

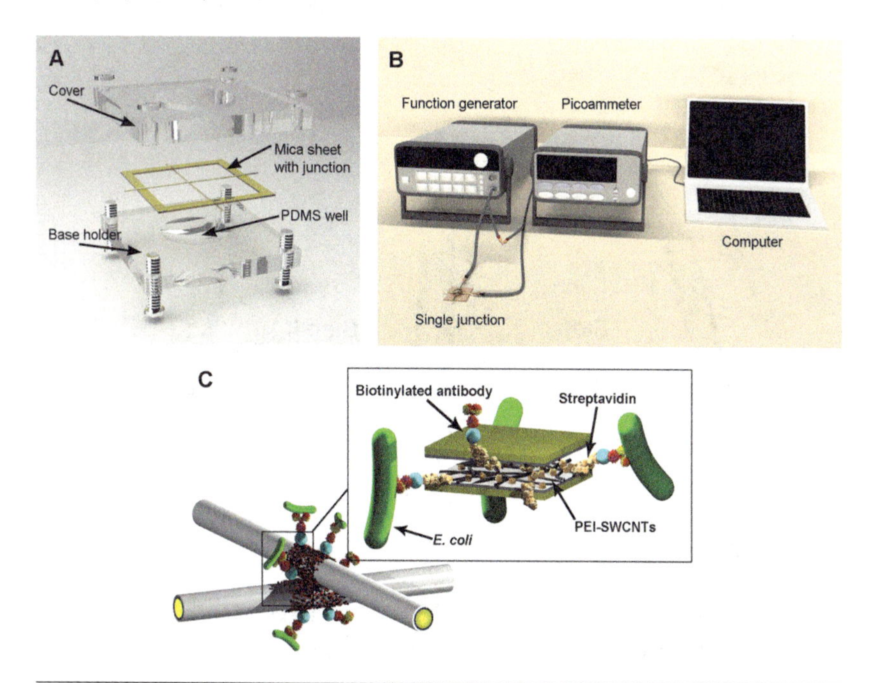

Figure 8.4 Design of a single-walled carbon nanotube (SWCNT) single junction biosensor for *E. coli* detection: (A) Setup of single junction nanosensor, (B) Experimental setup of SWCNT sensor for electrical current measurement for *E. coli* detection, and (C) Graphical illustration of *E. coli* captured on functionalized SWCNT junction. (Reproduced from Yamada, K.,. *PLoS One*, 9, e105767, 2014.)

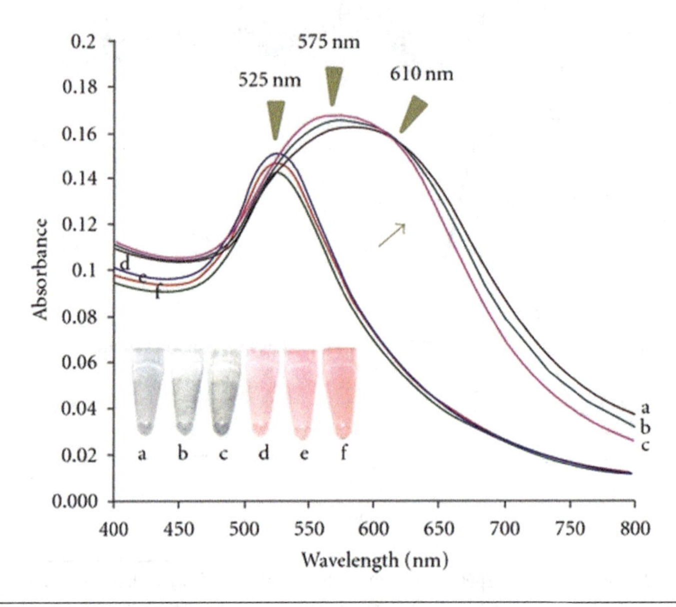

Figure 8.7 Absorbance spectra and vials of swine DNA in mixed meatball prepared from (a) pure pork, (b) mixtures of pork–beef, (c) pork–chicken, (d) chicken–beef, (e) pure beef, and (f) pure chicken with AuNPs and deionized solution. (Reproduced from Ali, M.E. et al., *J. Nanomater.*, 1–7, 2012.)

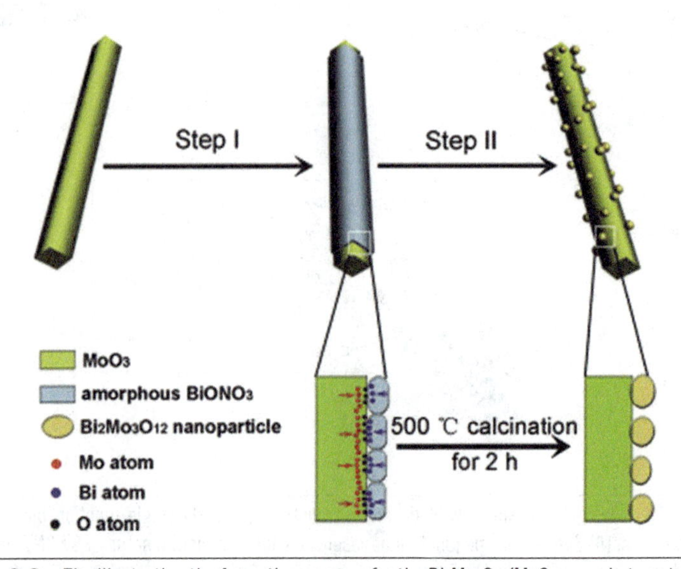

Figure 9.2 The illustration the formation process for the $Bi_2Mo_3O_{12}/MoO_3$ nano-heterostructures. (Reproduced from Liu, T. et al., *Chem. Eng. J.*, 244, 382–390, 2014.)

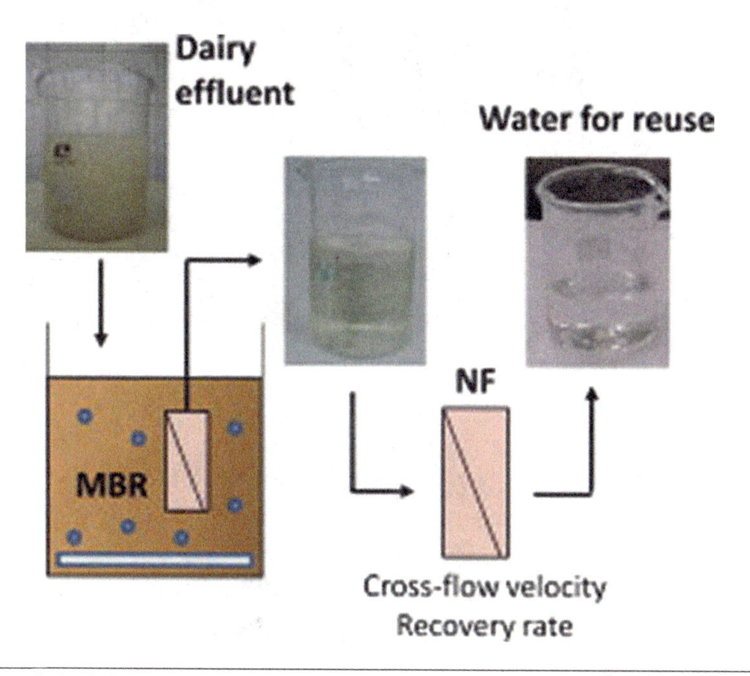

Figure 9.4 Dairy wastewater treatment. (Reproduced from Andrade, L. et al., *Sep. Purif. Technol.*, 126, 21–29, 2014.)

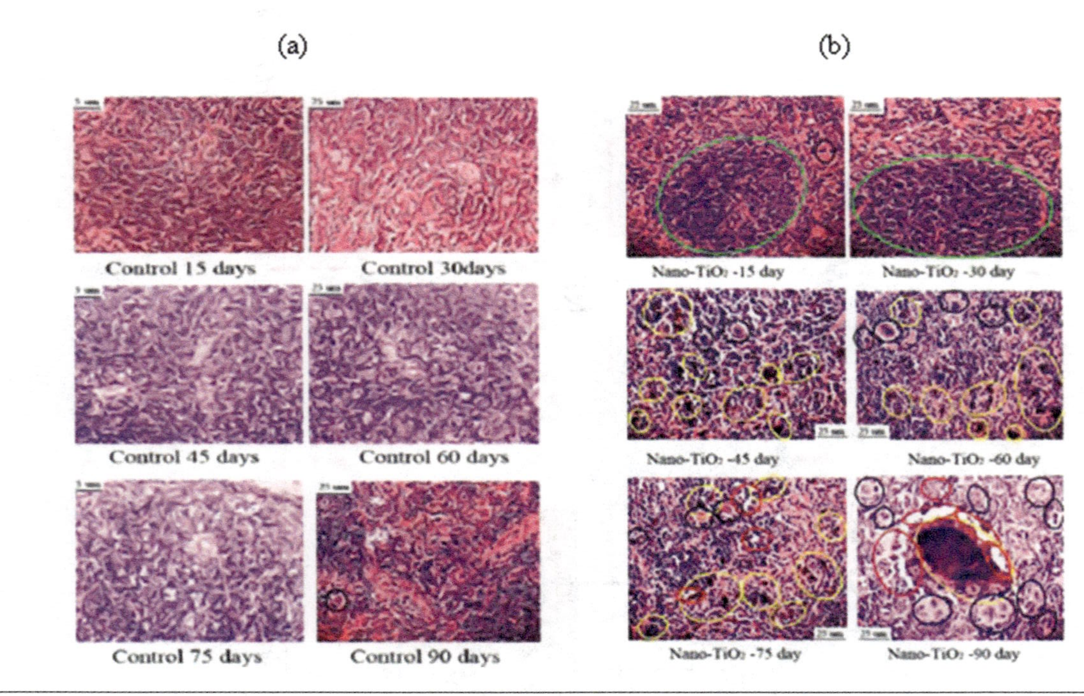

Figure 10.2 Histopathological changes of mice spleen with (a) control and (b) nano-TiO_2 exposure for different durations (Black circle: macrophage infiltration, green circle lymphocyte proliferation; yellow circle: nano-TiO_2 aggregation; red circle: fat degeneration and cell necrosis). (Reproduced from Sang, X. et al., *J. Agricult. Food Chem.*, 61, 5590–5599, 2013.)

6

Applications in the Bakery Industry

Introduction

Bakery products are staple food in most parts of the world. Bread has especially been considered more staple food for the Western world for decades. Developments in science and technology helped to evolve new products and processes in the bakery industry. Developments in nanoscience and nanotechnology are helping various sectors of the food industry from farm to fork, and the bakery industry is not an exception for this evolution (Neethirajan and Jayas 2011). The major application of nanotechnology is the development of active packaging materials in the bakery industry. Nanoencapsulation and nanoemulsion technologies can be used to add essential micro and macro nutrients to the bakery products during processing. Nanocomposites with nanoclay and other nanoparticles can be used for active packaging of bread and other bakery products to keep the bakery products for a long time with good quality (Alhendi and Choudhary 2013). Keeping the bakery products fresh is the most challenging task of the bakery industry due to quick physical and chemical changes occurring in the products. Active packaging techniques with nanomaterials can alter the mechanical, thermal, and gas transmission properties of the packaging polymers in order to keep the bakery products fresh and safe for a prolonged time. Nanotechnology also helped in the development of active packaging films with enhanced antimicrobial properties, which can control the growth of microorganisms like bacteria, mold, and yeast in the raw and processed products of the bakery industry. Nanoparticles of silver (Ag), gold (Au), and copper (Cu) can be incorporated with packaging polymers while casting, or they can apply the coating on the polymers to inhibit the growth of microbes.

The bakery industry also uses conventional analytical techniques, such as plate culturing, gas chromatography (GC), and liquid chromatography, for detecting microorganisms in raw and processed products, which are either time consuming or expensive. Nanosensors developed using nanotechnology can be used to detect mold and bacteria attack on baked products, and these can be used for quick detection of mycotoxins in the raw materials like flour and processed products like bread (Akbari et al. 2015; Baeumner 2004; Gomes et al. 2015).

Applications During Processing

Bakery products are the good source of carbohydrates, proteins, minerals, and vitamins. Nowadays the use of whole grains in bakery products is increasing due to the consumer's demand for nutrition-rich food. Use of whole grains in manufacturing bakery products such as bread increases the fiber content of the product as well as other micronutrients. Nutritional supplements are also added with the ingredients to improve the nutritional value of the bakery products. The advancement in nanotechnologies like nanoencapsulation, nanoemulsions, and nanofortification can help the bakery industry to add essential micro and macro nutrients with the ingredients of the bakery products without altering the textural, physical, and sensory characteristics of products (Chaudhry et al. 2008; Flora et al. 1998; Mozafari et al. 2008; Sekhon 2010). In particular, nanoencapsulation can help in adding nutrients like Omega-3 and other nutritional compounds to the bakery products to increase dietary nutritional value without altering consumer's acceptance of the product. Nanoencapsulation can also help in target delivery of the nutrients, which may increase the bioavailability of the nutritional compounds. Applications of nanoscience and nanotechnology in food processing are explained in several chapters (Chapters 1, 5, 7, and 8) of this book; nevertheless, applications focusing on the bakery industry are discussed in this chapter.

Omega-3 fatty acids are one of the main essential fats for the human body, which affect the receptors in cell membranes. Omega-3 fatty acid helps to prevent heart disease and heart stroke, as well as plays a major role in protecting the human body from cancers (de Deckere

2000; HPSH 2017). Tuna fish oil is the major source of Omega-3 fatty acids, but some people do not like to take tuna fish oil in their diet because of the unpleasant smell of the fish oil. George Western Foods ltd., Enfield, Western Australia, Australia, produced bread incorporated with tuna fish oil nanocapsules with the commercial name of "Tip-Top," which target delivers Omega-3 fatty acids. The tuna fish oil is capsulated at the nanoscale level and mixed with the bread during the bread manufacturing process. These fish oil nanocapsules break only after reaching the stomach and release the fish oil directly into the digestive system. Nanoencapsulation masks the unpleasant odor of fish oil so that people who have the dislike due to the smell can also include this bread in their diet to intake Omega-3 fatty oils (Neethirajan and Jayas 2011).

Flax seed oil is another source of Omega-3 fatty acids, and Mogol et al. (2013) developed white bread with high-amylose corn starch (HACS) nanoencapsulting Omega-3 fatty acids (from cold-pressed flax oil) and silymarin (derivative from the plant *Silybum marianum*, which is used as a natural antimicrobial agent). Deterioration during processing and storage due to oxidation of these polyunsaturated fatty acids (Omega-3 fatty acids) is the most common issue when using Omega-3 fatty acids as ingredients in the food products. The nanoencapsulation technique helps to keep the Omega-3 fatty acids stable for a long time as well as targets delivery of the nutrient. The use of silymarin in food products is limited due to its susceptibility to degradation during the thermal processing of foods. Nanoencapsulation also prevents the thermal degradation of silymarin. For encapsulating Omega-3 (Mogol et al. 2013), first 100 g of HACS was mixed with 1 L of KOH (0.1 M) and heated to 80°C for about 30 min, and then the solution was cooled to room temperature. The cooled HACS-KOH solution was kept on a magnetic stirrer and cold-pressed flax oil (10 g for 100 g HACS) was added to it, and then homogenized at 20 psi (137.89 kPa) through a high-pressure homogenizer (Model: Micro De Bee, BEE International, MA, USA). The collected suspension was spray dried at inlet and outlet temperatures of 200°C and 90°C, respectively, for manufacturing nanoencapsulated Omega-3 fatty acid particles. For making nanoencapsulated silymarin particles (Mogol et al. 2013), first, the silymarin was mixed with food-grade ethanol (10% w/w concentration) and then added to a HACS-KOH solution (1:10 ratio). Then similar to Omega-3 fatty acid nanoencapsulation, the solution was homogenized and then the suspension was made as nanosize particles using spray drying. The above-prepared nanoencapsulated HACS-Omega-3 fatty acid, and HACS-silymarin were added into bread dough at various concentrations (1.0%, 2.5%, 5.0%, and 10.0%), and the bread was baked according to the American Association of Cereal Chemists (AACC)-10B standard method (AACC 2000). Results showed that the bread density, color, and volume changed significantly when the HACS-Omega-3, or HACS-silymarin concentration was above 2.5% (Figure 6.1). The density of the bread increased, and the loaf volume decreased with the increase in

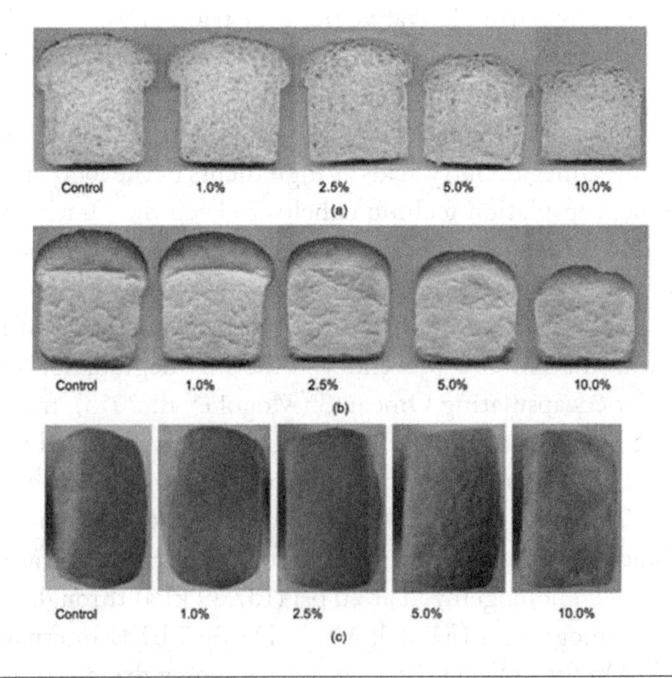

Figure 6.1 (See color insert.) (a) Crumb, (b) Crust, and (c) Top views of breads incorporated with different amounts of nanoencapsulated high-amylose corn starch (HACS) and Omega-3 particles. (Reproduced from Mogol, B.A. et al., *Agro Food Ind. Hi Tech.*, 24, 62–65, 2013.)

nanoencapsulated HACS-Omega-3 or silymarin concentrations. But the sensory test results showed that the bread quality was in the acceptable range when the HACS-Omega-3 or silymarin concentrations were ≤5.0%. The oxidation and thermal degradation test results proved nanoencapsulation significantly decreased the thermal degradation of silymarin (4% loss when nanoencapsulation, but 11% in regular form), and also oxidation of Omega-3 fatty acids (Mogol et al. 2013). Gökmen et al. (2011) developed similar kind of white bread as Mogol et al. (2013) with 1.0%, 2.5%, 5.0%, and 10.0% nanoencapsulated Omega-3 fatty acids using HACS (preparation of nanoencapsulated Omega-3 was same as Mogol et al. [2013]) and tested the texture, color, porosity, and oxidation of lipids using hexanal, nonanal, and thermal degradation (using the derivatives acrylamide and hydroxymethyl furfural [HMF]). The textural parameters like hardness and chewiness increased with HACS-Omega-3 fatty acid concentration, but the springiness index did not change with the increase in HACS-Omega-3 fatty acid concentration (remained at 0.9). The crust color became lighter and porosity of the bread slices lowered when the HACS-Omega-3 fatty acid concentration increased above 2.5%. The hexanal and nonanal concentrations were measured after seven days of storage at room temperature by analyzing the headspace of the vial containing a piece of bread slice using the gas chromatography-mass spectroscopy (GC-MS) system (model Agilent 6890 coupled with Agilent 5973, Agilent Technologies, Waldbronn, Germany). The HMF and acrylamide content of the bread slices were analyzed using an HPLC system (model: Agilent 1200, Agilent Technologies, Waldbronn, Germany). The oxidation test results showed that the encapsulation of Omega-3 fatty acids from flax seed oil significantly reduced the formation of hexanal and nonanal (derivatives of the poly unsaturated fatty acid oxidation process) after seven days of storage at room temperature (Figure 6.2). An acrylamide test also showed that the nanoencapsulation significantly reduced the formation of acrylamide due to thermal degradation during baking (Figure 6.3), and also the acrylamide formation in bread decreased with increase in HACS-Omega-3 fatty acid concentration. The HMF test also showed that

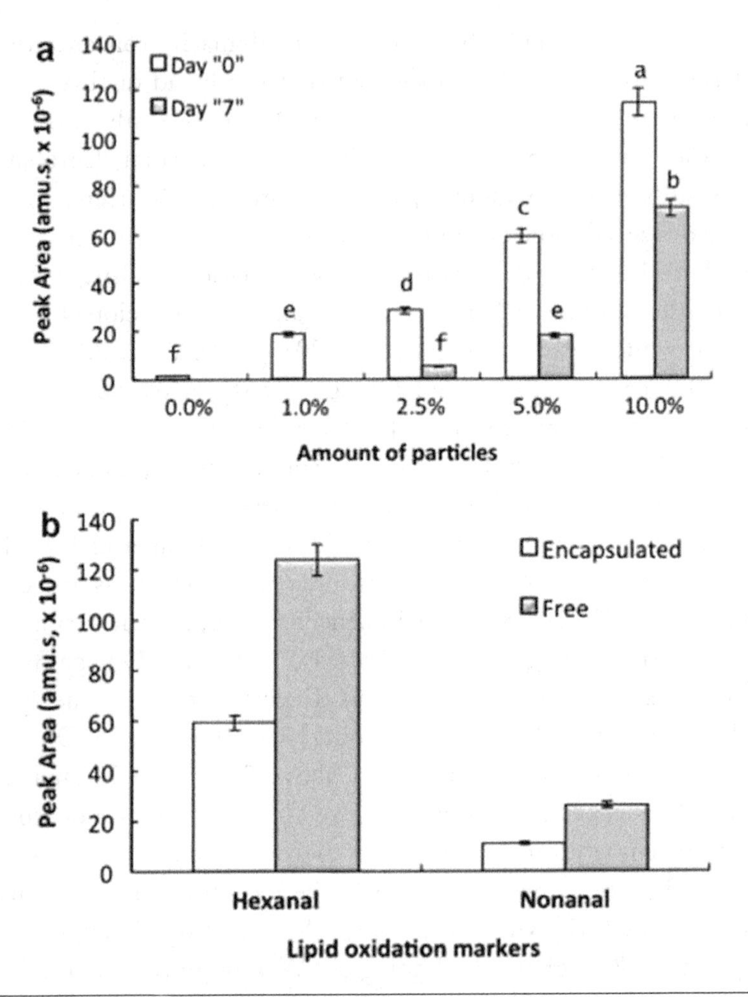

Figure 6.2 (a) Hexanal levels in the bread at different HACS-Omega-3 fatty acid levels on day 0 and day 7 of storage at room temperature. (b) Hexanal and nonanal levels of nanoencapsulated and free HACS-Omega-3 particles (0.5% concentration) in bread after 7 days of storage at room temperature. (Reproduced from Gökmen, V. et al., *J. Food Eng.*, 105, 585–591, 2011.)

HMF concentration was significantly reduced when the HACS-Omega-3 fatty acid concentration was above 5% (Figure 6.4). But nanoencapsulation of Omega-3 fatty acids increased the formation of HMF, and Gökmen et al. (2011) attributed that the high amylase content of HACS and the physico-chemical changes due to

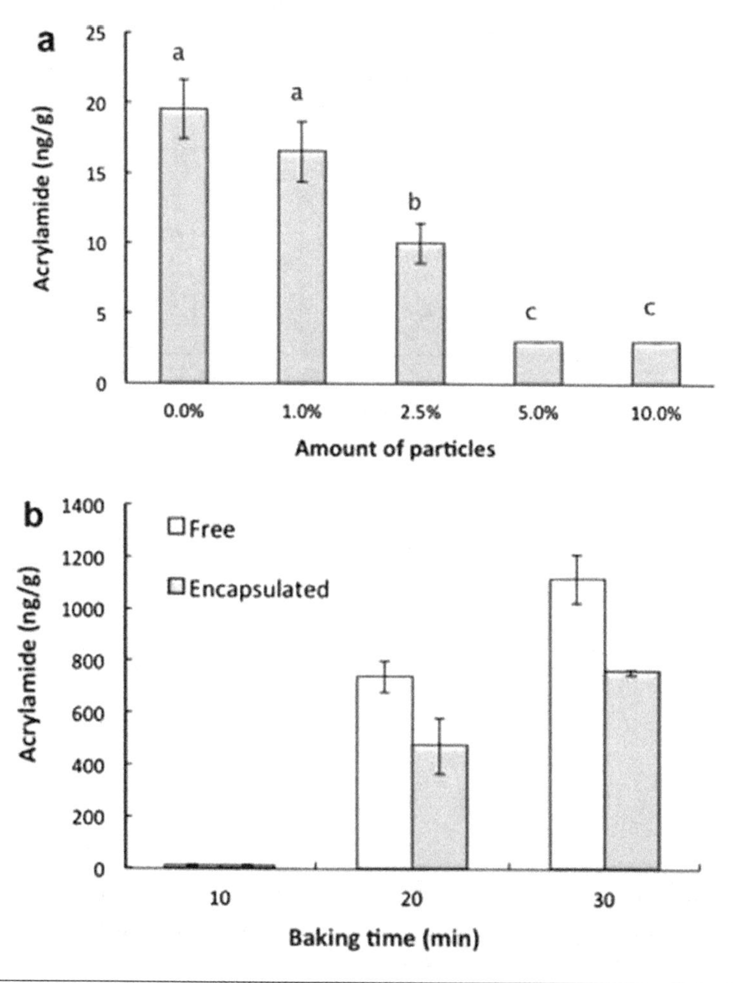

Figure 6.3 (a) Acrylamide levels in the bread at different HACS-Omega-3 fatty acid levels. (b) Acrylamide levels of nanoencapsulated and free HACS-Omega-3 particles (1.0% concentration). (Reproduced from Gökmen, V. et al., *J. Food Eng.*, 105, 585–591, 2011.)

the interaction between HACS-Omega-3 fatty acid particles and food matrix during baking might be the reason for the increase in HMF while using the nanoencapsulation technique. The sensory test of bread slices showed that the nanoencapsulation reduced the off flavor produced by the Omega-3 fatty acids.

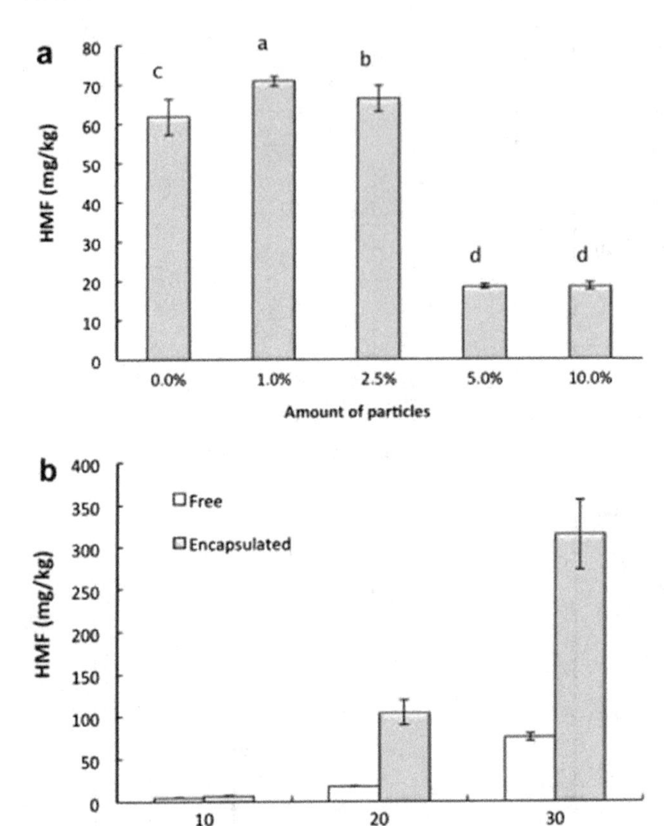

Figure 6.4 (a) HMF levels in the bread at different HACS-Omega-3 fatty acid levels. (b) HMF levels of nanoencapsulated and free HACS-Omega-3 particles (1.0% concentration). (Reproduced from Gökmen, V. et al., *J. Food Eng.*, 105, 585–591, 2011.)

Applications During Packaging

The shelf life of bakery products is very low (few hours to few days after baking) due to their high water activity. The freshness of the bakery products also lasts only for a very short time. The major microorganisms that spoil bakery products are mold, bacteria, and yeast. Their development on products affects the shelf life of the products. Staling (hardening of bread due to physical and chemical changes) is another factor affecting the quality (freshness) of the products. The microorganism growth is encouraged by the high water activity (0.90–0.95) of the baked product, but the loss of moisture in the package may lead to staling of bread and other bakery products. Thus, the packaging of bakery products plays a major role in the quality preservation of the bakery products. The packaging should keep the product fresh and soft as well as control microbial growth for extended shelf life and better quality. The bakery industry is using different techniques such as modified atmosphere packaging (packed with low oxygen/high carbon dioxide), irradiation of products with gamma or ultraviolet rays, and adding antimicrobial additives during production to achieve better shelf life and quality of the baked products (Degirmencioglu et al. 2011; Hussain and Jamil 2012; Marathe et al. 2002; Pateras 2007; Rasmussen and Hansen 2001). Developments in nanoscience and nanotechnology help the packaging industry to develop packaging polymers incorporating nanoparticles or nanotubes that enhance gas and moisture transmission and have excellent mechanical and thermal properties of packaging films for storing bakery products for an extended time without physical, chemical, and microbial quality losses (Alhendi and Choudhary 2013; Arora and Padua 2010; Han et al. 2011; Rhim et al. 2013; Sorrentino et al. 2007).

In modified atmosphere packaging (MAP), the oxygen concentration inside the packaging is reduced to inhibit the growth of aerobic microorganisms in the stored product. Generally, the gas concentration inside the package is artificially modified with higher amount of nitrogen and carbon dioxide (several mix ratios 70% N_2:30% CO_2; 30% N_2:70% CO_2; 60% N_2:40% CO_2; and 100% CO_2) to create low oxygen levels inside the package to increase the shelf life of the stored product. Degrimencioglu et al. (2011) tested various mixture ratios of N_2 and CO_2 and found that using pure CO_2 (100% CO_2) controlled the growth of all the aerobic microorganisms (mold, bacteria, and yeast) in food products. They also tested the oxygen-absorbing materials (oxygen scavengers) to control the mold growth in bakery products and compared the results with the bakery products stored in MAP at same storage conditions. The oxygen concentration of the package with oxygen scavenger was as low as 0.014%, but the MAP package had an oxygen concentration of 0.6%. Their results showed that oxygen scavengers could be used for increasing the shelf life of stored products better than the MAP package because mold can grow in an environment where the oxygen concentration is higher than 0.4%. Using nanotechnology, packaging films can be developed incorporating oxygen-scavenging material like titanium dioxide

(TiO_2) nanoparticles with plastic polymers (high-density polyethylene and low-density polyethylene) to use for packing bakery products. Cozmuta et al. (2015) developed a packaging film with TiO_2 and silver nanoparticles (AgNPs) for bread packaging, and it showed promising results in inhibiting microbial growth.

Active packaging is the recently developed concept, in which the packed product interacts with the components of the packaging to improve the shelf life of the product along with the quality (Ray et al. 2006). In regular active packaging, oxygen scavengers such as ascorbic acid, iron powder, or enzymes are used to reduce the oxygen concentration to less than 0.01% in order to avoid spoilage of stored products. Some of the researchers also used CO_2 scavengers (like CaO) to remove the CO_2 produced by the stored products for keeping the food safe for a longer time (Vermeiren et al. 1999). Application of ethylene-absorbing agents and moisture-regulating products helps to keep the moisture content of the package at a safe level in order to maintain the quality of the stored product (Alhendi and Choudhary 2013). Maintaining the moisture content of the product is a tricky process because high moisture leads to spoilage due to microbial growth, but the low moisture content dries the product, which affects the texture of bakery products as well as leads to staling. Packaging polymers with low permeation rates were used to maintain the moisture content of the packed products, and sometimes water-absorbing pads were used for control the excess moisture released by the products like meat during storage (Vermeiren et al. 1999). To control the microbial growth, packaging polymers were coated with antimicrobial agents like peptide, alcohol, bacteriocine, and sometimes metals (Ag, Au, and Cu) were used in active packaging of food products to enhance the shelf life (Suppakul et al. 2003). But sometimes some of these antimicrobial agents need to migrate to food from the packaging films to control the microbial growth, which may affect the consumers' acceptability of the products, and sometimes these may pose health risks to the consumer (Siegrist 2008). Application of organic materials such as natural essential oil as a coating of packaging film, or edible films, are other active packaging technique used for extending shelf life and maintain the quality of the baked products. Studies have been conducted to determine the effects of using cinnamon essential oil applied edible films, and cassava film with cinnamon and clove powder on mold (*Aspergillus niger, Penicillium expansum*) growth and shelf life of bread (Balaguer et al. 2013; Gutiérrez et al. 2011; Kechichian et al. 2010; Lopes et al. 2014). Even though the essential oils controlled mold and other microbial growth in baked products, the sensory tests found the aroma of the essential oil on the products after using these essential oil films in the package, and this negatively affected consumer acceptance of the baked product.

Recent developments in nanoscience and nanotechnology have also made an enormous growth in the area of active packaging of food products. Nanocomposites with nanoclay, AgNPs, gold nanoparticles (AuNPs), and TiO_2 nanoparticles have been tested for packaging of fruits, vegetables, dairy, meat, bakery, and other food products (raw and processed), and yielded positive results on enhancement of shelf life of the stored products (Alhendi and Choudhary 2013; Arora and Padua 2010; Avella et al. 2005; De Azeredo 2009; Rhim et al. 2013). Incorporation of nanoclay with the packaging polymers positively affects the physical, thermal, mechanical, and permeation (gas and moisture) properties of the polymers and results in the enhancement of the quality and shelf life of stored products. Nanocomposite packaging films with nanoclay have better moisture stability and lower gas permeation than regular polymer materials. The addition of nanoclay and other nanoparticles also give better resistance to ultraviolet light. This may also help to keep the stored product safe and good for a longer period. Ray et al. (2006) found that a random distribution of nanoclay particles in the film improved the thermal and gas permeation behavior more than when polymer material was mixed as intercalated or non-intercalated silicon sheets with nanoclay. Regular packaging films such as low-density polyethylene (LDPE) and high-density polyethylene (HDPE) when incorporated with AgNPs, AuNPs, CuNPs, and TiO_2NPs also enhanced antimicrobial, gas transmission, and mechanical properties and increased the shelf life of the bakery products. Incorporation of these nanoparticles with polymers reduced mold, bacteria, and yeast growth in a substantial manner. A detailed review of the applications of nanotechnology on the packaging of bakery products is available elsewhere (Alhendi and Choudhary 2013).

Essential oils like cinnamon oil and oregano oil have antimicrobial properties against molds and other foodborne pathogens. Otoni et al.

(2014) developed a methylcellulose edible film with cinnamon and oregano essential oil nanoemulsions to pack bread and tested along with methylcellulose edible film with coarse essential oil and methylcellulose edible film without any essential oil mix for *A. niger*, and *P. expansum* growth in bread stored at 25°C. A 4% (w/v) of coarse clove bud or oregano essential oil was mixed with Tween 80 (3% [w/v]) and double-distilled water and stirred for 5 min at 1000 rpm. For making nanoemulsions, this mixture was ultrasonicated at a frequency and power input of 20 kHz and 400 W for 10 min using a ultrasonicator (model DES500, Unique Group, Indaiatuba, Brazil). For making the film, methylcellulose (2% [w/v]) and polyethylene glycol (0.2% [w/v]) were mixed with the coarse and nanoemulsified clove bud or oregano essential oil solutions and homogenized at 6 rpm for 30 min followed by 2 h resting. The scanning electron microscopy and droplet analysis tests showed that the average size of nanoemulsified clove buds and oregano essential oil droplets were 180 and 250 nm, respectively. Then the prepared film-forming solution was casted on to metalized polypropylene sheets and dried at room temperature (25°C ± 2°C) overnight. Bread slices obtained from the local grocery store were sliced and placed between two films and stored at room temperature (25°C ± 2°C) for 15 days. Samples were collected from each treatment once in 5 days (0, 5, 10, and 15 days) for microbial analysis using a plate-culturing method. The complete procedure of development and testing of methylcellulose edible film with clove bud or oregano essential oil nanoemulsified droplets on mold and yeast growth in stored bread is graphically shown in Figure 6.5. The results showed that edible films with nano-modified clove buds showed more inhibition against molds and yeast growth in bread slices when compared with the other films with or without essential oils. The control treatment had more microbial count than other treatments, and the film with the synthetic commercial antimicrobial agent (solution comprising calcium propionate, ethanol, sorbic acid, and grain alcohol) had the similar effect as the control treatment on mold growth after 10 days of storage at room temperature (Figure 6.6). Statistical grouping test (Tukey's test) results showed that nanoemulsification of oregano oil did not affect the mold growth and showed similar inhibition pattern as the edible films with coarse oregano oil. But the nanoemulsification of clove buds showed better mold and yeast inhibition capacity

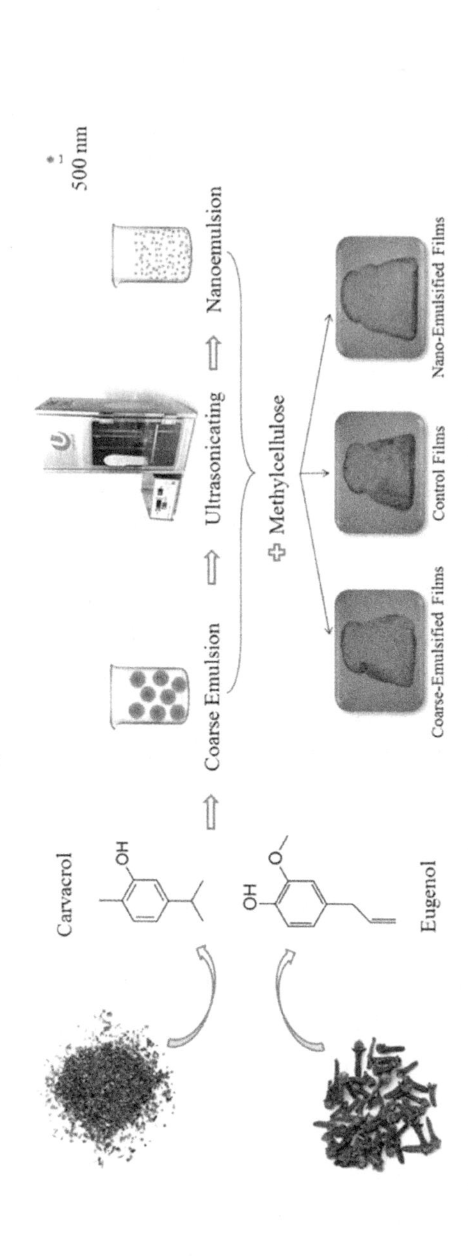

Figure 6.5 **(See color insert.)** Schematic representation of shelf life extension of sliced bread packed using methylcellulose edible film incorporated with clove bud and oregano essential oil nanoemulsions. (Reproduced from Otoni, C.G. et al., *J. Agric. Food Chem.*, 62, 5214–5219, 2014.)

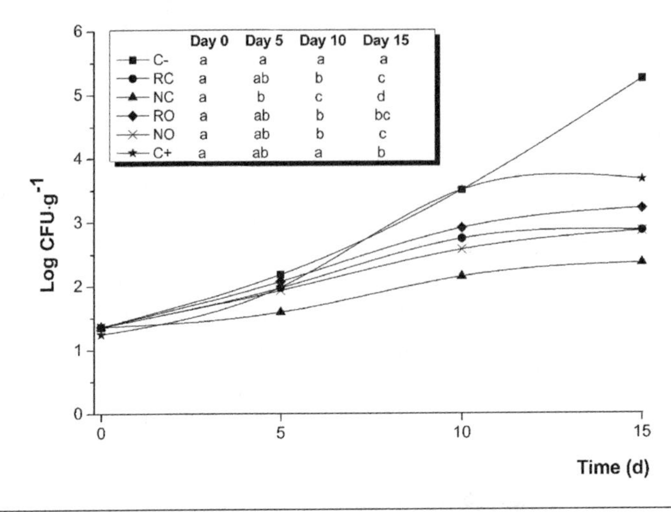

Figure 6.6 Microbial (mold and yeast) count on sliced bread packed with control (C−), commercial antimicrobial agents (C+), methylcellulose edible film with clove bud (coarse: RC, nanoemulsion: NC), and oregano (coarse: RO, nanoemulsion: NO) essential oils stored at 25°C. (Reproduced from Otoni, C.G. et al., *J. Agric. Food Chem.*, 62, 5214–5219, 2014.)

than the coarse clove buds when mixed with methylcellulose to make edible packaging films (Figure 6.6). The mechanical tests of edible films also proved the methylcellulose edible films casted with nano clove bud and oregano essential oil droplets showed better flexibility (had a higher elongation at breakage [>54%]) when compared with the regular films and films with coarse clove bud and oregano essential oil droplets.

Noshirvani et al. (2017a) developed an active packaging technology with chitosan-carboxymethyl cellulose-oleic acid (CMC-CH-OL) nanopolymer film incorporated with zinc oxide nanoparticles (ZnO NPs) for coating and packaging of bread to increase the shelf life of the product. For making the nanopolymer film, 0.2 g of chitosan was mixed with 50.0 mL of acidic water and stirred vigorously overnight and then a NaOH solution (3.0 M) was added to bring the pH of the solution to 6.3. Simultaneously, ZnO NPs (at 0.5%, 1% and 2% concentrations, obtained from Sigma Aldrich, average size 25 nm) were mixed with 50.0 mL of water and agitated with an ultrasonic bath for 30 min. After sonication, 0.4 g of carboxymethyl cellulose sodium salt (CMC) was mixed with the ZnO solution and stirred for 1 h. Then the chitosan and CMC with ZnO NP solutions were mixed and 0.2 mL of Tween 80 solutions was added for emulsification and stirred for 15 min. A 0.3 mL of oleic acid was added and homogenized using an ultrasonic probe for 15 min, and then 0.3 mL of glycerol was added and stirred again for 15 min. The final solution was poured in to 9.0 cm diameter Petri dishes and kept inside an environmental chamber for 72 h set at 25°C and 50% relative humidity (RH). The bread slices were placed in between two above prepared nanopolymer films and packed with polyethylene bags and kept at 25°C for two months. When the *A. niger* spores were artificially inoculated on the bread slices and packed using CMC-CH-OL-ZnO NPs nanopolymer films (ZnO NPs concentration 2.0%), fungal growth was visibly seen only after 22 days of storage whereas most of the other samples showed fungal growth after 15 days. For the bread slices without any inoculation, visible fungal growth was observed after 3 days of storage for the control (without active packaging), but the slices packed using 2% ZnO NPs CMC-CH-OL-ZnO NPs nanopolymer film showed fungal growth after 35 days of storage at 25°C (Figure 6.7). The water activity tests also showed the bread slices packed with nanopolymer film had the higher water-holding capacity, which helped to slow down the staling of bread slices and maintained the bread texture for

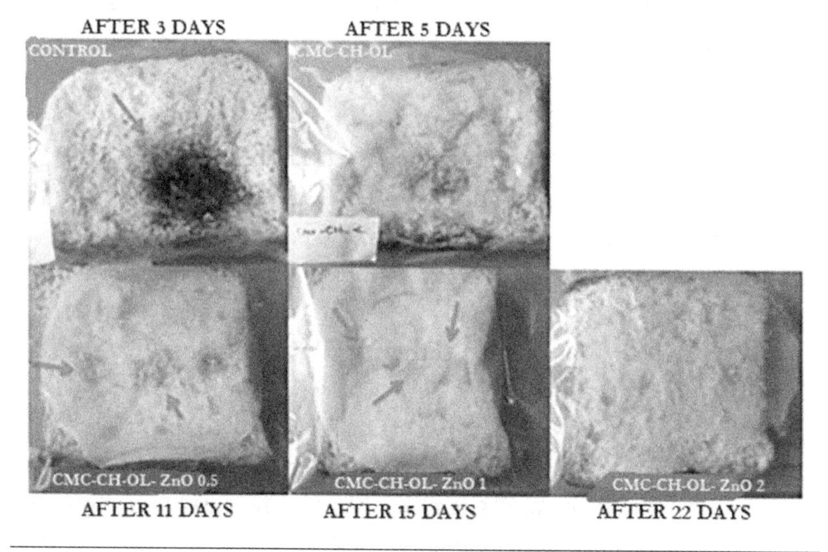

Figure 6.7 (See color insert.) Visual appearance of bread stored at 25°C inoculated with *Aspergillus niger* in control and active packaging film (coated with chitosan-carboxymethyl cellulose-oleic acid (CMC-CH-OL) incorporated with 0.5%, 1%, and 2% of zinc oxide nanoparticles). (Reproduced from Noshirvani, N. et al., *Food Packaging and Shelf Life*, 11, 106–114, 2017a.)

a long time when compared with control samples (Noshirvani et al. 2017a). Noshirvani et al. (2017b) tested CMC-CH-OL packaging films with and without ZnO NPs against *A. niger* using disc diffusion method. The above-prepared CMC-CH-OL films with and without ZnO NPs were placed in Petri dishes, and *A. niger* spores obtained from Iran Institute of Industrial and Scientific Research (Tehran, Iran) were dispersed on the potato dextrose agar media. These Petri dishes with media were incubated at 25°C for 6 days, and then colony-forming units were counted. The microbial count test results showed the inhibition rate of CMC-CH-OL film without ZnO NPs were 15.2%, and the zone of inhibition was 11.4 mm, but when 0.5% (weight basis) ZnO NPs were added with CMC-CH-OL films, the inhibition rate and zone increased to 13.3%, and 22.2 mm, respectively. The concentration of ZnO NPs in CMC-CH-OL films also had a significant effect on inhibition of *A. niger*. The inhibition rate and zone of CMC-CH-OL films with 2% ZnO NPs were 40.1%, and 30.1 mm, respectively. They attributed the presence of chitosan in the control films (CMC-CH-OL films without ZnO NPs) might be the reason for 15.2% of inhibition rate in control samples, whereas the Zn^{+2} ions released from the ZnO NPs damaged the cell membranes and increased the inhibition rate when ZnO NPs was added into the CMC-CH-OL packaging polymers.

Carbon nanotubes (CNT) are mixed with packaging films, and these modified packaging films are used for active packaging of meat, bakery, and processed food products. Salehifar et al. (2013) developed an active packaging material with LDPE incorporated with multi-walled carbon nanotubes (MWCNTs). The packaging films were prepared using a solvent casting method at 100°C for an hour using LDPE, Xylene, and MWCNTs. A solution with 95% Xylene, 5% LDPE, and MWCNTs with different concentrations (0.1%, 0.2%, and 0.3%) was prepared using an ultrasonic disruptor for 30 min. Then the mixture was casted in glass dishes and dried at 80°C for 8 h to evaporate Xylene. After drying, glass dishes were kept at room temperature for 48 h, and then the films were extracted from the dishes by immersing them in the distilled water. Iranian lavash bread was packed in the above-prepared packaging film and kept at room temperature. The microbial test was carried out by counting the number of colonies formed on 14th and 21st day of storage. The Colony Forming

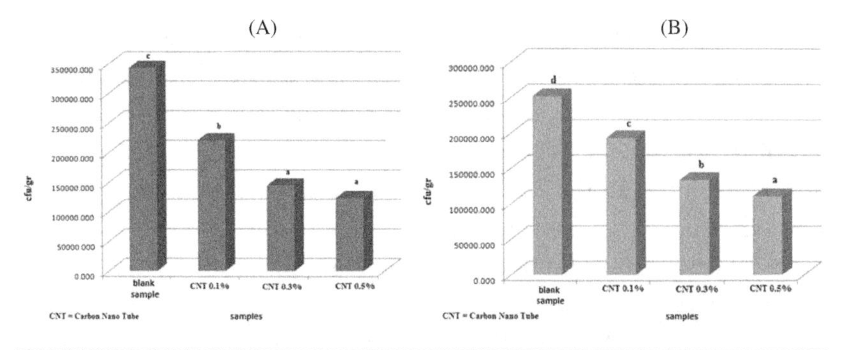

Figure 6.8 Microbial count on (A) 14th, and (B) 21st day of lavish bread stored in regular LDPE packaging film (blank sample), LDPE packaging film incorporated with 0.1%, 0.2%, and 0.5% of MWCNTs. (Reproduced from Salehifar, M. et al., *Euro. J. Exp. Bio.*, 3, 183–188, 2013.)

Units (CFU) count was higher for bread packed with regular LDPE film than the lavish bread packed in LDPE/MWCNT packaging film, and an increase in MWCNT concentration reduced the microbial growth further. They also found that there was no significant difference in the microbial count for bread packaged and stored in films with 0.3% MWCNT and 0.5% MWCNT (Figure 6.8). The texture analysis of bread proved that the staling effect on the bread was less in the samples stored in packaging film with 0.5% MWCNT than in the samples stored in regular LDPE films. The moisture permeability test of the nanocomposite films showed the addition of MWCNTs with LDPE decreased the moisture permeation rate, which may keep the bread softer for a long time, and the antimicrobial property of the CNTs inhibited the growth of molds on the bread (Salehifar et al. 2013).

Cozmuta et al. (2015) developed an active packaging film with (HDPE) incorporated with Ag and TiO_2 nanoparticles to improve the shelf life of wheat bread. Titanium dioxide has moisture and oxygen-scavenging properties, and the Ag has good antimicrobial properties. When these two materials are added with the regular packing films (HDPE or LDPE), the packaging material's gas and moisture transmission properties, and antimicrobial property improve, which can lead to increase in shelf life of the packed product. The sol-gel procedure was used to prepare TiO_2 solution (Yang et al. 2005). The titanium tetraisopropoxide, nitric acid (65%), ethanol, and ultrapure water were mixed and kept for two weeks for aging. After aging, the gel was immersed into $AgNO_3$ solution (0.005M, 0.4 mg gel/mL solution) and stirred for 24 h at a speed of 160 rpm. The obtained gel was washed with ultrapure water and ethanol mix 5 times and dried at 500°C for 2 h at a heating rate of 4°C/min, which helped the Ag^+ ions reduction onto the TiO_2 surface and titania structure crystallization. The above-prepared Ag/TiO_2 nanocomposite (5 g) and 2 mL of C_2H_5OH were mixed into a suspension and manually coated using a bar coater on the surface of the HDPE film at room temperature and then air dried for 10 min. The bread slices were sandwiched between two films and kept for 6 days at 30°C and analyzed for yeasts, mold, *Bacillus subtilis* and *Bacillus cereus.* For comparison, bread slices were also packed in regular HDPE film and kept unpacked under the same conditions for six days and a similar microbial analysis was carried out. The microbial count tests showed that there were no yeast and mold growth above maximum allowable limit (2 CFU/g) on bread packed with HDPE film incorporated with Ag/TiO_2 nanocomposite after 6 days of storage, whereas the bread slices unpacked and packed with regular HDPE films had yeast and mold growth after 2 and 4 days of storage, respectively. The *B. subtilis* count test showed that bread slices packed in Ag/TiO_2 + HDPE films had a lower microbial count of 4.87 log CFU/g (initial count was 2.89 log CFU/g), whereas unpacked and HDPE film packed slices had 11.2 and 10.1 log CFU/g, respectively after 6 days of storage. The minimum level for causing foodborne illness for *B. subtilis* is 6 log CFU/g, so the bread slices packed with regular HDPE film and unpacked ones crossed this limit after 3 and 4 days, respectively, but the Ag/TiO_2 + HDPE film kept the bread slices at safe level even after 6 days of storage. The *B. cereus* count tests also proved that HDPE film incorporated with Ag/TiO_2 nanocomposite had the lowest microbial count (1.75 log CFU/g) among all three treatments, and unpacked and packed with regular HDPE film bread slices had microbial counts of 2.76 log CFU/g, and 3.01 log CFU/g, respectively, after 6 days of storage (Figure 6.9). None of these samples reached the minimum level that causes foodborne illness (5 log CFU/g) after 6 days of storage (Cozmuta et al. 2015). The gas and moisture transmission rate tests also showed that the HDPE films incorporated with Ag/TiO_2 nanocomposites behaved same as regular HDPE packaging films. The moisture permeation rate for HDPE and Ag/TiO_2 + HDPE films were 6.5 and 6.3 $g/(m^2.day)$, respectively.

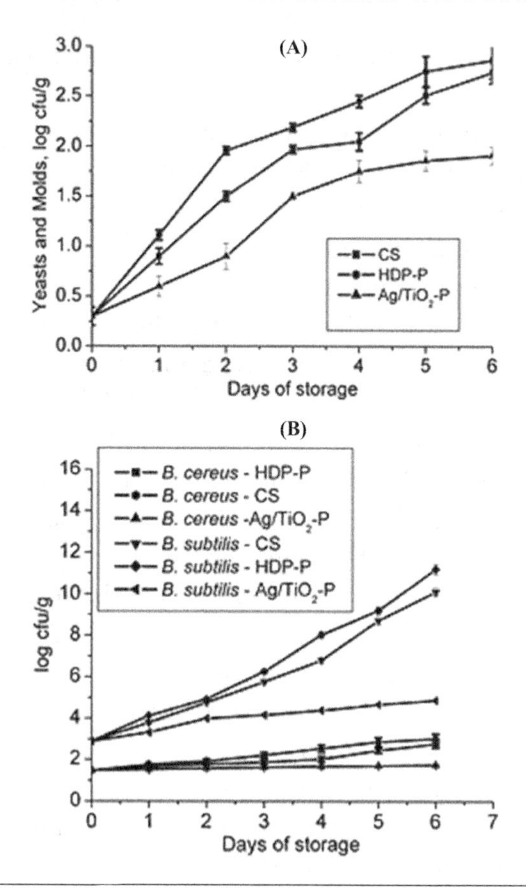

Figure 6.9 Microbial count of (A) yeast and mold and (B) *Bacillus cereus* and *Bacillus subtilis* on bread packed with HDPE film (HDP-P), HDPE film incorporated with Ag/TiO_2 nanocomposite (Ag/TiO_2-P), and unpacked (CS) during storage. (Reproduced from Cozmuta, A.M. et al., *Packaging Tech. Sci.*, 28, 271–284, 2015.)

Oxygen permeation rates through HDPE and Ag/TiO_2 + HDPE films were 1.6×10^2 $cm^3/(m^2.day.Pa)$ and 1.5×10^2 $cm^3/(m^2.day.Pa)$, respectively (Cozmuta et al. 2015). These results showed a great application potential of incorporation of Ag/TiO_2 nanocomposites into regular packaging polymers to increase the shelf life of bakery products.

The shelf life of sponge cake is very low (a day or two) due to its high water activity, which makes it susceptible to mold growth within a short time of storage. Use of oxygen scavengers and ethanol emitters can help to increase the shelf life of sponge cake. Janjarasskul et al. (2016) added oxygen-scavenging material BestKept® (Alpine Foods, Bangkok, Thailand), and ethanol-emitting material MaxxLive®

(Alpine Foods, Bangkok, Thailand) into the headspace of sponge cake packages. The microbial growth test showed that the total plate count in the control package (without any oxygen scavengers or ethanol enhancer) exceeded the safe level (104 CFU/g) within 3 days of storage at room temperature, whereas the packages containing oxygen scavenger or ethanol enhancer had a shelf life of 42 days. Since the titanium dioxide nanoparticles (TiO_2 NPs) and nanoclay composites have good oxygen-scavenging properties, they can also be tested for increasing the shelf life of sponge cake and other bakery items that have shorter storage time (Janjarasskul et al. 2016).

Packaging polymers of HDPE incorporated with the copper nanotubes (Cu nanotubes) were tested to inhibit *Staphylococcus aureus* growth (Bikiaris and Triantafyllidis 2013). Cu nanotubes were synthesized using a N,N,N′,N′-tetramethylethylenediamine (TMEDA), cyclohexane, and $CuCl_2$ solution (Cho and Huh 2009). A 20 mL of $CuCl_2$ (0.2 M) aqueous solution was taken in a beaker, and 8 mmol glucose+ 2.4 mL TMEDA were mixed and stirred for 15 min. Then cyclohexane (40 mL) and cetyltrimethylammonium bromide (CTAB) was added and autoclaved for 12 h at 120°C. After cooling to room temperature, Cu nanotubes were filtered out and washed with ethanol and dried for 12 h at 60°C. The HDPE polymer and the prepared Cu nanotubes (at 0.5, 1.0, 2.5, and 5.0% concentrations) were placed in a planetary ball mill (model: S100, RETSCH, Haan, Germany) and ground for 3 h at a speed of 500 rpm. Then the packaging film was prepared using a melt-mixing method using a reomixer (model: 600, Haake–Buchler Instruments ltd., Saddle Brooke, NJ) for 15 min at a temperature of 220°C and a speed of 30 rpm. The above-prepared HDPE film with Cu nanotubes was placed inside Petri dishes and 1.2×10^5 CFU/mL of *S. aureus* strain was applied on the sheet and incubated for 24 h at 37°C. The results showed there was no significant difference in microbial count on regular HDPE film and the HDPE film with 0.5% Cu nanotubes. *S. aureus* growth was decreased with increase in Cu nanotube concentration above 1.0% (Bikiaris and Triantafyllidis 2013). This result showed the potential of using Cu nanotubes and nanoparticles along with regular packaging polymers to inhibit the growth of major foodborne pathogens in raw ingredients and processed bakery products.

Applications in Quality Control and Food Safety

As discussed in the applications in packaging section, the major reason for the quality deterioration of bakery products and risk for food safety are mold, yeast, and bacteria. The packing film with nanocomposites of Ag, Au, Cu, or TiO_2 nanoparticles can be used for safe storage of bakery products. Mycotoxin infection in raw materials like flour is another safety concern in the bakery industry. Nanosensors based on spectral, electrical, impedance or resistance, colorimetric, or immunoassay techniques are used in the bakery industry to detect the major foodborne pathogens and mycotoxins in raw materials as well as in processed final products (Avella et al. 2010; Azhir et al. 2012; Baeumner 2004; Bernardes et al. 2014; Gomes et al. 2015; Kaittanis et al. 2010; Wang and Irudayaraj 2008; Wang et al. 2008; Yamada et al. 2014; Zhao et al. 2004). A detailed discussion about nanosensors used for quality control and food safety is given in Chapter 4. However, some of the quality control and food safety applications of nanotechnology mainly focusing on bakery industry are discussed briefly in this section.

A nanosensor was developed using CuO nanoparticles and nanolayers to detect *A. niger* growth in bakery products by sensing the odorous gases released by the fungi (Etefagh et al. 2013). The sol-gel method was used to synthesize of CuO nanoparticles. Copper nitrate ($Cu [NO_3]_2 \cdot 3H_2O$) was mixed with the ethanol and deionized water solution (1:1 v/v), and then ethylene glycol and citric acid were added to the solution and stirred for 1 h at 40°C. The solution was kept at 100°C for 4 h to evaporate excess solvents, and then the gel was desiccated at 600°C for 1 h to obtain the CuO powder, and the powder was further milled to nanoparticles. To get the nanosensors, the above-prepared CuO nanoparticles were applied on a thin glass plate at 400°C by the spray pyrolysis method and then cured with the solution containing $Cu (CH_3COO)_2.H_2O$ and distilled water. *A. niger* was cultured in a growth plate, and the nanosensors were exposed to the culture plate kept at varying temperature from 32°C to 42°C, and the resistance of the nanosensors was measured using a nano ammeter. The growth of *A. niger* was increased at the temperature range of 32°C–37°C, and it was noted that the resistance of the sensor decreased with the growth of *A. niger* on the culture plate (Figure 6.10). Etefagh et al.

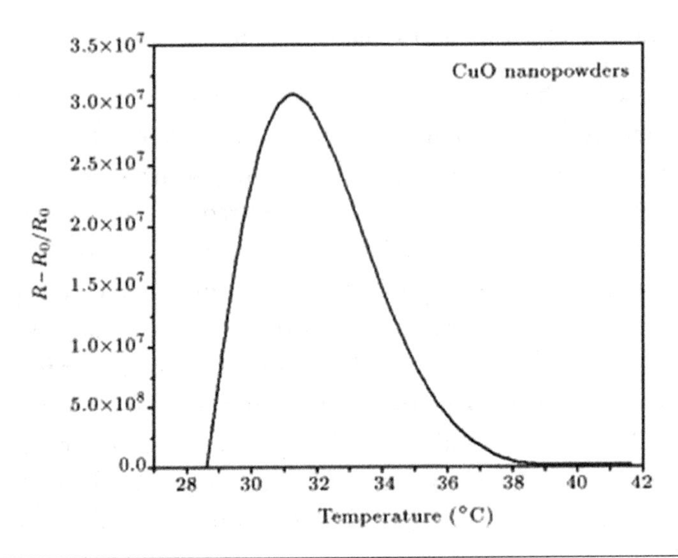

Figure 6.10 Changes of resistance of a CuO nanosensor when exposed to *Aspergillus niger* culture plate kept between 28°C and 42°C. (Reproduced from Etefagh, R. et al., *Sci. Iran.*, 20, 1055–1058, 2013.)

(2013) interpolated the gases produced by the *A. niger* fungi from the reaction with the oxygen ions in the surface and associated resistance changes in the semiconductor. They suggested this rapid change in resistance due to the presence of fungi could be used for detection of *A. niger* presence in bakery products quickly. Even though this sensor's reaction was quick, the sensitivity of the sensor was also affected by the ambient temperature, which may produce false positive results.

The shelf life of bread is low (3–4 days) due to its high water activity (0.96), and mold growth is the common reason for spoilage of bread and other bakery products (Cioban et al. 2010). *Penicillium* spp. and *A. niger* are the two most common fungi species that spoil the bakery products, and these two species itself cause around 60% spoilage (Alhendi and Choudhary 2013). Traditionally, the bakery industry uses propionic acid to control the mold growth in bakery items, but the recent trend of chemical preservative-free food among the consumers led the bakery industry to use active packaging techniques to inhibit mold growth without directly using any chemical preservatives in the manufacturing process.

The major fungal species that affect bakery products are: *A. niger, Aspergillus flavus, Europium amstelodami, Europium repens, Europium rubrum, Europium herbariorum,* and *Penicillium corylophillum*, and some of this fungal growth leads to mycotoxin production. Traditionally, microbial plating and counting (colony forming unit counting) is the most common method used for identification of fungal growth in bakery products, but this method is time-consuming (Schnürer et al. 1999). Electronic nose based on GC-MS is one of the nanosensing techniques used for rapid analysis of odorous gases produced by fungal and bacterial growth to detect fungal spoilage (Keshri et al. 2002). Early detection of fungal growth can help to avoid mycotoxin development in bakery products. Vinaixa et al. (2004) developed an e-nose-based detection technique to identify major fungal species that affect bakery products using analysis of odorous gases released into the head space by in vitro and in situ methods. They also tested static headspace (SH) optimization and solid phase microextraction (SPME) sampling techniques for rapid detection of fungal species. The all seven fungal species spores recovered from bakery items were artificially inoculated into vials and analyzed using an e-nose based on a GC-MS (model: QP 5000 GC/MS, Shimadzu Corp., Tokyo, Japan) system along with a control blank to analyze the volatile gases present in the nanoscale level at the headspace of these vials. An auto sampler coupled with the MS-e-nose collected the gas samples from the headspace of the vials, and GC-MS analyzed the samples. Principal component analysis (PCA), and discriminant function analysis (DFA) techniques were used for data analysis using MATLAB® software. The PCA models developed from the results had classification accuracies of 88%, 98%, and 100% after 24, 48, and 72 h, respectively, between blank and fungal spore inoculated samples. The two-dimensional DFA model results showed that, for the first 24 h after fungal inoculation, there were some mix groupings between blank and inoculated samples. After 48 h of inoculation, blank and inoculated samples were separated nicely, as well as the samples from same species clumped together and the distance between different species increased after 48 h (Figure 6.11). Vinaixa et al. (2004) also found that the fungal species released CO_2 for the first 24 h of inoculation, and the other volatile components were released after 48 h due to secondary metabolism. The SPME sampling method gave higher classification than the SH sampling method. This nanosensing technique with MS-e-nose could be useful for early detection (in 48 h of fungal development) of major fungal species, which spoil bakery products

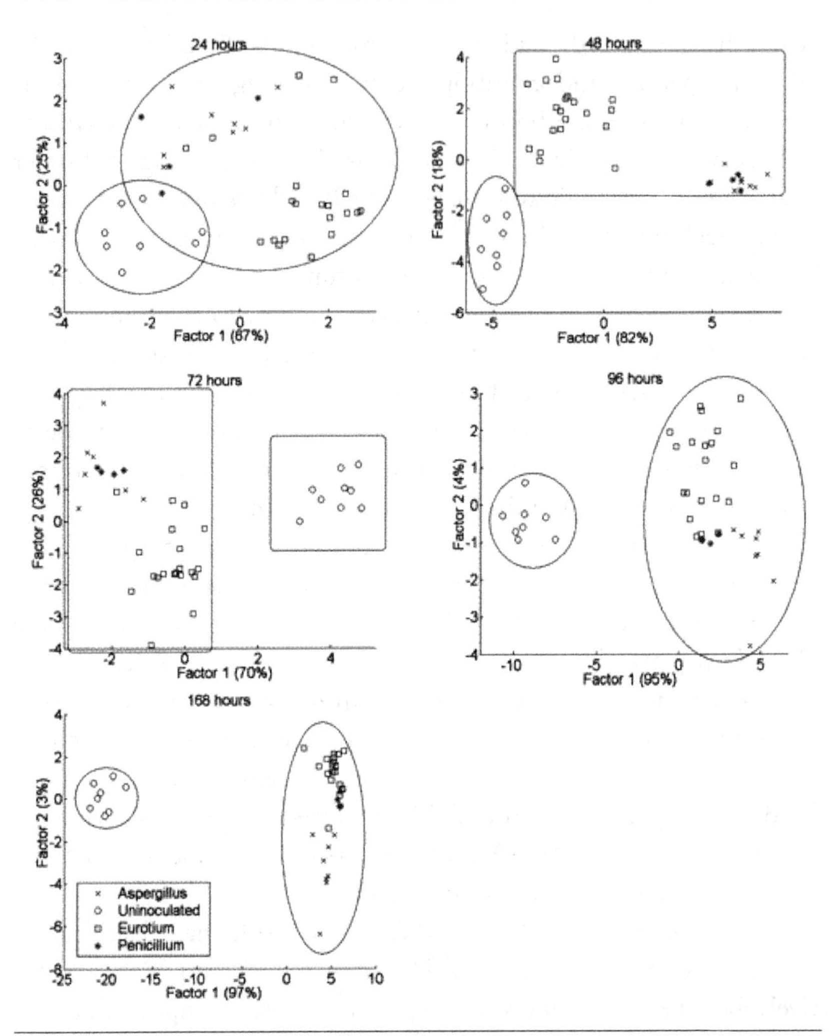

Figure 6.11 Two-dimensional DFA prediction plots of headspace analysis of 24, 48, 72, 96, and 168 h after fungal inoculation. (Reproduced from Vinaixa, M. et al., *J. Agric. Food Chem.*, 52, 6068–6074, 2004.)

but cannot be used for rapid detection since the total analysis time was 25 min. Vinaixa et al. (2004) suggested the use of multiple SPME fibers in parallel to get higher throughput from the unit, which can reduce analysis time to 5 min, and then the system could be used for rapid detection of fungal spoilage in bakery products.

All-purpose and whole wheat flour are the major ingredients in the bakery industry, and *Escherichia coli* contamination in flour is the major concern among industry and consumers. Bruna et al. (2012) developed a packaging film incorporated with copper-modified montmorillonite ($MtCu^{2+}$) and regular LDPE to inhibit the growth of *E. coli* in the packed product, which can be used to pack flour. Montmorillonite clay was stirred for 3 h at 60°C and then centrifuged at 4000 rpm after the stirring. During this reaction period, the ion exchange process modified the montmorillonite clay into $CuSO_4$. The sediment collected after centrifugation was washed with distilled water three times, and then dried over night at 80°C. After drying, the product was ground to the particle size less than 45 μm (325 mesh). The supernatant of the centrifuged solution was diluted with water and the required amount of copper in the solution was obtained in the $MtCu^{2+}$ (copper-modified montmorillonite clay). A microcompounder (15 mL capacity) with twin screws set at 200°C was used to mix LDPE and $MtCu^{2+}$, and both the solutions were mixed for 15 min to obtain a homogeneous mixture. A microcast film line (DSM Xplore, Sittard, The Netherlands) was connected with the compounder to obtain the packaging film (thickness of 50 μm, a width of 300 mm). To check the effect of adding $MtCu^{2+}$ with LDPE on *E. coli* growth, packaging film with different concentrations of $MtCu^{2+}$ (0, 1%, 2%, 3%, and 4%) were placed in Petri dishes, and pure *E. coli* culture was plated and kept in an incubator for 24 h set at 37°C. After 24 h, number of colonies formed was counted, and the results showed that regular LDPE packaging film without $MtCu^{2+}$ had higher amount of microbial count, and also the LDPE films incorporated with $MtCu^{2+}$ at less than 2% concentrations did not have significant effect on inhibition of *E. coli* growth. But the number of colonies formed was decreased with increase in $MtCu^{2+}$ concentrations above 2% (Figure 6.12). The LDPE packaging film incorporated with 4% $MtCu^{2+}$ concentration had lower microbial count (reduced about 94% of microbial load). The thermal properties test of the film showed that the LDPE incorporated with nanoclay composite had higher thermal barrier effect than the regular polymer films (Bruna et al. 2012). These LDPE films with nanoclay composite can be used as a packaging material for raw ingredients like flour as well as finished products in the bakery industry to avoid *E. coli* contamination.

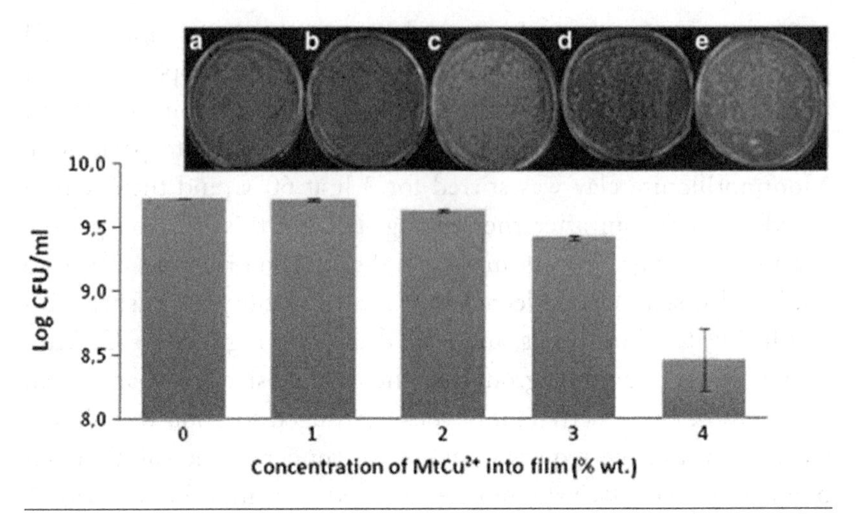

Figure 6.12 Culturing plates with *Escherichia coli* on packaging film with various concentrations of $MtCu^{2+}$ [(a) regular LDPE film without $MtCu^{2+}$, (b) LDPE film with 1% $MtCu^{2+}$, (c) LDPE film with 2% $MtCu^{2+}$, (d) LDPE film with 3% $MtCu^{2+}$, (e) LDPE film with 4% $MtCu^{2+}$]. (Reproduced from Bruna, J. et al., *Appl. Clay Sci.*, 58, 79–87, 2012.)

Conclusion

Nanotechnology is mainly used in packaging of bakery products in order to extend the shelf life of the product by controlling the mold, yeast, and bacterial growth during logistics and storage of bakery products. The demand for usage of nanoparticle-coated packaging polymer films and edible films are increasing due to the consumer's preference of chemical residue-free bakery and other food products. Even though the bakery industry is trying to expand active packaging techniques with nanocomposite films, still consumers are reluctant to accept the products that have direct contact with the Ag or other nanoparticles. The surveys clearly show that consumer acceptance level is high when nanotechnology is applied as a non-contact material for food preservation, but the acceptance level is low if nanoengineered products are used as ingredients or have direct contact with food. So the bakery industry, as well as scientific community, should work together to gain confidence of consumers by proving the potential benefits of nanotechnology in food as well as clearing the doubts about the effects of using nanotechnology on human health. Nanoencapsulation of active bio-ingredients like Omega-3 fatty acids helps to develop functional bread and other bakery products with improved nutritional and health benefits. Particularly the encapsulation helps to mask the unpleasant odor and taste of Omega-3-fatty-acid-rich products, such as fish oil or flax seed oil, which may be the major constraint for some people to avoid the Omega-3 fatty acid products. Since the baking process is a high-temperature unit operation, the thermal degradation of functional components like Omega-3 is common, and the nanoencapsulation prevents this thermal degradation and also the development of unpleasant components. Nanoencapsulation also reduces the risk of oxidation of unsaturated free fatty acid components like Omega-3 during storage of bakery products. Some industries like George Western Foods in Australia are taking the advantage of nanotechnology and are commercially producing bakery products with nanoencapsulated fish oil for target delivery of Omega-3 fatty acid directly to the human digestion system. Nanotechnology also can be applied to enhance quality and safety of bakery products through the use of nanosensors and active packaging components.

References

AACC. 2000. *Approved Methods of Analysis.* 10ed. St Paul, MN: AACC international: 08–12.

Akbari E., Buntat Z., Afroozeh A., Zeinalinezhad A., Nikoukar A. (2015) *Escherichia coli* bacteria detection by using graphene-based biosensor. *IET Nanobiotechnology* 9:273–279.

Alhendi A., Choudhary R. (2013) Current practices in bread packaging and possibility of improving bread shelf-life by nano-technology. *International Journal of Food Science and Nutrition* 3:55–60.

Arora A., Padua G. (2010) Nanocomposites in food packaging. *Journal of Food Science* 75(1):43–49.

Avella M., De Vlieger J.J., Errico M.E., Fischer S., Vacca P., Volpe M.G. (2005) Biodegradable starch/clay nanocomposite films for food packaging applications. *Food Chemistry* 93:467–474.

Avella M., Errico M.E., Gentile G., Volpe M.G. (2010) Nanocomposite sensors for food packaging. *Nanotechnological Basis for Advanced Sensors* 501–510. doi:10.1007/978-94-007-0903-4_53.

Azhir E., Etefagh R., Shahtahmasebi N., Mohammadi M., Amiri D., Sarhaddi R. (2012) *Aspergillus niger* biosensor based on tin oxide (SnO_2) nanostructures: Nanopowder and thin film. *Indian Journal of Science and Technology* 5:3010–3012.

Baeumner A. (2004) Nanosensors identify pathogens in food. *Food Technology* 58:51–55.

Balaguer M.P., Lopez-Carballo G., Catala R., Gavara R., Hernandez-Munoz P. (2013) Antifungal properties of gliadin films incorporating cinnamaldehyde and application in active food packaging of bread and cheese spread foodstuffs. *International Journal of Food Microbiology* 166:369–377.

Bernardes P.C., de Andrade N.J., Soares N.D.F. (2014) Nanotechnology in the food industry. *Bioscience Journal* 30:1919–1932.

Bikiaris D.N., Triantafyllidis K.S. (2013) HDPE/Cu-nanofiber nanocomposites with enhanced antibacterial and oxygen barrier properties appropriate for food packaging applications. *Materials Letters* 93:1–4.

Bruna J., Peñaloza A., Guarda A., Rodríguez F., Galotto M. (2012) Development of $MtCu^{2+}$/LDPE nanocomposites with antimicrobial activity for potential use in food packaging. *Applied Clay Science* 58:79–87.

Chaudhry Q., Scotter M., Blackburn J., Ross B., Boxall A., Castle L., Aitken R., Watkins R. (2008) Applications and implications of nanotechnologies for the food sector. *Food Additives & Contaminants: Part A* 25:241–258. doi:10.1080/02652030701744538.

Cho Y.S., Huh Y.D. (2009) Synthesis of ultralong copper nanowires by reduction of copper-amine complexes. *Materials Letters* 63(2):227–229.

Cioban C., Alexa E., Sumalan R., Merce I. (2010) Impact of packaging on bread physical and chemical properties. *Bulletin of University of Agricultural Sciences and Veterinary Medicine Cluj-Napoca. Agriculture* 67.

Cozmuta A.M., Peter A., Cozmuta L.M., Nicula C., Crisan L., Baia L., Turila A. (2015) Active packaging system based on Ag/TiO_2 nanocomposite used for extending the shelf life of bread. Chemical and microbiological investigations. *Packaging Technology and Science* 28:271–284.

De Azeredo H.M. (2009) Nanocomposites for food packaging applications. *Food Research International* 42:1240–1253.

de Deckere E.A. (2001) Health aspects of fish and n-3 polyunsaturated fatty acids from plant and marine origin. Nutritional Health, Humana Press, Totowa, NJ. pp. 195–206.

Degirmencioglu N., Göcmen D., Inkaya A.N., Aydin E., Guldas M., Gonenc S. (2011) Influence of modified atmosphere packaging and potassium sorbate on microbiological characteristics of sliced bread. *Journal of Food Science and Technology* 48:236–241.

Etefagh R., Azhir E., Shahtahmasebi N. (2013) Synthesis of CuO nanoparticles and fabrication of nanostructural layer biosensors for detecting *Aspergillus niger* fungi. *Scientia Iranica* 20:1055–1058.

Flora K., Hahn M., Rosen H., Benner K. (1998) Milk thistle (*Silybum marianum*) for the therapy of liver disease. *The American Journal of Gastroenterology* 93:139–143.

Gökmen V., Mogol B.A., Lumaga R.B., Fogliano V., Kaplun Z., Shimoni E. (2011) Development of functional bread containing nanoencapsulated Omega-3 fatty acids. *Journal of Food Engineering* 105:585–591. doi:10.1016/j.jfoodeng.2011.03.021.

Gomes R.C., Pastore V.A.A., Martins O.A., Biondi G.F. (2015) Nanotechnology applications in the food industry. A review. *Brazilian Journal of Hygiene and Animal Sanity* 9:1–8. doi:10.5935/1981-2965.20150001.

Gutiérrez L., Batlle R., Andújar S., Sánchez C., Nerín C. (2011) Evaluation of antimicrobial active packaging to increase shelf life of gluten-free sliced bread. *Packaging Technology and Science* 24:485–494.

Han W., Yu Y., Li N., Wang L. (2011) Application and safety assessment for nano-composite materials in food packaging. *Chinese Science Bulletin* 56:1216–1225.

HPSH. (2017) Omega-3 fatty acids: An essential contribution, Harvard D.H. Chan School of Public Health, Boston, MA.

Hussain S., Jamil K. (2012) Studies on the shelf life enhancement of traditional leavened bread. *Pakistan Journal of Biochemistry and Molecular Biology* 45:81–84.

Janjarasskul T., Tananuwong K., Kongpensook V., Tantratian S., Kokpol S. (2016) Shelf life extension of sponge cake by active packaging as an alternative to direct addition of chemical preservatives. *LWT-Food Science and Technology* 72:166–174.

Kaittanis C., Santra S., Perez J.M. (2010) Emerging nanotechnology-based strategies for the identification of microbial pathogenesis. *Advanced Drug Delivery Reviews* 62:408–423.

Kechichian V., Ditchfield C., Veiga-Santos P., Tadini C.C. (2010) Natural antimicrobial ingredients incorporated in biodegradable films based on cassava starch. *LWT-Food Science and Technology* 43:1088–1094.

Keshri G., Voysey P., Magan N. (2002) Early detection of spoilage moulds in bread using volatile production patterns and quantitative enzyme assays. *Journal of Applied Microbiology* 92:165–172.

Lopes F.A., Soares N.F.F., Lopes C.C.P., Silva W.A., Júnior J.C.B., Medeiros E.A.A. (2014) Conservation of bakery products through cinnamaldehyde antimicrobial films. *Packaging Technology and Science* 27:293–302.

Marathe S., Machaiah J., Rao B., Pednekar M., Rao S.V. (2002) Extension of shelf-life of whole-wheat flour by gamma radiation. *International Journal of Food Science and Technology* 37:163–168.

Mogol B.A., Gokmen V., Shimoni E. (2013) Nano-encapsulation improves thermal stability of bioactive compounds Omega fatty acids and silymarin in bread. *Agro Food Industry Hi-Tech* 24:62–65.

Mozafari M.R., Khosravi-Darani K., Borazan G.G., Cui J., Pardakhty A., Yurdugul S. (2008) Encapsulation of food ingredients using nanoliposome technology. *International Journal of Food Properties* 11:833–844.

Neethirajan S., Jayas D.S. (2011) Nanotechnology for the food and bioprocessing industries. *Food and Bioprocess Technology* 4:39–47.

Noshirvani N., Ghanbarzadeh B., Mokarram R.R., Hashemi M. (2017a) Novel active packaging based on carboxymethyl cellulose-chitosan-ZnO NPs nanocomposite for increasing the shelf life of bread. *Food Packaging and Shelf Life* 11:106–114.

Noshirvani N., Ghanbarzadeh B., Mokarram R.R., Hashemi M., Coma V. (2017b) Preparation and characterization of active emulsified films based on chitosan-carboxymethyl cellulose containing zinc oxide nano particles. *International Journal of Biological Macromolecules* 99:530–538. doi:10.1016/j.ijbiomac.2017.03.007.

Otoni C.G., Pontes S.F.O., Medeiros E.A.A., Soares N.F.F. (2014) Edible films from methylcellulose and nanoemulsions of clove bud (*Syzygium aromaticum*) and oregano (*Origanum vulgare*) essential oils as shelf life extenders for sliced bread. *Journal of Agricultural and Food Chemistry* 62:5214–5219. doi:10.1021/jf501055f.

Pateras I.M. (2007) Bread spoilage and staling. Technology of Breadmaking, Springer, Boston, MA. pp. 275–298.

Rasmussen P.H., Hansen A. (2001) Staling of wheat bread stored in modified atmosphere. *LWT-Food Science and Technology* 34:487–491.

Ray S., Quek S.Y., Easteal A., Chen X.D. (2006) The potential use of polymer-clay nanocomposites in food packaging. *International Journal of Food Engineering* 2(4):1–11. doi:10.2202/1556-3758.

Rhim J.W., Park H.M., Ha C.S. (2013) Bio-nanocomposites for food packaging applications. *Progress in Polymer Science* 38:1629–1652.

Salehifar M., Beladi N.M., Alizadeh R., Azizi M.H. (2013) Effect of LDPE/MWCNT films on the shelf life of Iranian Lavash bread. *European Journal of Experimental Biology* 3(6):183–188.

Schnürer J., Olsson J., Börjesson T. (1999) Fungal volatiles as indicators of food and feeds spoilage. *Fungal Genetics and Biology* 27:209–217.

Sekhon B.S. (2010) Food nanotechnology–An overview. *Nanotechnology, Science and Applications* 3:1–15.

Siegrist M. (2008) Factors influencing public acceptance of innovative food technologies and products. *Trends in Food Science and Technology* 19:603–608.

Sorrentino A., Gorrasi G., Vittoria V. (2007) Potential perspectives of bio-nanocomposites for food packaging applications. *Trends in Food Science and Technology* 18:84–95.

Suppakul P., Miltz J., Sonneveld K., Bigger S.W. (2003) Active packaging technologies with an emphasis on antimicrobial packaging and its applications. *Journal of Food Science* 68:408–420.

Vermeiren L., Devlieghere F., Van Beest M., De Kruijf N., Debevere J. (1999) Developments in the active packaging of foods. *Trends in Food Science and Technology* 10:77–86.

Vinaixa M., Marín S., Brezmes J., Llobet E., Vilanova X., Correig X., Ramos A., Sanchis V. (2004) Early detection of fungal growth in bakery products by use of an electronic nose based on mass spectrometry. *Journal of Agricultural and Food Chemistry* 52:6068–6074.

Wang C., Irudayaraj J. (2008) Gold nanorod probes for the detection of multiple pathogens. *Small* 4:2204–2208.

Wang R., Ruan C., Kanayeva D., Lassiter K., Li Y. (2008) TiO_2 nanowire bundle microelectrode based impedance immunosensor for rapid and sensitive detection of *Listeria monocytogenes*. *Nano Letters* 8:2625–2631.

Yamada K., Kim C.T., Kim J.H., Chung J.H., Lee H.G., Jun S. (2014) Single-walled carbon nanotube-based junction biosensor for detection of *Escherichia coli*. *PLoS One* 9:e105767.

Yang Q., Sha J., Ma X., Yang D. (2005) Synthesis of NiO nanowires by a sol-gel process. *Materials Letters* 59(14–15):1967–1970.

Zhao X., Hilliard L.R., Mechery S.J., Wang Y., Bagwe R.P., Jin S., Tan W. (2004) A rapid bioassay for single bacterial cell quantitation using bioconjugated nanoparticles. *Proceedings of the National Academy of Sciences of the United States of America* 101:15027–15032.

7

Applications in the Dairy Industry

Introduction

Applications of nanotechnology and nanomaterials in the dairy industry are on the rise and have varied applications in production, processing, quality control, and packaging of dairy products. Skim milk is a naturally nanoengineered product, which contains nanoscale-level protein and nutritional compound particles. Production of low-fat and zero-fat dairy products is gaining consumers' attention due to their health benefits, and the dairy industry is investing a lot on producing these low-fat products with similar rheological, textural, and sensory properties as regular-fat level products. Nanotechnology is a great tool for this option, which can help to produce low-fat or zero-fat products like yogurt, ice cream, or milk with the same functionality and physicochemical properties (Lee et al. 2013; Mohammadi et al. 2015; Park et al. 2007; Relkin et al. 2008; Seo et al. 2009). The dairy products can be fortified using nanoscale nutritional compounds like zinc and iron without changing the rheological and sensory properties. Nanotechnology also helped to add supplementary calcium at nanoscale level in yogurt resulting in improved nutritional intake and digestibility (Seo et al. 2009).

Nanotechnology is mainly applied in quality control and safety aspects during manufacturing and logistics of dairy products. Nanosensors developed with the help of silver and gold nanoparticles can be used to detect major foodborne pathogens affecting dairy products quickly and accurately, resulting in savings of millions of dollars to the dairy industry. Silver and gold have natural antimicrobial properties, which can be used in the nanoscale level to control the pathogenic bacterial cultures in the dairy products and functionalized as nanosensors with pathogen-specific antibodies to selectively attract microorganisms. The regular analytical tools like fluorescence, spectroscopy, mass spectroscopy (MS), liquid chromatography (LC), or high-pressure liquid chromatography coupled with these nanosensors can be used for quantification of the microorganisms (Cressey 2013; Font et al. 2008; Ganmaa and Sato 2005; González-Sálamo et al. 2017; GoodFood 2008; Lee et al. 2012, 2013).

Adulteration of milk and milk products with organic and inorganic substances is the major concern among the consumers nowadays, and melamine adulteration is the most recently reported milk adulteration (Gossner et al. 2009; Nag 2010; Schoder 2012). Sometimes milk producers and processors dilute the milk with water to increase the quantity, but it has a negative effect on the quality of the milk since this practice reduces the protein level, which is a major quality parameter in the dairy industry throughout the world. Melamine is an organic chemical substance commonly used in plastics and adhesives. It contains a high amount of nitrogen substances and when added to milk it artificially increases protein in diluted milk by cheating protein analyzers, since most of the common protein analyzing techniques use measured nitrogen level for predicting protein content in a product. Melamine has the serious health effects in humans, and especially in infants and children; it may lead to death due to renal failure (Ni et al. 2014; Zhang et al. 2009). In 2008, melamine adulteration in milk powder and baby formulas resulted in six infant causalities and more than 300,000 sick children in China (Zhang et al. 2009). Nanotechnology-based sensors developed using gold and silver nanoparticles can identify melamine adulteration in milk and milk products. Most of these melamine detection sensors are colorimetric sensors, so melamine presence in dairy products can be rapidly detected using the visible color change in the sensor solution. These sensors can also be coupled with spectroscopic or LC instruments to quantify the melamine level in milk (Guo et al. 2014; Ping et al. 2012; Xin et al. 2015). Due to the nature of present-day dairy farming, milk producers are using antibiotics for the cows, which end up in milk. Residues of these antibiotics have some ill effects on human health, and nanotechnology-based sensors can be used to detect these unwanted antibiotic residues in milk, even if these are present in very low levels. Magnetic nanoparticles (MNPs) can be used to selectively attract and separate the residues of the antibiotics, and then common analytical tools can be used for quantification. The MNP-based sensing techniques have also been tested for detecting mycotoxins in dairy products (Font et al. 2008; Ganmaa and Sato 2005; González-Sálamo et al. 2017; GoodFood 2008; Lee et al. 2012; Sekhon 2010).

Another major application of nanotechnology in the dairy industry is in packaging of dairy products. Nanotechnology helps to develop packaging polymers and materials with specific mechanical and gas transmission properties to increase shelf life of the dairy products. Packaging materials developed using nanoclay composites can act as oxygen scavengers, which can inhibit the growth of foodborne pathogens in dairy and other food products. Milk and infant nursing (feeding) containers have been developed using silver nanoparticles (AgNPs) to increase the storage time and also to maintain the quality of the dairy products. Some of these technologies have been successfully tested and commercially implemented to store milk. Few examples are: milk containers with AgNPs by Nanox Technologies, Brazil; double-layer ultraviolet (UV)-protected milk containers with titanium dioxide nanoparticles by Fonterra Ltd., New Zealand; and the Baby Dream feeding bottles with AgNPs, by Baby Dream ltd., South Korea. The Nanox containers have increased the shelf life of milk to 15 days from a maximum of 7 days in southern parts of Brazil, where the safe storage time of milk is short (maximum 7 days) due to the lack of safe storage facilities and structures at the domestic level (Alisson 2015; BabyDream 2017; Dickinson 2017; Sekhon 2010).

Applications During Processing

Nanotechnology has a huge potential for applications in the dairy industry during processing of dairy products. The nutritional quality and functionality of milk and milk products can be enhanced using nanotechnology by adding nanoscale nutritional supplements and minerals without altering the physicochemical, textural, and sensory properties of products. Researchers also found that the nutritional intake and digestibility of the dairy products can be improved while using nanomaterials for fortification of milk and yogurt (Park et al. 2007;

Santillán-Urquiza et al. 2017; Seo et al. 2009). Nanoencapsulation of nutritional compounds can deliver the functional compounds of the dairy product directly to the target receptors, which can also improve the nutritional benefits of the dairy products. Nanotechnology can also be used to develop low-fat and cholesterol-free (or zero-fat) dairy products with the same physicochemical properties as regular products.

Park et al. (2007) added nano calcium (Ca) (average size of 30–900 nm) to milk and tested the effect of nano-Ca-enhanced milk on bone loss in ovariectomized rats. Daily 1 mL of milk with 20 mg nano Ca was fed to rats and the body weight gain, food efficiency ratio, serum calcium concentration, and femoral bone mineral density of the rats were analyzed after 18 weeks. They reported significant decrease in body weight gain and serum Ca and significant increase in bone mineral density, stiffness, and energy for nano-Ca-milk-fed rats than controls (Figure 7.1). The trabecular bone area was also higher in the rats fed with nano-Ca containing milk (Park et al. 2007). Based on these results, Park et al. (2007) suggested consumption of milk supplemented with nano Ca could reduce the bone loss as well as increase the bone calcium metabolism.

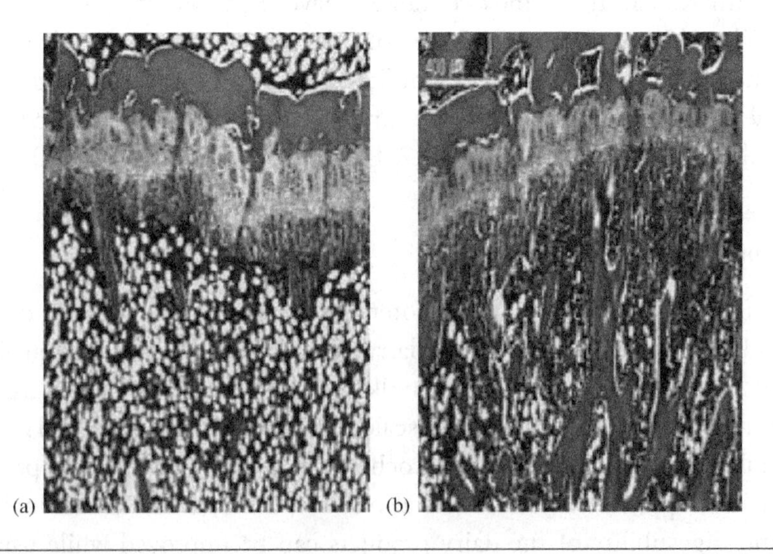

Figure 7.1 (See color insert.) Micrographs of trabecular bones of ovariectomized rats after 18 weeks feeding with (a) control and (b) nano calcium supplemented milk. (Reproduced from Park, H.S. et al., *Asian Aust. J. Anim. Sci.*, 20, 1266–1271, 2007.)

Yogurt naturally has high protein and calcium content, but minerals such as iron and zinc are low in yogurt. In recent times, fortification of food products is gaining lots of attention of industry and consumers. Santillán-Urquiza et al. (2017) prepared a fortified yogurt with calcium, zinc, and iron nanoparticles and compared it with fortified yogurt having microsize mineral (calcium, zinc, and iron) particles. Yogurt was prepared using pasteurized milk and the micro and nano mineral particles were added into yogurt at various concentrations using a stirrer for 20 min at a stirring rate of 120 agitations/min. Then the yogurt was incubated for 5 h at 45°C, then stored at refrigeration temperature (4°C) for 28 days, and samples were analyzed for rheological, textural, and sensory properties once a week (at 7, 14, 21, and 28 days of storage). The pH values of all the samples decreased (4.65–4.30), and acidity values increased (0.86–0.90 g/100 mL) after 28 days of storage. Syneresis tests showed that the yogurt fortified with nano calcium, zinc, and iron particles showed more stability when compared with control samples with no added mineral particles, and also the textural analysis proved nanoparticle-fortified yogurt had a stronger gel structure than control samples, as well as micromineral added samples (the firmness of control, nano Ca, and micro-Ca-fortified samples were 0.60, 0.93, and 0.60 N, respectively). Yogurt fortified with nanosize minerals had more digestibility than the control and fortified with microsize mineral particles, and the nanomineral-fortified yogurt had the best sensory scores when compared with the other samples (Santillán-Urquiza et al. 2017).

Low-fat and cholesterol-free dairy products are gaining more attention of the consumers due to their health benefits, and dairy industry is responding by producing low cholesterol and cholesterol-free products with the same taste as regular fat content dairy products. Seo et al. (2009) added chitosan (derivative of chitin) nanoparticle powder into low cholesterol yogurts to maintain the natural taste and to increase shelf storage time and tested the effect of chitosan NPs on sensory, microbial, and physicochemical properties of the yogurt. Various concentrations of chitosan NPs (0.1%, 0.3%, 0.5%, and 0.7%, w/v) (536 nm diameter) were added into low cholesterol yogurt samples and allowed to stabilize for 24 h at 10°C. Then the samples were stored at refrigeration temperature (4°C) for 0, 5, 10, 15, and 20 days to test the effect on sensory and physicochemical properties.

The scanning electron microscope images of the samples showed that the average diameter of chitosan NPs was 562 nm, and the physicochemical tests showed no significant changes in color of yogurt after 20 days of storage at 4°C when compared with the control samples. The pH value of the control samples decreased from 4.21 to 3.94 after 20 days of storage at 4°C, whereas no significant change in pH for the chitosan NP-added samples was noticed. The lactic acid bacterial culture (*Lactobacillus bulgaricus* and *Streptococcus thermophiles*) counts were 4.75×10^8 to 9.70×10^8 cfu/mL after 20 days of storage at 4°C, which proved the increased shelf storage time for low cholesterol yogurts, and the sensory evaluation test also showed no significant changes in overall sensory scores between control and chitosan NP-added low cholesterol yogurt (Seo et al. 2009).

Probiotics deliver the beneficial living bacterial cells to humans directly into gut ecosystem. Nanotechnology helped to develop nanoencapsulation of probiotic bacterial cultures, which increase the shelf life of the probiotic bacteria as well as improve delivery and release of the probiotic by target delivery into the certain part of gastrointestinal system where it reacts only with certain receptors. Many researchers proved this encapsulation and target delivery improved the efficiency of the food and nutraceutical products (Huang et al. 2009; Yu and Huang 2010). Probiotic nanofoods were developed by Dr. Kim in combination with calcium to help to reduce the risk of osteoporosis (a disease-causing fragility and fracture of bones due to deterioration of bone tissues) in humans. Dr. Kim developed probiotic nanofoods in the forms of powder, capsules, and liquids, which can be directly mixed with food (at room temperature) or yogurt to enhance daily diet (Sekhon 2010).

Liposomes are one of the microencapsulation techniques used in food industry to encapsulate flavors, vitamins, microorganisms, antioxidants, enzymes, and antimicrobial components of food in order to improve taste, flavor, texture, and shelf life of the product (Mohammadi et al. 2015). The major components of liposomes are lipid or phospholipids and aqueous medium. Nanoliposomes are similar to liposomes and manufactured using high energy to get the particle size of lipids or phospholipids from 30 nm to 1 μm, dispersed in the aqueous medium. Nanoliposomes have been extensively tested in cheese production process to achieve specific taste and flavor as well as fastening the cheese ripening process (Mozafari et al. 2008). Nanoliposomes have the advantage of higher surface area due to their submicron size over the liposomes, which improves the bioavailability of the target substance for the cheese ripening process. The major processing step of cheese manufacturing is proteolysis during ripening, which affects the taste and smell of the end product. Proteolysis during cheese ripening process and using nanoliposome techniques develop more flavors in a short time, which is economically beneficial for cheese manufacturers (Mohammadi et al. 2015; Mozafari et al. 2008). Reducing ripening time also helps to reduce the risk of bitterness development and texture defects. The enzymes encapsulated using nanoliposome technology helped to reduce the loss of enzymes in whey and increased the enzyme distribution in the curd, which decreased protein loss and bitterness as well as improved the texture of cheese through the regulated ripening process (Mohammadi et al. 2015). Encapsulation of cheese enzymes also protected casein from the early proteolysis process. Release of the enzymes from nanoliposome encapsulation depends on the process temperature, pH, and the ionic concentration during cheese ripening process, which plays a major role in the development of more flavors in short time. The encapsulation of flavor enzymes (protease) using a heating method for formation of protease encapsulation reduced the cheese ripening process time (Vafabakhsh et al. 2013).

Nanotechnology also helped to develop nanoencapsulated lipids and food ingredients such as curcumin using dairy products like stearin-rich milk, which restricted the oxidation of lipids and improved the stability of the curcumin. The sodium caseinate-stabilized oil in water nanoemulsions prepared using stearin-rich milk reduced the rate of degradation of lipids (Relkin et al. 2008; Sekhon 2010; Wang et al. 2009). Development of new types of membranes with nanosize sieves helped to filter milk for cheese manufacturing, and these nanosieves can also be used for preparation of fat colloids filled with water, which can be used for low-fat dairy products like ice cream and milk with the similar taste as high-fat products (GoodFood 2008). The food manufacturing giant "Unilever" is researching about the development of low-fat ice cream with the use of nanotechnology by reducing the emulsion's particle size to nano level, which helps to make the texture of ice cream smoother and reduces the use of emulsion for ice-cream manufacturing by 90%. Unilever also projects that use of nanoscale emulsions in ice-cream manufacturing will reduce the fat content to less than 1%, where the ice-cream manufactured using traditional method has a fat content of 8%–16% (Farhang 2007).

Applications in Quality Control and Safety

Researchers around the world are working on the development of nanosensors for applications in food quality control and food safety, and the food industry is also investing millions of dollars into such research programs. In the dairy industry, presence of chemical and antibiotics residues in milk, adulteration of milk and milk products with melamine, and deterioration of dairy products due to the pathogenic microorganisms are the major quality and safety issues. Nanotechnology-based colorimetric sensors have been tested successfully for detection as well as quantification of melamine present in the milk and milk products, and the results showed that these nanotechnology-based sensing techniques are quick and reliable as well as very easy to implement in the production plants. Antibiotic residues can be identified using the nanosensors, and mycotoxin presence in milk products can be sensed also using these sensing tools developed using nanomaterials and nanotechnology. Silver and gold nanoparticle (AgNP and AuNP)-based sensors and magnetic nanoparticle-based sensing techniques have been successfully demonstrated by the researchers to detect and quantify pathogenic microorganisms in the dairy products.

Detection of Adulterants

Melamine is an organic chemical most commonly used in plastics, countertops, dishware, adhesives, and whiteboards. Milk producers and processors sometimes add water into raw milk to increase the volume, which decreases the protein content of milk. The protein content of milk and other food products is measured by standard analytical techniques using the nitrogen content of the product. Addition of melamine into the diluted milk increases the nitrogen content thus artificially inflating the protein content of the milk. Use of melamine in dairy or other food products is not approved by World Health Organization and also poses some health risks like kidney stone formation. Melamine presence in milk is being detected using regular analytical techniques, such as high-performance liquid chromatography (HPLC), gas chromatography/mass spectrometry (GC/MS), MS, surface-enhanced Raman spectroscopy (SERS), and HPLC coupled with MS (HPLC/MS). Most of these methods need a long time for detection, expensive equipment, and highly qualified personnel (Xin et al. 2015). Some of these methods also need pretreatments like extraction and derivatization. Nanotechnology-based sensors and sensing techniques can be used to easily detect melamine adulteration in milk by using colorimetric changes (Ai et al. 2009; Cai et al. 2014; Giovannozzi et al. 2014; Guo et al. 2014; Kumar et al. 2014; Ni et al. 2014; Song et al. 2015; Xin et al. 2015).

AuNPs (approximate diameter of 12 nm) stabilized with a cyanuric acid (CA) derivative 1-(2-mercaptoethyl)-1,3,5-triazinane-2,2,4-trione (MTT) were used to detect melamine adulteration in raw milk and milk products (infant formulas) (Ai et al. 2009). CA and its derivatives have the capability of producing strong triple hydrogen bonding with melamine (Sherrington and Taskinen 2001), and MTT-stabilized AuNPs dispersed well in the distilled water and produced a wine red color. When the raw milk was added into the solution, there were no significant changes in the color of the solution. But when solution (8 μL) with different concentration of melamine (1.0, 2.5, and 5 ppm) was added into the MTT-stabilized AuNPs sensor solution (1 mL), the sensor solution's color changed from red to violet blue (Figure 7.2). This color change was clearly seen by the naked eye, as well as a clear change in using surface Plasmon resonance (SPR) spectra at 519 nm wavelength (Ai et al. 2009). MTT-stabilized AuNPs bind with melamine and produce hydrogen bonding, which results in a reproducible color change depending on the concentration of the melamine content. This nanosensor-based method can be used to determine the melamine content in dairy products and pet foods even at small quantities (up to 2.5 ppb).

Xin et al. (2015) developed a colorimetric sensor using AuNPs to detect melamine presence in the pure raw milk using methonobactine (Mb)-mediated AuNP formation. The sensor solution was in wine red color without and with raw milk, but when melamine was added into the solution at different concentrations (0 to 5.6×10^{-6} M),

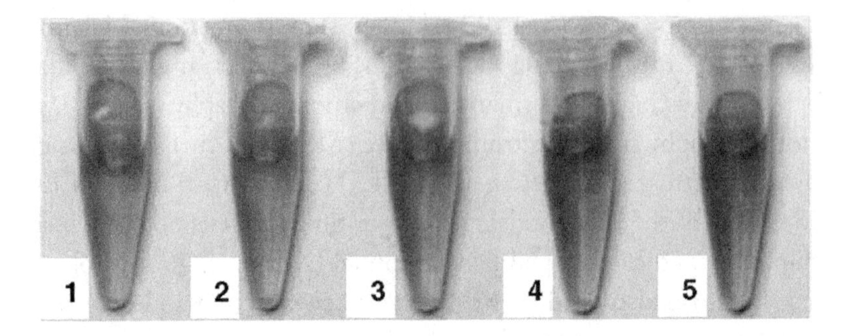

Figure 7.2 **(See color insert Figure 4.3.)** Colorimetric nanosensors for melamine detection in milk: (1) without any addition; (2) with the addition of the extract from blank raw milk; (3) with the addition of the extract containing 1 ppm (final concentration: 8 ppb); (4) with the addition of the extract containing 2.5 ppm (final concentration: 20 ppb); and (5) with the addition of the extract containing 5 ppm. (final concentration: 40 ppb). (Reproduced from Ai, K. et al., *J. Am. Chem. Soc.*, 131, 9496–9497, 2009.)

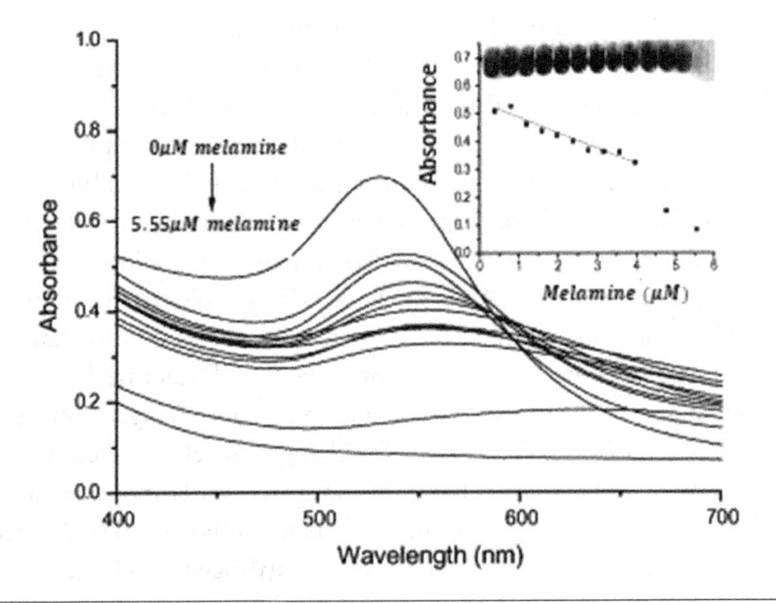

Figure 7.3 UV-absorbance spectra of the AuNPs sensor at different concentrations of melamine. Inset: Visual color change and the UV-absorbance at 539 nm of gold nanoparticle sensor solution at different concentrations of melamine (from left to right: 0, 0.4, 0.8, 1.2, 1.6, 2.0, 2.4, 2.8, 3.2, 3.6, 4.0, 4.8, and 5.6×10^{-6} M). (Reproduced from Xin, J.Y. et al., *Food Chem.*, 174, 473–479, 2015.)

the color of the sensor liquid was slowly changed from red to blue and finally ended up at pale yellow color at the melamine concentration of 5.6×10^{-6} M (Figure 7.3). Melamine presence in the solution affected the aggregation of the AuNPs, which caused the visible color changes in the sensor solution (Xin et al. 2015).

Song et al. (2015) developed a colorimetric melamine detection technique by using AgNPs modified with sulfanilic acid (SAA). They made the sensing solution by adding $AgNO_3$ (2.0 mL of 0.01 M) and SAA (4.0 mL of 0.01 M) with double-distilled water (90 mL). Then $NaBH_4$ (8.8 mg) was added and stirred vigorously for 2 h at room temperature. The average diameter of the AgNPs used in this sensor solution was 6.7 nm, and a 3 mL of sensor solution (SAA-AgNPs) was added to the 50 mL of test solutions (raw milk and milk with various concentrations of melamine) and incubated at room temperature for 5 min. The color of the sensing solution changed from yellow to blue-green whenever melamine was present in the milk solution (Figure 7.4), and the detection limit of this sensor was 10.6 nM (Song et al. 2015). Melamine in the solution produced hydrogen bonding similar to AuNPs and induced the aggregation of SAA-modified AgNPs, which resulted in visible color changes in the sensor solution. Song et al. (2015) proposed that this simple colorimetric sensing technique is rapid (needs only 5 min incubation time) and very economical (costs $0.5/sample) to detect melamine adulteration in milk products.

Guo et al. (2014) developed hollow gold chip nanosensors to detect and quantify the melamine in milk with the use of the SERS

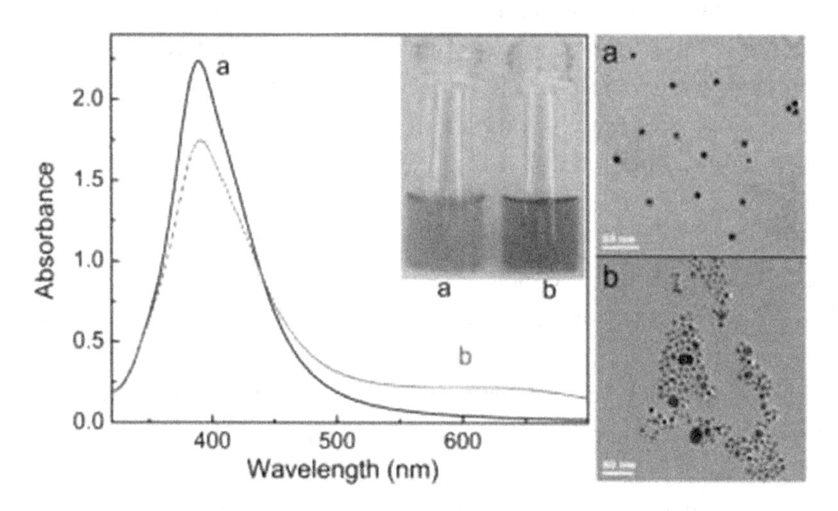

Figure 7.4 UV-Vis spectra (left) (insert image: colorimetric response of sensing solution) and transmission electron microscopy (TEM) images (right) of SAA-AgNPs sensor solution in the absence (a) and presence (b) of 0.9 μM melamine. (Reproduced from Song, J. et al., *Food Control.*, 50, 356–361, 2015.)

technique. AuNPs (average diameter of 30 nm) were prepared using modified Schwartzberg's method (Guo et al. 2014). A sodium citrate solution (500 μL, 0.1 M) and cobalt chloride solution (100 μL, 0.4 M) were mixed with 50 mL of distilled water and deoxygenated using ultrapure nitrogen (N_2) for 1 h, and sodium borohydride solution (150 μL, 1 M) was added using a rapid magnetic stirrer, then the solution was kept under constant N_2 for 45 min. After this reaction time, gold precursor (50 μL, 0.1 M) was added, and the nitrogen supply was stopped for oxidization. Then the AuNPs were synthesized using 0.01% $HAuCl_4$ (100 mL) and 1% sodium citrate (2 mL) by boiling for 30 min. Glass wafers with the size of 1 cm × 1 cm were coated with polydiallydimethylammonium chloride (PDDA) solution by immersing them for 1 h followed by a rinse and dry with water, and N_2, respectively. To create gold chip nanosensors, these PDDA-coated glass wafers were soaked in the prepared AuNP colloid for 6 h and then rinsed and dried using water and N_2, respectively. For testing this developed sensor for detecting melamine in raw milk samples, melamine was added into milk at different concentrations (1, 10, 50, and 100 ppm), and then the sample was centrifuged for 20 min at a speed of 5000 rpm. The AuNP nanosensor was immersed in the supernatant solution after the centrifugation for 2 min, and then washed and dried using water and N_2, respectively. Then the SERS spectra was obtained using a Raman spectrophotometer (Model: JobinYvon/HORIBA labRam, Irvine, California, USA), and the results showed that significant differences in spectra of milk with and without melamine. The complete sensor fabrication and detection method is graphically shown in Figure 7.5. Guo et al. (2014) also found a positive correlation between Raman peak intensity at 712 cm^{-1} and the melamine concentration, and they obtained a linear equation to quantify the melamine level in milk with a correlation coefficient of 0.9912. Guo et al. (2014) showed that this developed sensor can be used to detect and quantify melamine without any sample pretreatment in 1 h total procedure time with a detection limit of 1 ppm.

Kumar et al. (2014) developed AuNP-based nanosensors stabilized with citrate to detect melamine in milk. They obtained AuNPs with an average diameter of 21 nm by mixing trisodium citrate (5 mL, 38.8 nM) into a boiling 100 mL chloroauric acid (1 mM) on a hot

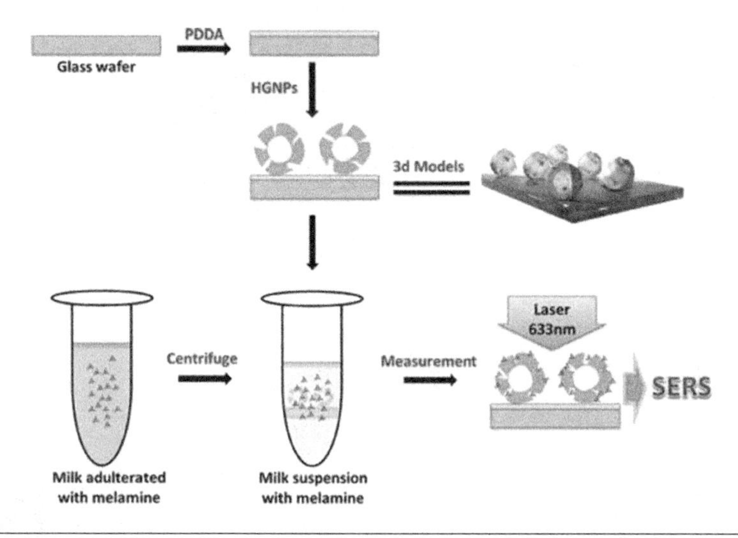

Figure 7.5 Schematic diagram of AuNP nanosensors chip fabrication and melamine detection procedure. (Reproduced from Guo, Z. et al., *Talanta.*, 122, 80–84, 2014.)

plate with magnetic stirrer and stirred for 15 min. Raw milk samples were centrifuged at 10,000 rpm for 5 min to remove milk proteins, which affect the interaction between AuNPs and melamine. Similar to previous nanosensing techniques, the melamine presence in the milk creates aggregation of AuNPs, which were stable under aqueous samples, and resulted in a color change of wine red to blue. The developed sensor was tested against raw milk samples with different concentrations of melamine and found the melamine concentration of 1 mg/L produced rapid color change from red to blue, and at lower melamine concentrations the color change was slow (Figure 7.6). The spectra obtained using UV-Vis spectrometer (Model: SPECORD 200, Analytikjena, Germany) also showed a visible peak at 640 nm, which was used to quantify melamine concentration in a raw milk sample. This sensor took 15 min for detection and quantification of melamine, and this sensor can be used to detect melamine in milk as low as 0.05 mg/L (Kumar et al. 2014).

Cai et al. (2014) developed a AuNP-based nanosensor functionalized with 3-mercaptopriopionic acid (MA) to detect melamine in raw milk and infant formulas. One milliliter of AuNPs stabilized with citrate were prepared and stored at 4°C using the similar techniques as followed by Guo et al. (2014) and Kumar et al. (2014). The AuNPs

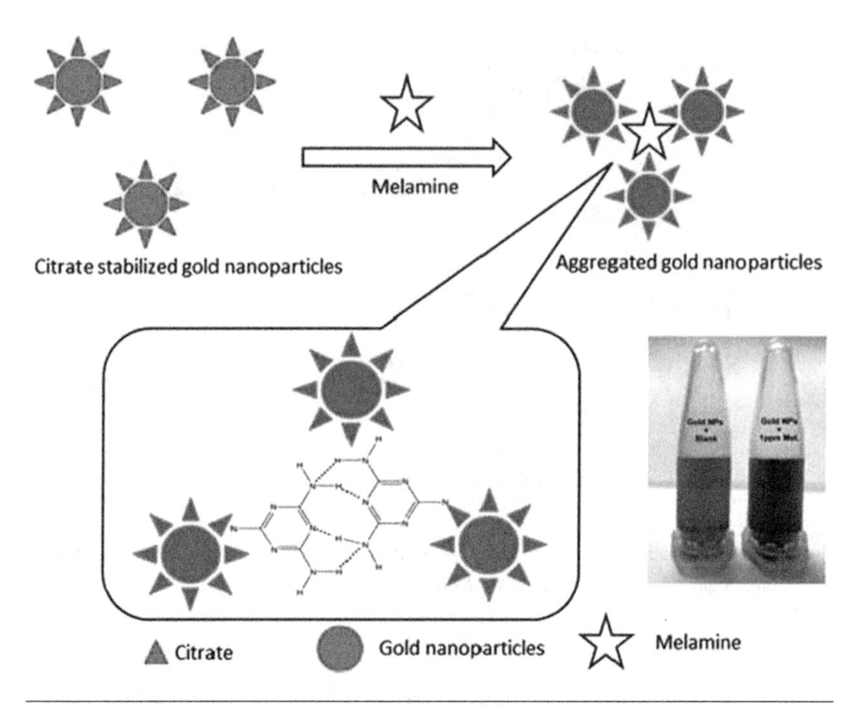

Figure 7.6 (See color insert.) AuNPs nanosensors for melamine detection in milk (inset: AuNPs sensing solution in absence and presence of 1 mg/L melamine). (Reproduced from Kumar, N. et al., *Anal Biochem.*, 456, 43–49, 2014.)

were mixed with 200 μL of MA (0.5 M) for 6 h and centrifuged at 5000 rpm for 10 min to separate MA-modified AuNPs. When the milk and infant formula solutions with different concentrations melamine were tested with these MA-modified AuNPs (which acted as nanoprobes), visible color change was observed by the naked eye over the melamine concentration of 30 μg/mL (Figure 7.7). But when the absorbance of UV-Vis spectra was used to quantify the melamine concentration in milk and milk products, this developed sensor was able to detect melamine concentration as low as 0.4 μg/mL (Cai et al. 2014). They also found this developed sensor had good selectivity for melamine over the milk proteins, which reduced melamine detectability of other similar AuNPs nanosensors developed by other researchers (Guo et al. 2014; Kumar et al. 2014).

Ni et al. (2014) developed a colorimetric nanosensor using bare AuNPs (160 μL, 2.1×10^{-8} M) mixed with 3,3′,5,5′-tetramethlybenzidine (TMB) (200 μL, 8.3 mM) and H_2O_2 (200 μL, 1.4 M) to detect melamine in milk and milk products. Melamine with different

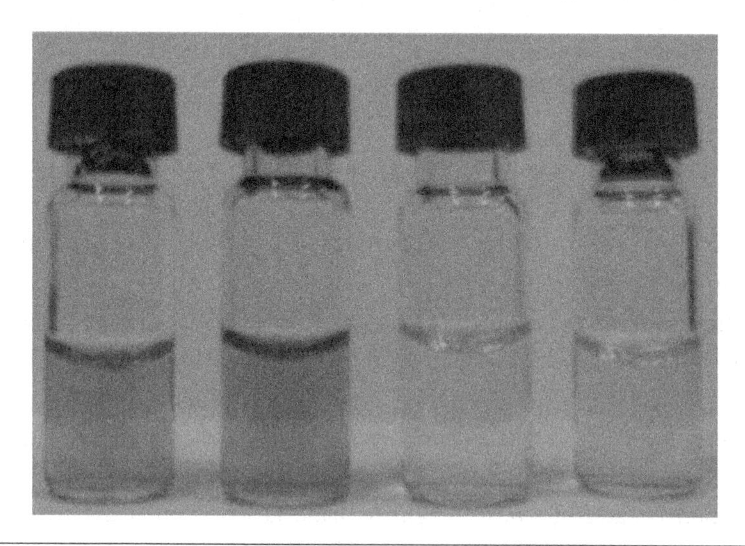

Figure 7.7 (See color insert.) MA-functionalized AuNPs nanosensors reaction with addition of melamine at different concentrations (from left to right: 0, 10, 20, and 30 μg/mL). (Reproduced from Cai, H.H. et al., *J. Food Eng.*, 142, 163–169, 2014.)

concentrations in water or milk and milk powder were tested with this nanosensor, and a clear color change from red to blue was noticed when melamine was present in the sample. For milk and milk products, supernatants of the solution obtained through centrifugation of 5.0 mL milk (or 5.0 mg of milk powder) with 1.5 mL trichloroacetic acid (2 M) for 10 min at a speed of 10,000 rpm was used for detection using this TMB-H_2O_2–AuNPs colorimetric nanosensor solution. The results proved that this sensor detected melamine in milk samples as low as 0.5 μM concentration with the naked eye, and when the UV-Vis spectrophotometer was used the detection limit was 0.2 μM, which is lower than the melamine safety limit set by US Food and Drug Administration (20 μM) (Ni et al. 2014).

Giovannozzi et al. (2014) developed an AuNP nanosensor with SERS for rapid detection of melamine in raw milk. Citrate-synthesized AuNPs (40 nm size) were prepared following the other reported methods (Cai et al. 2014; Guo et al. 2014; Kumar et al. 2014). Milk samples (4 mL) were mixed with 200 μL HCl (1 M) under vortex for 10 s and then centrifuged at a speed of 14,000 rpm for 30 min. Supernatant of the solution was mixed with 60 μL of NaOH (1 M) to adjust solution's pH to 4.7 and then centrifuged for 30 min at a speed of 14,000 rpm. Previously prepared AuNPs were added with the supernatant solution and analyzed immediately with SERS spectroscopy. From the SERS spectral curves, a regression equation was developed using the normalized Raman band at 715 cm^{-1} for quantification of melamine in raw milk samples ($R^2 = 0.99$). Giovannozzi et al. (2014) found that the detection limit of this AuNPs sensing was 0.17 mg/L and had higher quantification accuracy when the melamine concentration range was between 0.57 and 5.0 mg/L. This method took about 30 min for quantification, but this sensing technique requires a SERS spectroscopy unit, which increases the cost of testing as well as requires highly qualified personnel.

Ping et al. (2012) also developed a probe using label-free AgNPs to detect the melamine concentration in raw milk. AgNPs were obtained by adding sodium borohydride (10 mL, 10 mM) drop by drop into 20 mL of silver nitrate (0.25 M) + sodium citrate (0.25 M) solution and stirred for 10 min followed by cooling to room temperature and stewing for 8 h. A 2.0 mL of chloroform and trichloroacetic acid (10%) solution was added to raw milk (4 mL diluted to 10 mL with water) and centrifuged at a speed of 13,000 rpm for 10 min to separate milk proteins. The $NaHCO_3$ (10%) solution was added with the supernatant of the centrifuged solution to adjust the pH to 8.0, then centrifuged for 10 min at a speed of 10,000 rpm. Finally, 800 μL of AgNP solution was added with 600 μL of supernatant solution and kept under room temperature for 20 min. During this reaction time, melamine bound with the AgNP solution and produced concentration-dependent colorimetric changes (yellow to light red), which was noticed by the naked eye. The spectra obtained by Vis-NIR spectrophotometer (Figure 7.8) also showed that melamine concentration can be found using ratio of absorption at 402, and 500 nm (A_{500}/A_{420}) with the use of a linear equation (correlation coefficient = 0.9989). Ping et al. (2012) proved that this nanoparticle-based probe can detect melamine in milk and milk products with a concentration as low as 2.32 mM. Similar kinds of sensors with silver and gold nanoparticles were developed by various researchers (Cao et al. 2009, 2010; Chi et al. 2010; Guo et al. 2010; Han and Li 2010; Li et al. 2010; Liang et al. 2011; Qi et al. 2010; Su et al. 2011; Wei et al. 2010) using nanotechnology for rapid detection of melamine adulteration in milk, which can help the regulatory agencies as well as

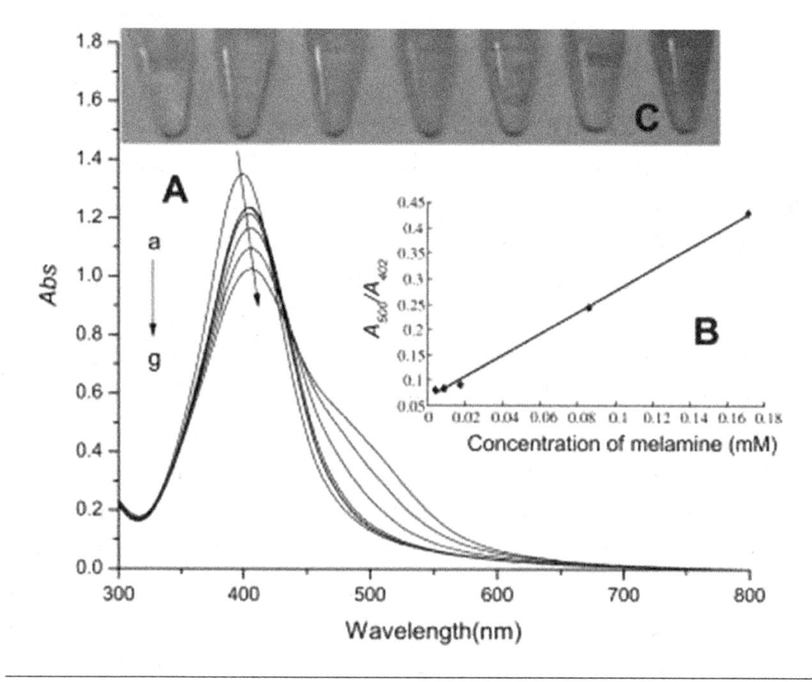

Figure 7.8 **(See color insert.)** (A) Absorption spectra, (B) absorption ratio (A_{500}/A_{402}) vs melamine concentration, and (C) using AgNPs mixed in raw milk and various concentrations of melamine (from left to right 0, 0.002, 0.004, 0.008, 0.017, 0.085, 0.1700 mM, respectively). (Reproduced from Ping, H. et al., *Food Control*, 23, 191–197, 2012.)

food safety personnel to identify and quantify the melamine in milk and milk products quickly and accurately.

Detection of Chemical Residues

Micro and nanosensors have been developed through GoodFood (2008) European projects for food safety analysis as well as for quality control of food products during production. These sensors can be used directly in the production line to control the quality of dairy products. Researchers in the GoodFood project developed sensors to detect residues of antibiotics in milk as well as aflatoxin M_1 in milk. They developed four types of sensors (Fluidic-Fluorescence system using magnetic beads, wavelength-integrated optimal system [WIOS], dip-stick sensors, and multi-analyte enzyme-linked immunosorbent assay [ELISA] sensors) for detecting antibody residues in the milk. In the Fluidic-Fluorescence system, milk was mixed with magnetic beads (functionalized with antibiotic specific antibodies) in the capture chamber of microfluidic chip, and then the magnetic beads were captured and analyzed using a fluorescence optical sensor in the detection chamber. The presence of antibodies in the milk was detected using the differences in the optical signal due the antibody–antigen reaction. The WIOS sensors used the change of the refractive index at the wavelength guide of the sensor for antibiotic residues in the milk. The dipstick sensors were developed using "antibiotic-binding molecules" (ABMs) with the nitrocellulose membrane and AuNPs for detection of residues of antibiotics. The milk was mixed with an ABM solution, and then dipstick sensor was placed in the solution. The capillary action of the solution with ABM and nitrocellulose created a color intensity change on the dipstick, which was used to detect the antibiotic residues by comparing with the standard (control) samples. All the major antibiotics residues can be detected from a single sample using the multi-analyte ELISA sensor (Figure 7.9). All four types of sensors were successfully tested for all major antibiotics residues

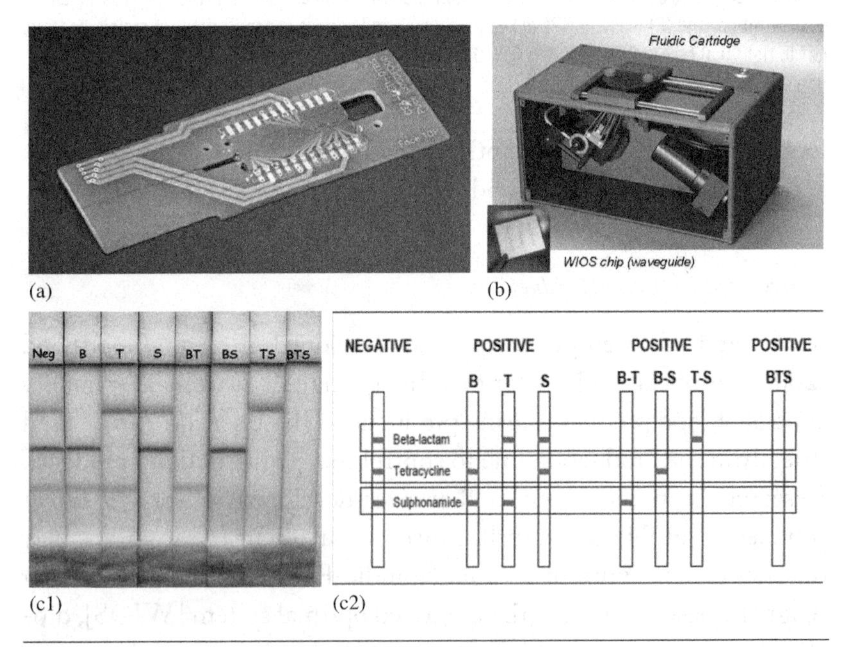

Figure 7.9 (a) Fluidic-Fluorescence microfluidic sensor, (b) WIOS sensor, and (c) dipstick sensor [(c1) sensor with positive and negative spiked milk with various antibiotic residues, (c2) schematic of obtained results with dipstick sensor (c1) tested with milk with antibiotic residues] for antibiotic residue detection in milk. (Reproduced from GoodFood, *Food Safety and Quality Monitoring with Microsystems*, European Council, Brussels, Belgium, 2008.)

(sulphonamide, tetracycline, and betalactam) used in cattle farming by the GoodFood project researchers as well as validated by real samples from Nestle (GoodFood 2008).

An ELISA-based detection system with MNPs was developed to detect antibiotic (sulfonamide) residues in raw milk samples (Font et al. 2008). The antibody-functionalized MNPs were used for separation and enrichment of antibiotic residues, and the ELISA technique with spectrophotometer was used for quantification of the residues. The commercially available carboxyl group (MP-COOH)-modified MNPs (diameter 196 nm, active chemical functionality 0.155 mmol g^{-1}) from Estapor (Product No. 00–39, Merck Millpore Ltd., Burlington, MA, United States) were obtained and then sulfonamide-specific antibody (As 167)-functionalized MNPs (Ab-NPs) were prepared by coupling nanomagnetic beads with the purified immunoglobulin fractions (made by ammonium sulfate precipitation of As 167). In a microtiter (96 wells) plate, an Ab-MNP suspension (0.25 mg mL^{-1} concentration, 50 μL/well) was added and then standard solutions SPY (25 μL/well, concentrations from 50 μM to 0.125 nM and zero) and SA1-HRP (25 μL/well, concentration 0.025 μmL^{-1}) of SA1 (5-[6-(4-amino-benzenesulfonylamino)-pyridin-3-yl]-2-methyl-pentanoic acid) + HRP (horseradish peroxidase) were added and incubated at room temperature for 30 min. The plate was then washed and the milk sample solution was added to the plate (100 μL/well) and incubated at room temperature for 30 min, and then H_2SO_4 4N (50 μL/well) was added to stop the reaction. The absorbance of the solution at 450 nm was measured by a spectrometer (Model: Varian Inova 500, Varian Inc. Palo Alto, California, USA). The absorbance curves showed significant differences among milk samples with and without sulfonamide residues, and the results proved this ELISA technique with Ab-MNPs could be able to detect sulfonamide residues as low as 0.5 μg L^{-1}, which is lower than the European Union's maximum residue allowable limit for milk (100 μg L^{-1}) (Font et al. 2008). They also successfully tested the same Ab-MNP-ELISA technique to detect the sulfonamide residues in cow using the cow's hair samples, and the complete detection procedure is graphically shown in Figure 7.10.

Estrogen is the one of the major steroid hormones found in food products, and it has the ability to cause carcinogenic cancers in

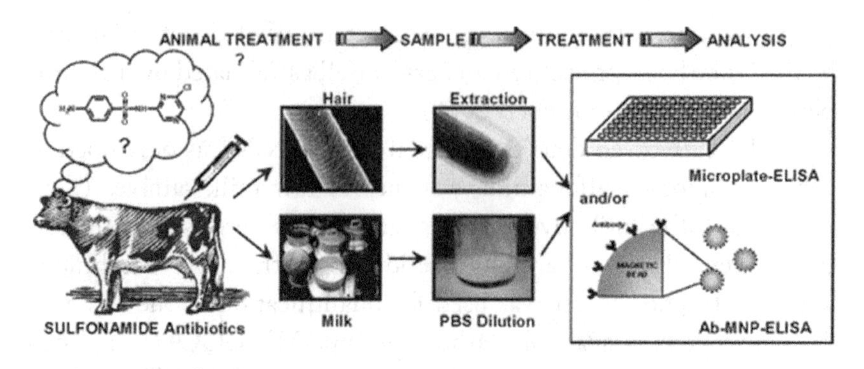

Figure 7.10 Schematic representation of Ab-MNPs-ELISA-based sulfonamide residue detection procedure. (Reproduced from Font, H. et al., *J. Agric. Food Chem.*, 56, 736–743, 2008.)

humans, like ovarian, breast, and prostate cancers (Ganmaa and Sato 2005). Milk is the one of the main sources of estrogen intake by humans, and estrogen levels in raw milk is increasing day by day because nearly 75% of the milk is produced from pregnant cows in modern dairy farms (Gao et al. 2011; Tso and Aga 2010). In general, LC, MS, and solid phase extraction (SPE) techniques have been used for detection and quantification of estrogen in milk, but these are time-consuming methods, as well as samples have to be purified from other substances in milk, which interfere with the detection of estrogen. Gao et al. (2011) developed a detection method using a polypyrrole (PPy)-coated magnetic nanoparticle (PPy/MNP) sensor coupled with LC-MS to quantify the estrogen levels in raw milk. The PPy/MNPs used to separate estrogen from the milk without protein precipitation and LC-MS system was used to quantify the extracted estrogen. A 5 g of ferric trichloride hexahydrate ($FeCl_3.6H_2O$) was mixed with 100 mL of ethylene glycol (EG), and then 15 g of sodium acetate (NaAc) and 50 mL of ethylene diamine (ED) were mixed and stirred for 30 min. Then the solution was placed in an autoclave and heated at 200°C for 8 h, and then cooled to room temperature, and MNPs were collected using a magnet. Collected MNPs were washed with water and ethanol and dried using a vacuum drier for 6 h at 60°C. After drying, 1.0 g of prepared MNPs were mixed with 9.1 g of ferric chloride ($FeCl_3$), added to 100 mL of deionized water in a 250 mL flask, and shook at a speed of 150 shakes/min in a water bath for 3 h set at a temperature of 25°C. For PPy coating of MNPs, 0.5 mL of PPy monomers and 20 mL of SDS solution was added to the solution, and then the mixture was shaken for another 12 h in the same water bath. The PPy-coated MNPs (i.e., MNPs/PPy) were collected by magnetic separation, washed by ethanol and water, and then dried by vacuum drier (60°C) for 6 h. For analyzing estrogen in the milk, 5 mg of the above prepared MNPs/PPy were placed in a glass vial (15 mL) and activated using methanol and Milli-Q water, and then the diluted milk sample (10 mL) was poured into the vial and vortexed for 3 min. The MNPs/PPy absorbed the estrogen in the milk sample during this vortexing procedure and then separated from the solution using a magnet, and the MNPs/PPy were washed with water (2 mL). Washed MNPs/PPy were placed in a vial, and 1 mL of acetone was added and vortexed for 0.5 min for dissolving the estrogen. Then the MNPs/PPy were separated using a magnetic method and collected residue was used to quantify the estrogen using a LC-MS. This rapid magnetic solid phase extraction (MSPE) technique had an estrogen detection limit of 5.1–66.7 ng/L and needed only 3 min for analyzing a milk sample (Gao et al. 2011).

Bisphenol A (BPA) is the component used in the manufacturing of food and beverage containers, and the industry is reducing use of this chemical due to its ill effects on human health. The common techniques like liquid chromatography (HPLC), GC/MS, and ELISA methods are used for detection of BPA residues in food and beverages (Basheer and Lee 2004; Inoue et al. 2000; Moreno et al. 2011). An aptamer biosensor with AuNPs was developed to detect the BPA residues in milk samples (Zhou et al. 2014). They polished a glassy carbon electrode (GCE) with Al_2O_3 slurry and washed with double-distilled water and ethanol, then dried with nitrogen at room temperature. A graphene oxide (GO) (5.0 μL of 1.0 mg mL^{-1} concentration) suspension was poured uniformly on to the prepared GCE and then rinsed with double-distilled water followed by immersing in the 0.1 mM chloroauric acid (0.1 mM) and sulphuric acid (0.5 M) to electrodeposit the AuNPs on to the electrode. Then the anti-BPA aptamer was assembled on the electrode and immobilized with various concentrations of anti-BPA at 4°C for 16 h. The prepared electrode was connected with an electrochemical workstation (Model: CHI 440A, Shanghai CH Instrument Inc., Shanghai, China) to measure the BPA residues using the change in electrochemical properties. The Tris–HCl buffer solution (20 mL) was added with 5.0 mL of raw milk and sonicated and shaken for

15 min and 10 min, respectively. After centrifuging the mixture for 10 min, a supernatant of the solution was collected using a membrane filter (0.22 μm) and spiked with different concentrations of BPA for testing the aptamer biosensor's detectability. Their results showed that the developed aptamer-based biosensors with AuNPs had average BPA recovery of 105% and the relative standard deviation <5.2% when tested with various levels of BPA in milk and also showed strong reproducibility of results (Zhou et al. 2014).

Detection and Control of Pathogenic Microorganisms

Due to their antimicrobial properties, AgNPs have been tested for controlling foodborne pathogens and other microorganisms like fungi. AgNPs ruptured the cell wall of pathogenic bacteria, which resulted in the damage of the cell membrane (Figure 7.11). Synthesis of AgNPs from silver nitrate using biological materials like plant proteins has been researched by scientists around the world, and Lee et al. (2013) tested cow milk using synthesized AgNPs to control fungal infection in agricultural materials. They mixed 4 mL of cow milk with 96 mL of commercially available silver nitrate (1 mM) and

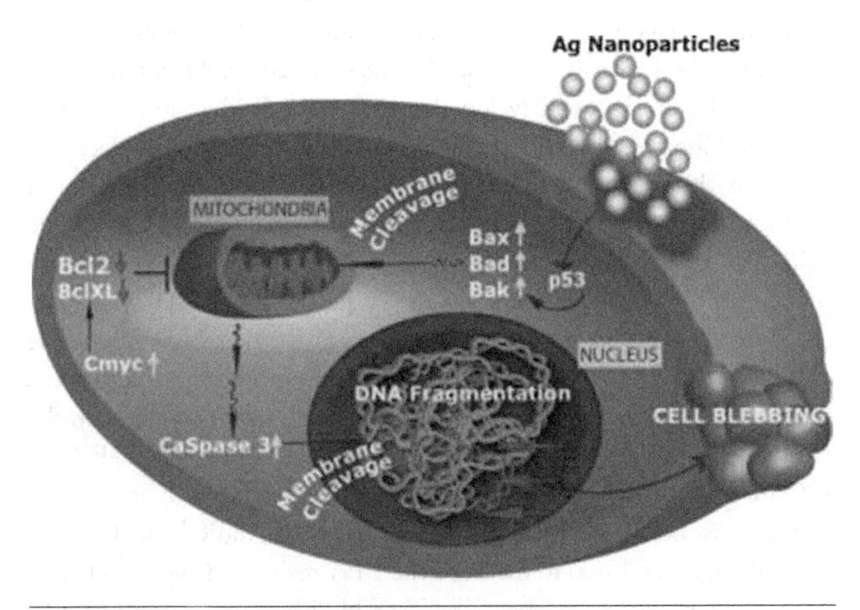

Figure 7.11 Schematic representation of AgNPs causing cell damage. (Reproduced from Gopinath, P. et al., *Colloids Surf. B Biointerfaces*, 77, 240–245, 2010.)

mixed in a rotary shaker at a speed of 180 rpm for 8 h. After the reaction time, synthesized AgNPs were obtained by centrifugation at a speed of 13,000 rpm for 15 min and then purified using autoclaved water. The effect of prepared AgNPs (average size 30–90 nm diameter) against the growth of fungal species *Colletotrichumcoccodes*, *Monilinia* sp., and *Pyricularia* sp.) was assessed by culturing the fungal samples using a potato agar plating method with and without AgNPs (Lee et al. 2013). Results of their study showed that the synthesized AgNPs reduced the growth of *Colletotrichumcoccodes*, *Monilinia* sp., and *Pyricularia* sp. by 87.1%, 86.5%, and 83.5%, respectively. Results also showed the growth of fungal species reduced with increase in AgNPs concentration. Lee et al. (2013) tested AgNPs synthesized with cow milk only with pure fungal cultures, and they suggested for more testing to implement the AgNPs as fungicides for applying on agricultural products.

Lee et al. (2012) tested AgNPs for selective detection of major bacterial cultures affecting yogurt (*Bifidobacterium lactis* (Bb12), *S. thermophiles, L. bulgaricus, Bifidobacterium longum*, and *Lactobacillus acidophilus*) using Matrix-assisted laser desorption/ionization mass spectrometry (MALDI-MS) technique. Fresh yogurt samples obtained from the store, samples stored in the refrigerator for 1 wk, 2 wk, and 1 month after expiry date, and samples stored at room temperature for 2, 5, and 7 days were tested for feasibility of using AgNPs for rapid detection of pathogenic bacterial cultures. Two different concentrations (0.035 mg and 0.35 mg) of AgNPs were mixed with 100 μL of yogurt samples and incubated at 37°C for 10 min, and then spotted on a MALDI plate for quantification using MALDI-MS. The results showed that the use of AgNPs (at a concentration of 0.035 mg) for selective capturing of pathogenic bacterial cultures increased the detectability of the MALDI-MS technique by 2 to 8 folds in yogurt, but a higher concentration of AgNPs (0.35 mg) reduced the bacterial colony counts due to the AgNPs antimicrobial properties (Lee et al. 2012).

Zhang et al. (2016) tested AuNPs encapsulated using poly (amidoamine) dendrimer and enhanced double-wall carbon nanotubes (CNT) for detecting *Escherichia coli* in dairy products (milk, infant formula, milk powder, and yogurt). They polished GCEs with 0.3 μm alumina slurry and then rinsed by acetone and water.

The chitosan membrane was formed on the GCE by casting 5 μL of chitosan (1%) dispersed in acetic acid, and then the poly (amidoamine) dendrimer-gold nanoparticles were bonded to the chitosan-GCE. The electrode was activated by casting 0.10 M N-hydroxysuccinimide (NHS) in phosphate-buffered saline (PBS) and then incubated with 1.0 mg mL^{-1} antibody (*E. coli* specific) for 50 min at 37°C. Different concentrations of *E. coli* (5.0×10^2, 1.0×10^3 and 5.0×10^3 cfu mL) were mixed with the solution containing 1 mg of dairy product (milk, infant formula, milk powder, or yogurt) dispersed in 10 mL of PBS (0.01 M, pH 7.4) solution. The prepared electrode was placed inside the sample solution, and the peak current was measured using differential pulse voltammetry (DPV). Their results showed that the peak current increased with the increase in *E. coli* concentration in pure cultures (Figure 7.12), and the recoveries using this sensor were between 996.8% and 108.7% when tested on real dairy samples (Zhang et al. 2016).

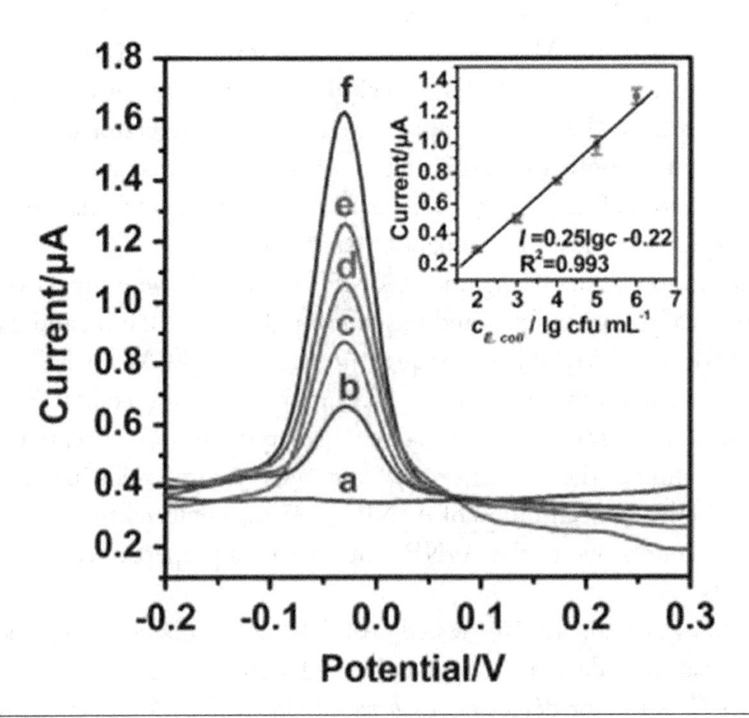

Figure 7.12 Peak DPV current of immune electro sensor when tested with different *E. coli* concentrations. ((a) 0, (b) 1.0×10^2, (c) 1.0×10^3, (d) 1.0×10^4, (e) 1.0×10^5, (f) 1.0×10^6 cfu mL^{-1}). (Reproduced from Zhang, X. et al., *Biosens. Bioelectron.*, 80, 666–673, 2016.)

González-Sálamo et al. (2017) used MNPs coated with polydopamine (pDA) for the extraction of estrogenic mycotoxin compounds from milk and yogurt followed by LC-MS for quantification of estrogenic compounds. The MNPs were prepared by dissolving 2.78 g of $FeSO_4$ and a 5.41 g of $FeCl_3 \cdot 6H_2O$ into 200 mL of HCl (0.5 M) and thoroughly mixed by magnetic stirring. Then this solution was mixed slowly into 300 mL of a NaOH (1.25 M) solution at room temperature using rapid stirring at 850 rpm for 30 min and neutralized with HCl (25%, w/w). For pDA coating, they dispersed the above-prepared MNPs (2.75 g/L concentration) into a solution of dopamine (15 mM) in PBS at a pH of 8.3 and stirred at 850 rpm for 6 h under room temperature. Then the pDA-coated MNPs (pDA/MNPs) were washed 6 times with the mixture solution of ACN:H_2O 50:50 (v/v), centrifuged at 4000 rpm for 15 min, and dried at 40°C. The protein components from the milk and yogurt samples were removed by centrifugation and spiked with estrogenic compounds at different concentrations to test the extraction ability of prepared pDA/MNPs. Filtered deproteinized milk and yogurt samples were mixed with pDA/MNPs in a beaker and manually shaken for 30 s, and then the sorbent was separated using a magnet. The separated sorbent was dried in a nitrogen stream, and the estrogenic compounds were separated from pDA/MNPs using 8 mL of MeOH under a manual agitation for 30 s. This extracted solution was then analyzed using LC-MS for quantification of estrogenic compounds. The optimization tests showed that 1.5 mL of milk, 75 μL of AcOH, and 3 mL of ACN; and 1.5 g of yogurt, 75 μL of AcOH, and 4.5 mL of ACN mixture combinations gave higher estrogenic compounds extraction rates using the pDA/MNPs for milk and yogurt samples, respectively. González-Sálamo et al. (2017) found that this MSPE technique using pDA/MNPs had recovery rates of 70%–120% and could detect estrogenic mycotoxin compounds at concentrations in the range of 0.21–4.77 μg/L and 0.29–4.54 μg/kg for milk and yogurt samples, respectively.

Applications in Packaging

Application of nanotechnology and nanomaterials in the food packaging sector is increasing in a rapid manner since the nanotechnology helps to improve the packaging polymer's thermal and mechanical properties, durability, as well as gas transmission properties (Chaudhry et al. 2008). Recent development in packaging materials for food and beverages are discussed in the Chapter 3 in detail, and the applications of nanotechnology-based packaging in dairy products are discussed in this chapter. The development of nanocomposites using nanoparticles like nanoclay AgNPs (up to 5% w/w) help the food manufacturers and distributors to safely store raw and processed dairy products for a long time. Nanocomposite polymers positively modified the gas transmission as well as moisture rates of the packaging film, and in the dairy industry, nanoclay-based packaging polymers have been successfully tested for cheese packaging (Akbari et al. 2006). Packaging materials with AgNPs also have improved antimicrobial properties, and milk containers with AgNPs were successfully adopted in Brazil to increase the storage time (Alisson 2015; Brody et al. 2008).

Shelf life of fresh milk is short, so the dairy industry uses various technologies such as sterilization, ultra-high temperature heating, to reduce the microbial load in order to increase the shelf life. Silver has good antimicrobial properties and researchers are using AgNPs for packaging of food products to inhibit the microbial growth and increase the shelf life. In Brazil, plastic milk bottles with polythene and AgNPs have been developed by a company, Nanox (Nanox Technologies, São Carlos, São Paulo state, Brazil) (Alisson 2015), and these bottles increased the shelf life of milk by >200% (from 7 days to 15 days) in Brazil (Figure 7.13). The AgNPs cause damage to the cell walls and also alter the membrane by which bacterial activity is inhibited (Figure 7.11). AgNPs were added with silica to make a larger nanoparticle center core (to restrict the seepage of silver into milk) and mixed with polythene pellets during the manufacturing process of these milk bottles. Nanox also used the same technology to produce flexible plastic milk bags (up to10 days shelf life) to use in the southern part of Brazil. Nanox Technologies also obtained FDA and EPA approvals for these technologies for commercial distribution. Double-layer lightproof plastic milk bottles also developed with titanium dioxide nanoparticles to save the UV sensitive components in the fresh milk (Vitamin B2) in New Zealand by the company named Fonterra (Fonterra Cooperative Group, Auckland, New Zealand) (Dickinson 2017).

Figure 7.13 Plastic milk bottle with AgNPs. (Reproduced from Alisson, E., Brazilian company doubles shelf life of pasteurized fresh milk, Agência FAPESP, http://agencia.fapesp.br/brazilian_company_doubles_shelf_life_of_ pasteurized_fresh_milk/21432, 2015.)

Due to its antimicrobial properties, silver is used as a preservative and packaging material for many food products. AgNPs are used in the food industry as a preservative and used to store food products, since it has the ability to inhibit the growth of major foodborne pathogens. Baby Dream® Co. Ltd. (South Korea) developed baby milk bottles with nanosilver particles and marketing with the commercial name of "Silver-nano noble nursing bottle" throughout the world (BabyDream 2017).

Conclusion

Nanotechnology and nanomaterials have been applied in the dairy industry mainly during processing by means of enhancement, fortification, and nanoencapsulation of micro and macro nutrient compounds into dairy products. Enhancing the nutritional properties of milk, yogurt, cheese, and ice cream using nanosize materials in dairy products without changing their functional properties is the major advantage of using nanotechnology in the dairy industry. Fortification of dairy products with supplementary minerals like calcium, zinc, and iron improves the health benefits of the products as well as nutritional intake. Nanoencapsulation of lipids and other nutritional compounds using nanotechnology helps targeted delivery of nutritional compounds to the specific receptors in the human gastrointestinal system. Nanosensors with Ag, Au, and

MNPs have been developed to selectively extract, detect, and quantify the chemical residues (BPA, antibiotic steroids), adulterants (melamine), and pathogenic microorganisms and their derivatives (bacteria, fungus, and mycotoxins). Most of these nanosensors can rapidly detect the melamine or other residues by visual color changes in the solution but need some pretreatment like removal of milk proteins, which may interfere with the results. These nanosensors also need to be coupled with expensive standard analytical tools like spectroscopy and chromatography instruments for quantification of chemical residues and adulterants, which may increase the cost and time for testing. Application of Ag and Au nanoparticles for detection and control of microorganisms, which grow in milk and milk products, is another major potential area of application for nanotechnology in the dairy industry, but similar to chemical residue detection and quantification by these nanosensors has to be linked with other expensive regular food analysis tools. AgNPs and nanoclay-based packaging materials can help the dairy industry to increase the shelf life of milk and milk products. But leaching of these nanomaterials into food products (milk, yogurt, or cheese) is the major concern among consumers. Even though nanotechnology helps to improve nutritional benefits and shelf life of dairy products in a substantial manner, the reservations about the impact of nanomaterials on the human health in the consumer's mind is the major obstacle in implementing most of these nanotechnology-based products and techniques in the market. More research is needed to develop regulatory guidelines for applying the nanotechnology and nanomaterials in the dairy industry as well as their effects on human health.

References

Ai K., Liu Y., Lu L. (2009) Hydrogen-bonding recognition-induced color change of gold nanoparticles for visual detection of melamine in raw milk and infant formula. *Journal of the American Chemical Society* 131:9496–9497. doi:10.1021/ja9037017.

Akbari Z., Ghomashchi T., Aroujalian A. (2006) *Potential of Nanotechnology for Food Packaging Industry*. Nano and Micro Technologies in the Food and HealthFood Industries, Amsterdam, the Netherlands. pp. 25–26.

Alisson E. (2015) Brazilian company doubles shelf life of pasteurized fresh milk, Agência FAPESP. Available from http://agencia.fapesp.br/brazilian_company_doubles_shelf_life_of_pasteurized_fresh_milk/21432.

BabyDream. (2017) Silver nano products. Available from http://www.nanotechproject.org/cpi/products/silver-nano-noble-ns-nursing-bottle. Accessed on June 18, 2017.

Basheer C., Lee H.K. (2004) Analysis of endocrine disrupting alkylphenols, chlorophenols and bisphenol-A using hollow fiber-protected liquid-phase microextraction coupled with injection port-derivatization gas chromatography–mass spectrometry. *Journal of Chromatography A* 1057:163–169.

Brody A.L., Bugusu B., Han J.H., Sand C.K., Mchugh T.H. (2008) Innovative food packaging solutions. *Journal of Food Science* 73:R107–R116.

Cai H.H., Yu X., Dong H., Cai J., Yang P.H. (2014) Visual and absorption spectroscopic detections of melamine with 3-mercaptopriopionic acid-functionalized gold nanoparticles: A synergistic strategy induced nanoparticle aggregates. *Journal of Food Engineering* 142:163–169. doi:10.1016/j.jfoodeng.2014.04.018.

Cao Q., Zhao H., He Y., Li X., Zeng L., Ding N., Wang J., Yang J., Wang G. (2010) Hydrogen-bonding-induced colorimetric detection of melamine by nonaggregation-based Au-NPs as a probe. *Biosensors and Bioelectronics* 25:2680–2685.

Cao Q., Zhao H., Zeng L., Wang J., Wang R., Qiu X., He Y. (2009) Electrochemical determination of melamine using oligonucleotides modified gold electrodes. *Talanta* 80:484–488.

Chaudhry Q., Scotter M., Blackburn J., Ross B., Boxall A., Castle L., Aitken R., Watkins R. (2008) Applications and implications of nanotechnologies for the food sector. *Food Additives & Contaminants: Part A* 25:241–258. doi:10.1080/02652030701744538.

Chi H., Liu B., Guan G., Zhang Z., Han M.Y. (2010) A simple, reliable and sensitive colorimetric visualization of melamine in milk by unmodified gold nanoparticles. *Analyst* 135:1070–1075.

Cressey D. (2013) Nanotechnology offers small food for thought. *The Guardian*, Guardian News & Media, London, UK.

Dickinson M. (2017) Lightproof bottle for the dairy industry. Available from https://matterchatter.wordpress.com/2013/03/21/light-proof-bottle-for-the-dairy-industry.

Farhang B. (2007) Nanotechnology and lipids. *Lipid Technology* 19:132–135.

Font H., Adrian J., Galve R., Estévez M.C., Castellari M., Gratacós-Cubarsí M., Sánchez-Baeza F., Marco M.P. (2008) Immunochemical assays for direct sulfonamide antibiotic detection in milk and hair samples using antibody derivatized magnetic nanoparticles. *Journal of Agricultural and Food Chemistry* 56:736–743. doi:10.1021/jf072550n.

Ganmaa D., Sato A. (2005) The possible role of female sex hormones in milk from pregnant cows in the development of breast, ovarian and corpus uteri cancers. *Medical Hypotheses* 65:1028–1037.

Gao Q., Luo D., Bai M., Chen Z.W., Feng Y.Q. (2011) Rapid Determination of estrogens in milk samples based on magnetite nanoparticles/polypyrrole magnetic solid-phase extraction coupled with liquid chromatography–tandem mass spectrometry. *Journal of Agricultural and Food Chemistry* 59:8543–8549. doi:10.1021/jf201372r.

Giovannozzi A.M., Rolle F., Sega M., Abete M.C., Marchis D., Rossi A.M. (2014) Rapid and sensitive detection of melamine in milk with gold nanoparticles by surface enhanced Raman scattering. *Food Chemistry* 159:250–256. doi:10.1016/j.foodchem.2014.03.013.

González-Sálamo J., Socas-Rodríguez B., Hernández-Borges J., Rodríguez-Delgado M.Á. (2017) Core-shell poly(dopamine) magnetic nanoparticles for the extraction of estrogenic mycotoxins from milk and yogurt prior to LC–MS analysis. *Food Chemistry* 215:362–368. doi:10.1016/j.foodchem.2016.07.154.

GoodFood. (2008) *Food Safety and Quality Monitoring with Microsystems*. European Council, Brussels, Belgium.

Gopinath P., Gogoi S.K., Sanpui P., Paul A., Chattopadhyay A., Ghosh S.S. (2010) Signaling gene cascade in silver nanoparticle induced apoptosis. *Colloids and Surfaces B: Biointerfaces* 77:240–245.

Gossner C.M.E., Schlundt J., Ben Embarek P., Hird S., Lo-Fo-Wong D., Beltran J.J.O., Tritscher A. 2009. The melamine incident: Implications for international food and feed safety. *Environmental Health Perspectives* 117(12), 1803–1808.

Guo L., Zhong J., Wu J., Fu F., Chen G., Zheng X., Lin S. (2010) Visual detection of melamine in milk products by label-free gold nanoparticles. *Talanta* 82:1654–1658.

Guo Z., Cheng Z., Li R., Chen L., Lv H., Zhao B., Choo J. (2014) One-step detection of melamine in milk by hollow gold chip based on surface-enhanced Raman scattering. *Talanta* 122:80–84. doi:10.1016/j.talanta.2014.01.043.

Han C., Li H. (2010) Visual detection of melamine in infant formula at 0.1 ppm level based on silver nanoparticles. *Analyst* 135:583–588.

Huang S., Chen J.C., Hsu C.W., Chang W.H. (2009) Effects of nano calcium carbonate and nano calcium citrate on toxicity in ICR mice and on bone mineral density in an ovariectomized mice model. *Nanotechnology* 20:375102.

Inoue K., Kato K., Yoshimura Y., Makino T., Nakazawa H. (2000) Determination of bisphenol A in human serum by high-performance liquid chromatography with multi-electrode electrochemical detection. *Journal of Chromatography B: Biomedical Sciences and Applications* 749:17–23.

Kumar N., Seth R., Kumar H. (2014) Colorimetric detection of melamine in milk by citrate-stabilized gold nanoparticles. *Analytical Biochemistry* 456:43–49. doi:10.1016/j.ab.2014.04.002.

Lee C.H., Gopal J., Wu H.F. (2012) Ionic solution and nanoparticle assisted MALDI-MS as bacterial biosensors for rapid analysis of yogurt. *Biosensors and Bioelectronics* 31:77–83. doi:10.1016/j.bios.2011.09.041.

Lee K.J., Park S.H., Govarthanan M., Hwang P.H., Seo Y.S., Cho M., Lee W.H., Lee J.Y., Kamala-Kannan S., Oh B.T. (2013) Synthesis of silver nanoparticles using cow milk and their antifungal activity against phytopathogens. *Materials Letters* 105:128–131. doi:10.1016/j.matlet.2013.04.076.

Li L., Li B., Cheng D., Mao L. (2010) Visual detection of melamine in raw milk using gold nanoparticles as colorimetric probe. *Food Chemistry* 122:895–900.

Liang X., Wei H., Cui Z., Deng J., Zhang Z., You X., Zhang X.E. (2011) Colorimetric detection of melamine in complex matrices based on cysteamine-modified gold nanoparticles. *Analyst* 136:179–183.

Mohammadi R., Mahmoudzade M., Atefi M., Khosravi-Darani K., Mozafari M.R. (2015) Applications of nanoliposomes in cheese technology. *International Journal of Dairy Technology* 68:11–23. doi:10.1111/1471-0307.12174.

Moreno M.J., D'Arienzo P., Manclús J.J., Montoya Á. (2011) Development of monoclonal antibody-based immunoassays for the analysis of bisphenol A in canned vegetables. *Journal of Environmental Science and Health, Part B* 46:509–517.

Mozafari M.R., Johnson C., Hatziantoniou S., Demetzos C. (2008) Nanoliposomes and their applications in food nanotechnology. *Journal of Liposome Research* 18:309–327.

Nag S.K. (2010) 6–Contaminants in milk: Routes of contamination, analytical techniques and methods of control. In: Griffiths M.W. (Ed.) *Improving the Safety and Quality of Milk*. Woodhead Publishing, Sawston, UK. pp. 146–178. doi:10.1533/9781845699420.2.146.

Ni P., Dai H., Wang Y., Sun Y., Shi Y., Hu J., Li Z. (2014) Visual detection of melamine based on the peroxidase-like activity enhancement of bare gold nanoparticles. *Biosensors and Bioelectronics* 60:286–291. doi:10.1016/j.bios.2014.04.029.

Park H.S., Jeon B.J., Ahn J., Kwak H.S. (2007) Effects of nanocalcium supplemented milk on bone calcium metabolism in ovariectomized rats. *Asian-Australia Journal of Animal Science* 20:1266–1271. doi:10.5713/ajas.2007.1266.

Ping H., Zhang M., Li H., Li S., Chen Q., Sun C., Zhang T. (2012) Visual detection of melamine in raw milk by label-free silver nanoparticles. *Food Control* 23:191–197. doi:10.1016/j.foodcont.2011.07.009.

Qi W.J., Wu D., Ling J., Huang C.Z. (2010) Visual and light scattering spectrometric detections of melamine with polythymine-stabilized gold nanoparticles through specific triple hydrogen-bonding recognition. *Chemical Communications* 46:4893–4895.

Relkin P., Yung J.M., Kalnin D., Ollivon M. (2008) Structural behaviour of lipid droplets in protein-stabilized nano-emulsions and stability of α-tocopherol. *Food Biophysics* 3:163–168.

Santillán-Urquiza E., Méndez-Rojas M.Á., Vélez-Ruiz J.F. (2017) Fortification of yogurt with nano and micro sized calcium, iron and zinc, effect on the physicochemical and rheological properties. *LWT-Food Science and Technology* 80:462–469. doi:10.1016/j.lwt.2017.03.025.

Schoder D. (2012) 34-Food adulteration with melamine on an international scale: Field work and troubleshooting in Africa. In: Hoorúr J. (Ed.) *Case Studies in Food Safety and Authenticity*. Woodhead Publishing, Sawston, UK. pp. 309–318. doi:10.1533/9780857096937.6.308.

Sekhon B.S. (2010) Food nanotechnology–An overview. *Nanotechnology, Science and Applications* 3:1–15.

Seo M.H., Lee S.Y., Chang Y.H., Kwak H.S. (2009) Physicochemical, microbial, and sensory properties of yogurt supplemented with nanopowdered chitosan during storage. *Journal of Dairy Science* 92:5907–5916. doi:10.3168/jds.2009-2520.

Sherrington D.C., Taskinen K.A. (2001) Self-assembly in synthetic macromolecular systems via multiple hydrogen bonding interactions. *Chemical Society Reviews* 30:83–93.

Song J., Wu F., Wan Y., Ma L. (2015) Colorimetric detection of melamine in pretreated milk using silver nanoparticles functionalized with sulfanilic acid. *Food Control* 50:356–361. doi:10.1016/j.foodcont.2014.08.049.

Su H., Fan H., Ai S., Wu N., Fan H., Bian P., Liu J. (2011) Selective determination of melamine in milk samples using 3-mercapto-1-propanesulfonate-modified gold nanoparticles as colorimetric probe. *Talanta* 85:1338–1343.

Tso J., Aga D.S. (2010) A systematic investigation to optimize simultaneous extraction and liquid chromatography tandem mass spectrometry analysis of estrogens and their conjugated metabolites in milk. *Journal of Chromatography A* 1217:4784–4795.

Vafabakhsh Z., Khosravi-Darani K., Khajeh K., Jahadi M., Komeili R., Mortazavian A.M. (2013) Stability and catalytic kinetics of protease loaded liposomes. *Biochemical Engineering Journal* 72:11–17.

Wang X., Wang Y.W., Huang Q. (2009) *Enhancing Stability and Oral Bioavailability of Polyphenols Using Nanoemulsions*. ACS Publications, Washington, DC.

Wei F., Lam R., Cheng S., Lu S., Ho D., Li N. (2010) Rapid detection of melamine in whole milk mediated by unmodified gold nanoparticles. *Applied Physics Letters* 96:133702.

Xin J.Y., Zhang L.X., Chen D.D., Lin K., Fan H.C., Wang Y., Xia C.G. (2015) Colorimetric detection of melamine based on methanobactin-mediated synthesis of gold nanoparticles. *Food Chemistry* 174:473–479. doi:10.1016/j.foodchem.2014.11.098.

Yu H., Huang Q. (2010) Enhanced in vitro anti-cancer activity of curcumin encapsulated in hydrophobically modified starch. *Food Chemistry* 119:669–674.

Zhang L., Wu L.L., Wang Y.P., Liu A.M., Zou C.C., Zhao Z.Y. (2009) Melamine-contaminated milk products induced urinary tract calculi in children. *World Journal of Pediatrics* 5:31–35.

Zhang X., Shen J., Ma H., Jiang Y., Huang C., Han E., Yao B., He Y. (2016) Optimized dendrimer-encapsulated gold nanoparticles and enhanced carbon nanotube nanoprobes for amplified electrochemical immunoassay of *E. coli* in dairy product based on enzymatically induced deposition of polyaniline. *Biosensors and Bioelectronics* 80:666–673. doi:10.1016/j.bios.2016.02.043.

Zhou L., Wang J., Li D., Li Y. (2014) An electrochemical aptasensor based on gold nanoparticles dotted graphene modified glassy carbon electrode for label-free detection of bisphenol A in milk samples. *Food Chemistry* 162:34–40. doi:10.1016/j.foodchem.2014.04.058.

8

Applications in the Meat Industry

Introduction

Growing interest for healthy, nutritious, chemical residue-free food at an affordable price among consumers is driving the whole food industry to make changes in product ingredient selection and processing steps. The meat industry is not an exception in this case, and increased food safety regulations around the world further require the meat industry to research more on innovative technologies to make meat products safe for human consumption (Young et al. 2013). Recent research and development in nanotechnology related to biological sciences can help the meat industry in product and process development (Singh et al. 2016). The major applications of nanotechnology in the meat industry are improving functionality and nutritional availability of products (by developing nanosize carriers (ingredients/additives) for improved nutritional delivery, flavor, and security), detection of foodborne pathogens, and improved packaging (Figure 8.1).

The meat industry has been using sodium salts as curing agents for a long time, but recent scientific studies found sodium could cause cardiovascular diseases and hypertension (Bošković et al. 2013). So the meat industry is now reducing the use of sodium, and nanotechnology can help to reduce the salt level through nanostructuring the materials (Singh et al. 2016). Similarly, nanotechnology can help the meat industry in supplementing or reducing the use of additives during processing by using nanostructuring, nanoencapsulation, and nanoemulsification of additives (Ramachandraiah et al. 2015).

Another major application of nanotechnology in the meat industry is in meat packaging. Packaging films manufactured using nanotechnology have excellent physical (strength, toughness, resistance to cracking) and chemical (resistance to chemicals) properties

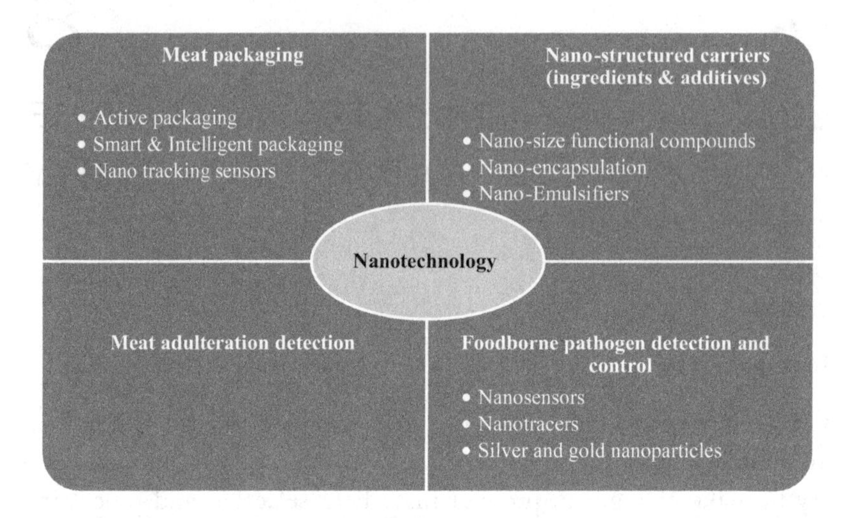

Figure 8.1 Application matrix of nanotechnology in meat processing.

(Ramachandraiah et al. 2015; Weiss et al. 2006). Plastic packaging films with nanoparticles (silver or titanium dioxide) have antimicrobial properties and prevent spoilage of meat. Packaging films with nanoparticles of clay also help to keep the meat fresh during storage by blocking entry of oxygen and moisture into the package as well as the escape of carbon dioxide from packaging material. Bayer commercially produces transparent plastic (polyamide 6 and polyamide 66) packaging films (commercial name: Durethan) with clay nanoparticles for meat and fish packaging, and the clay nanoparticles spread throughout the packaging film to prevent the permeation of gases and moisture through packaging material and keep meat and fish fresh during transportation and storage (Brody et al. 2008; Weiss et al. 2006).

The meat industry is spending millions of dollars every year to ensure food safety. Sensors made using nanotechnology can help to detect foodborne pathogens, which cause food poisoning and other health hazards, rapidly and accurately from meat products without compensating integrity of the product and package (Ramachandraiah et al. 2015). Some of the nanosensors function as tracing devices (nanotracers), which help to monitor the condition and integrity of the meat package from processing plant to the consumer's kitchen. Some meat-packing polymers made with silver or gold nanoparticles (AuNPs) help to control the growth of foodborne pathogens and increase the shelf life of the meat product. Sometimes nanotechnology-based sensors are coupled with other pathogen detection techniques (such as fluorescence, polymerase chain reaction [PCR] techniques) to reduce detection time or magnification of pathogen detection (Luo and Stutzenberger 2008).

Ingredients and Emulsifiers

Researchers are investigating on applying nanotechnology in meat processing for reducing fat content; minimizing the use of sodium, phosphate, and nitrate during production and storage; and also adding materials like probiotics and prebiotics to increase nutritional value as well as the taste of the product (Chaudhry et al. 2008; Weiss et al. 2006; Yusop et al. 2012; Zhao et al. 2009). Researchers around the world are also working on the formulation of functional components, which increase bioavailability to reduce the use of unhealthy compounds and processes, to promote health benefits using nanotechnology during meat processing and storage (Chaudhry et al. 2008; Olmedilla-Alonso et al. 2013). Ginger, commonly used as a tenderizer in the meat industry, showed improved penetration into meat products when it was applied as fine powder (microsize) when compared with raw ginger (Zhao et al. 2009). Ginger's functionality could be further improved, if ground into fine powder to nanoscale level and applied for meat tenderizing (Zhao et al. 2010).

Yusop et al. (2012) used nanoparticles of paprika oleoresin (average diameter 30 nm) for marinating chicken breast fillets and found the application of nanoparticles improved the marinating performance (higher color penetration, improved surface color values [high yellow and red values]) and sensory acceptance. They tested two levels of paprika concentration (1 and 3 g/100 mL) and two ingredient carriers (water and homogenized milk) for this test. Application of nanoparticle paprika-marinating solution in a concentration of 3 g/100 mL on chicken breasts resulted in greater absorption (9.11 ± 1.22 g/100 g) of the marinade and less losses (22.07 ± 2.52 g/100 g) during cooking. Chicken breasts marinated with 3 g/100 mL paprika level also showed better color values (higher CIE a^* [red] and b^* [yellow] values) after cooking when compared with the other level and control. Using milk as an ingredient carrier increased absorption of the marinade but had no significant effect on cooking losses. Application of nanoparticles at a higher concentration (3 g/100 mL) using the tumbling method improved the penetration and distribution efficiency of the marinade on chicken breasts and resulted in better sensory properties and less cooking loss (Yusop et al. 2012).

Functional components can be added to the food matrix during processing using nanoemulsions to improve antioxidant and antimicrobial properties of meat products (Weiss et al. 2006). Use of sunflower-oil-based nanoemulsion (AUSN-4) for storage of Indo-Pacific king mackerel fish steaks increased the shelf life of the steaks by 48 h when compared to control and antibiotic-treated steak samples, and also reduced the heterotrophic bacterial population up to 12 h (Joe et al. 2012). Nanoemulsion application also slows down the reduction of quality parameters (protein content, fat content, and carbohydrate) of steaks and Joe et al. (2012) found nanoemulsion could be used for short time storage of fish steaks where refrigeration facilities are not in place to store the captured fish. But sometimes the addition of nanoemulsion can lead to some adverse effects in raw and processed meat products during long-time storage. For example, adding oil-in-water nanoemulsion to pork sausages increased oxidation of the product (Salminen et al. 2013). When polylactic glycolic acid (PGLA) nanoparticles were used with phenolic compounds (benzoic acid, vanillic acid, syringic acid) for packaging of cooked and uncooked chicken meat, delivery of the phenolic compounds improved due to the larger surface area of the nanoparticles. Due to this effective delivery system, foodborne pathogens like *Escherichia coli* O157:H7 and *Listeria monocytogenes* were inhibited (6.0–6.5 log CFU/mL) at lower phenolics concentrations (1100 μg/mL) when compared to packing with phenolic compounds alone (5000 μg/mL) (Ravichandran et al. 2011).

The meat industry uses additives to improve the product texture, color, and taste, as well as increase productivity of the processing line, and nanotechnology helps the industry to incorporate these additives with meat products. Nonencapsulated functional components like vitamin C, vitamin E, and fatty acids were added to cured meat and sausages using 30 nm micelles (Commercial name: Novasol, Aquanova, Darmstadt, Germany), which reduced the ingredients' cost and improved the stability of the sausage color and processing speed (Alfadul and Elneshwy 2010). A nanoencapsulated system (nanospheres), with an average particle diameter of 0.01–10 μ, was developed for targeted delivery of sensory markers (flavor enhancers), proteins, and antioxidants for improving health and nutritional benefits of processed meat products (Shefer and Shefer 2003, 2004).

Nanotechnology helped to make edible coatings for meat products (such as sausage casings) with biopolymers like polysaccharides, lipids, and proteins (Weiss et al. 2006). Biopolymers with nanoparticles provide different functionalities like moisture and gas barriers to improve shelf life and maintain the quality of the meat products. Polysaccharide-based polymers provide excellent gas barrier properties (against oxygen, and carbon dioxide) but have poor moisture-barrier properties. Protein-based edible coatings also exhibit similar functionalities as polysaccharide-based polymers. On the same note, lipid-based polymers have greater moisture-barrier properties but poor gas-barrier properties. These biopolymers with nanoparticles serve as organic edible coatings, as well as increase the nutritional values (Weiss et al. 2006). Table 8.1 summarizes the applications of nanotechnology in the meat industry for functional ingredients and additives.

Table 8.1 Applications of Nanoparticle Materials as Additives and Functional Ingredients in Meat Processing

APPLICATION	NANOMATERIAL	PRODUCT	REFERENCES
Improve marinating performance	Paprika oleoresin	Chicken breast fillets	Yusop et al. (2012)
Delivery of the phenolic compounds to inhibit *Escherichia coli*	Polylactic glycolic acid (PGLA)	Cooked and uncooked chicken meat	Ravichandran et al. (2011)
Increase shelf life	Sunflower-oil-based nanoemulsion (AUSN-4)	Indo-Pacific king mackerel fish steaks	Joe et al. (2012)
Improve stability of the sausage color and processing speed	Nanocapsules (vitamin C, vitamin E, and fatty acids)	Cured meat and sausages	Alfadul and Elneshwy (2010)
Targeted delivery to improve health and nutritional benefits	Proteins and antioxidants nanospheres	Processed meat products	Shefer and Shefer (2003)

Pathogen Detection and Control

E. coli O157:H7, *Clostridium botulinum*, *Staphylococcus aureus*, Campylobacter, Norovirus, and *Toxoplasma gondii* are the most common foodborne pathogens in meat and meat products. These pathogens can cause various health issues from vomiting to death. Detection of foodborne pathogens in early stages is the major challenge for the meat industry in order to avoid serious economical and product loss (Luo and Stutzenberger 2008). Silver nanoparticles are commonly used for inactivation of common foodborne pathogens and were successfully tested for controlling gram-negative foodborne bacterium like *E. coli*, and results from studies showed that growth of gram-negative bacteria (number of colonies formed) had negative correlation with the concentration of silver nanoparticles (Figure 8.2) in the solution (Sondi and Salopek-Sondi 2004).

The conventional culturing (plate count method) technique is the most common method used for identification of foodborne pathogens, but it takes more than 72 h for detection of target pathogen (Kaittanis et al. 2010; Valdés et al. 2009). In the traditional culturing method, pathogen-specific solid or liquid media is used to detect a

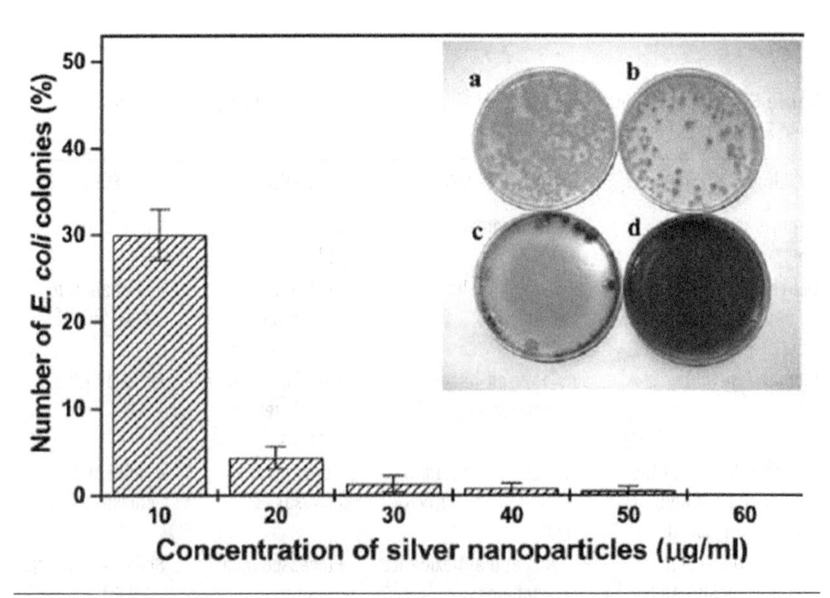

Figure 8.2 Effect of silver nanoparticle concentration in LB agar plate on growth of *E. coli* colonies formed. (Reproduced from Sondi, I., and Salopek-Sondi, B., *J. Colloid Interf. Sci.*, 275, 177–182, 2004.)

specific pathogen, so detection of multiple pathogens from a sample is not possible with the qualitative plate count technique. In the quantitative plate count method, a culturing layer is placed on an agar layer inside a petri dish, and visible colonies of grown microorganisms are counted after 24–72 h based on the target pathogens and media. Nowadays antibody-based immunological assay techniques, like the enzyme-linked immunosorbent assay (ELISA), and sequencing-based assay techniques (PCR, real-time PCR, and fluorescence in situ hybridization methods to detect the DNAs of foodborne pathogens) are used for rapid detection of foodborne pathogens (Duran and Marcato 2013; Inbaraj and Chen 2016).

Lysozyme is a type of monomeric protein that has been used as a preservative for meat (both raw and processed) and dairy products due to its antimicrobial properties. Chitosan (CS) is a polysaccharide produced by chitin deacetylation and widely applied as a vaccine, protein, and enzyme delivery system due to its adhesive capability. Wu et al. (2017) integrated lysozyme with CS nanoparticles (CS-NP) and tested its antibacterial activities against *E. coli* O157:H7 and *Bacillus subtilis*. They prepared nanoparticles by making CS and sodium tripolyphosphate (TPP) solution at 1:3 ratio (6 mL CS: 18 mL TPP), stirring using a magnetic stirrer at room temperature for 1 h and centrifuging nanoparticles at a pH of 5. The CS-Lys-NPs were prepared by adding lysozymes at a concentration of 1.25 mg/mL with the CS-NPs. Wu et al. (2017) tested the effect of prepared nanoparticle solutions against *E. coli* (Gram-negative bacterium) and *B. subtilis* (Gram-positive bacterium). Their results proved that CS-Lys-NP and CS-NP significantly prevented the growth of both bacteria and improved the antibacterial activities than CS alone. Transmission electron microscopic images of control and CS-NP/CS-Lys-NP-applied *E. coli* showed that CS-NP/CS-Lys-NP penetrated the *E. coli* cell membrane and resulted in cytoplasm leakage, which led to cell death. Researchers encouraged further research for use of these by the meat industry for processing meat, as well as on the incorporation of CS-Lys-NPs into the packaging of meat to increase the shelf life of products during storage and transportation. Abdou et al. (2012) tested CS-NPs for active coating of fish fingers (Carp fish) to increase shelf life. They prepared a CS-NP solution using a method (ionic gelation technique) described earlier (Wu et al. 2017) and tested for coliform bacteria, proteolytic bacteria, psychrophilic bacteria, yeast, and mold counts on fish fingers. Their results showed that fish fingers coated with CS-NPs had the lowest (2.87 log CFU/g) total bacterial count while storing at −18°C for 6 months, when compared with the commercial edible coating agents (5.27 log CFU/g). Rheological properties (shear stress and apparent viscosity) of fish fingers coated with a commercial edible coating and CS-NP coating were the same (Abdou et al. 2012).

The CS-NPs loaded with various metal ions (Ag^+, Zn^{2+}, Mn^{2+}, Cu^{2+}, and Fe^{2+}) were tested for their antibacterial activities against *E. coli, Salmonella choleraesuis,* and *S. aureus* (Du et al. 2009). They prepared CS-NPs by adding 1 mL of TPP to 25 mL of CS solution and stirred with a magnetic stirrer for 20 min. Then the solution was treated with ultrasonic waves for 30 min at 1.5 kW. The suspension of this solution was centrifuged at 12,000 × g (G force) for 10 min, and freeze dried. The metal ion solutions were added to freeze dried CS-NPs at a rate of 0.3% (w/v) and stirred for 12 h to attain a final concentration of 120 µg/mL, and then purified to get CS-NP-loaded metal ions. They found that loading CS-NPs with metal ions enhanced their antibacterial activity, and CS-NPs loaded with Ag^+ had the highest antibacterial activity (3, 3, and 6 minimum inhibitory concentration (MIC) for *E. coli*, *S. choleraesuis*, and *S. aureus*, respectively). CS-NPs loaded with Fe^{2+} had the lowest antibacterial activity (18, 18, and 36 MIC against *E. coli*, *S. choleraesuis*, and *S. aureus*, respectively).

A bionanosensor was developed with single-walled carbon nanotube (SWCNT) network to detect *Salmonella infantis* in chicken meat and eggs (Villamizar et al. 2008). They used SWCNT to construct a field effect transistor, and tween-20 solution was used along with CNT to restrict binding of other pathogens with the antibody. The antigen–antibody reaction produced a significant reduction in the electric current of the circuit and was used to detect the *S. infantis* presence in the meat sample. They reported that the detection limit of this biosensor was 100 CFU/mL, and detection time was 1 h. Applications of nanotechnology for pathogen detection and control are summarized in Table 8.2.

European researchers under the "Good Food" project developed microsensor technology- and nanotechnology-based sensors (using DNA biochips) for rapid detection of the harmful bacterium

Table 8.2 Applications of Nanotechnology for Foodborne Pathogen Detection and Control in Meat Industry

NANOPARTICLE/SENSOR	PATHOGENS	APPLICATION	REFERENCES
Lysozyme with CS nanoparticles	*Escherichia coli* O157:H7 and *Bacillus subtilis*	Antimicrobial agent	Wu et al. (2017)
CS-NPs	Coliform bacteria, proteolytic bacteria, psychrophilic bacteria, yeast, and mold count	Active coating of fish fingers to increase shelf life	Abdou et al. (2012)
CS-NPs with metal ions (Ag^{+}, Zn^{2+}, Mn^{2+}, Cu^{2+}, and Fe^{2+})	*E. coli*, *Salmonella choleraesuis*, and *Staphylococcus aureus*	Antibacterial agent	Du et al. (2009)
Bionanosensor with single-walled carbon nanotube (SWCNT)	*Salmonella infantis*	Pathogen detection in chicken meat and egg	Villamizar et al. (2008)
Volatile sensing system with (SnO_2) and indium (III) oxide (In_2O_3) sensors	*E. coli*, *Listeria monocytogenes*, and *S. infantis*	Pathogens detection using ammonia (NH_3), trimethylamine (TMA), and dimethylamine (DMA) volatiles	GoodFood (2008)
Silicon-gold nanorod array	*S. typhimurium*	Pathogen detection using fluorescence technique	Fu et al. (2008)
Water soluble quantum dots	*E. coli* and *S. aureus*	Rapid detection of pathogens	Xue et al. (2009)
Gold nanoparticles	*E. coli* O157:H7 and ovalbumin	Pathogen detection using colorimetric technique	Betala et al. (2008)
Electrical impedance based bio nanosensors with magnetic nanoparticles (MNPs)	*E. coli* O157:H7	Pathogen detection in ground beef	Varshney and Li (2007)
Quantum dots	*E. coli* and *S. typhimurium*	Pathogen detection using fluorescence sensor	Yang and Li (2006)
Carbon nanoparticle polyaniline sensor	*Salmonella* spp., *Bacillus cereus*, and *Vibrio parahaemolyticus*	Pathogen detection	Arshak et al. (2007)
SWCNT sensor	*E. coli*		Yamada et al. (2014)

(*E. coli*, *L. monocytogenes*, and *S. infantis*) in meat and fish products (GoodFood 2008). They also developed sensors to measure freshness of the fish samples using volatile components (ammonia [NH_3], trimethylamine [TMA], and dimethylamine [DMA]) in the package (Figure 8.3). This sensor was referred to as the "multisensory miniaturized gas chromatographic system (MMGCS)" (GoodFood 2008)

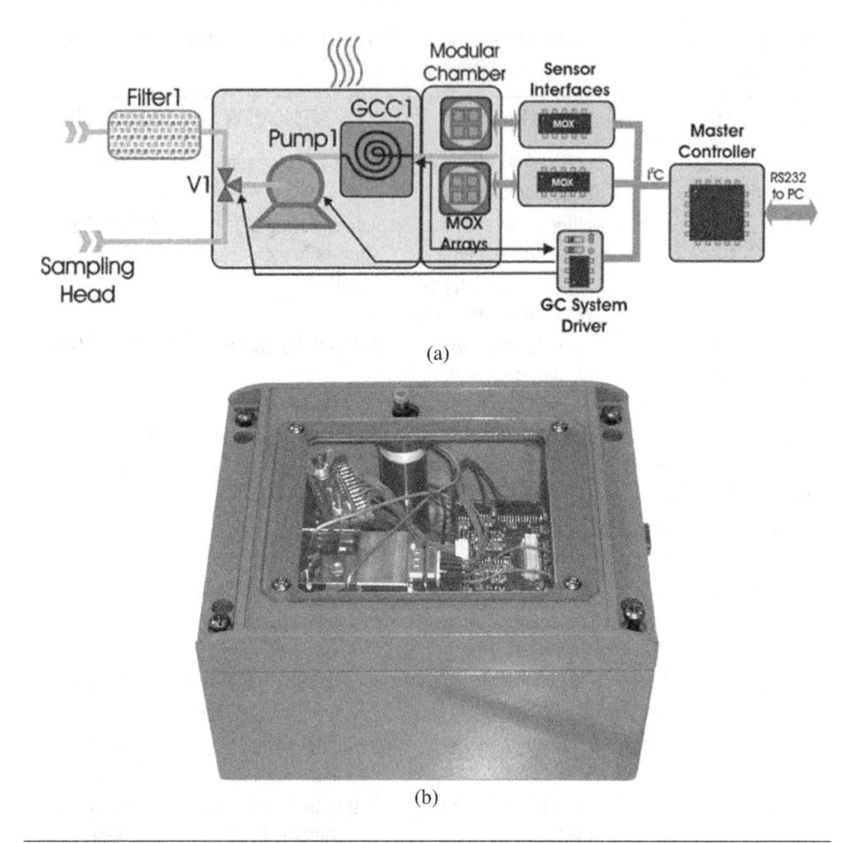

Figure 8.3 (a) Schematic view of the fish volatile measurement system and (b) Prototype of fish volatile measurement sensor. (Reproduced from GoodFood, *Food Safety and Quality Monitoring with Microsystems*, European Council, Brussels, Belgium, 2008.)

and contained two tin dioxide (SnO_2) and two indium (III) oxide (In_2O_3) sensors. This system is similar to a gas chromatography system, and the headspace air from the fish package was sampled and injected into the MMGCS for detection of the freshness of fish using TMA, DMA, and NH_3. The detection column temperature was set at 120°C to release the volatiles. The lower detection limit of this system was 0.5, 5.0, and 5.0 ppm for TMA, DMA, and NH_3, respectively.

Varshney and Li (2007) developed an electrical impedance-based bionanosensor to detect the *E. coli* O157:H7 presence in the ground beef with magnetic nanoparticles (MNPs), which were made up of Fe_3O_4 (more than 85%) and Fe (approximately 4×10^{11} particles/mg). The average diameter of MNPs used in this sensor was 135 nm. These MNPs coated with streptavidin, which acted as antibody conjugates, were coupled with a gold microelectrode array for measurement of impedance. When the ground beef with *E. coli* O157:H7 was introduced into the sensor, the MNP antibody conjugates separated *E. coli* O157:H7 from the sample in the presence of a magnetic field and changed the impedance value to aid in the detection of the *E. coli* O157:H7. Varshney and Li (2007) attributed the coupling of MNP antibody conjugates with the gold microelectrode for improving the detection potential of the microelectrode array. They found the detection limit of this sensor was 8×10^5 CFU/mL, and total time taken to detect *E. coli* O157:H7 from the ground beef was 35 min (from sample preparation to detection).

A SWCNT sensor was fabricated to detect *E. coli* (Yamada et al. 2014). They aligned SWCNTs and 50 μm diameter golden tungsten wires coated with polyethylenimine to form a junction and functionalized the junction with streptavidin and *E. coli* specific antibodies (Figure 8.4). Whenever the sample with *E. coli* entered into the junction, the antibody and bacterial cells changed the electrical current

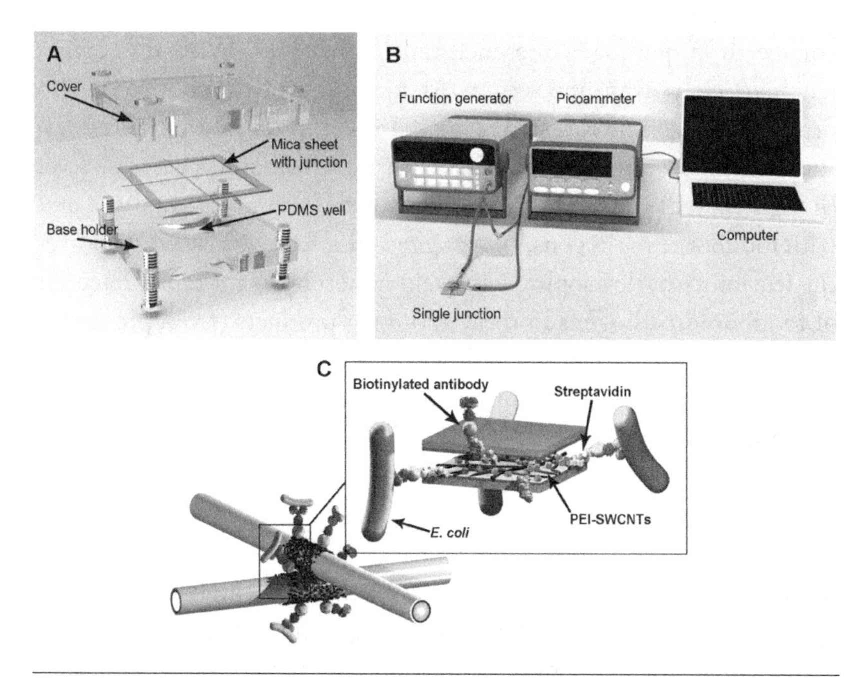

Figure 8.4 **(See color insert.)** Design of a single-walled carbon nanotube (SWCNT) single junction biosensor for *E. coli* detection: (A) Setup of single junction nanosensor, (B) Experimental setup of SWCNT sensor for electrical current measurement for *E. coli* detection, and (C) Graphical illustration of *E. coli* captured on functionalized SWCNT junction. (Reproduced from Yamada, K.,. *PLoS One*, 9, e105767, 2014.)

value at the junction, by which *E. coli* presence was detected. It was observed that the electrical current at the junction after functionalized with streptavidin and antibody was 0.59 μA, and dropped to 0.29 μA after the introduction of *E. coli* (at 10^8 CFU/mL concentration). They also noticed the change in electrical current (ΔI) was 290.90 nA, and 33.13 nA with and without SWCNT, respectively. Yamada et al. (2014) demonstrated that this developed SWCNT sensor could detect *E. coli* in 5 min with a detection limit of 10^2 CFU/mL.

Odors released by the foodborne pathogens can be used to detect the presence of those pathogens in food matrix by using polymer nanocomposite sensors. Arshak et al. (2007) developed a nanocomposite sensor with carbon nanoparticles and polyaniline to detect *Salmonella* spp., *Bacillus cereus* and *Vibrio parahaemolyticus* using gases produced from the pathogens. They monitored the voltage across the sensor and found that the voltage drop pattern was different for different types of foodborne pathogens. The basic principle of these types of nanocomposite sensors is, when the odor is absorbed by the polymer, it changes the electrical conductivity of the polymer material. By monitoring the voltage drop, pathogen presence can be detected. With this carbon nanoparticle polyaniline sensor, Arshak et al. (2007) noticed a 70% drop in voltage when tested with the odors produced by *Salmonella* spp. This sensor was tested only with pure bacterial cultures. Even though they did not test this sensor for food samples contaminated with foodborne pathogens, they suggested results of these tests proved the feasibility of developing handheld nanosensors for rapid detection of foodborne pathogens in meat and meat products (from processing line to consumer market) (Valdés et al. 2009). More detailed discussion about nanosensors used to detect foodborne pathogens is available in Chapter 5.

Packaging

Developments in nanotechnology have helped to improve meat packaging in recent times. Smart packaging and intelligent packaging materials developed using nanotechnology improve safety and shelf life of meat products during storage and transportation. Regular conventional food and meat packaging films are passive systems, which just play a role of protection barrier between the food packed and the surrounding. But some of the nanotechnology-based packing films serve as active packaging systems, which protect food as well as actively interact with food by removing undesirable factors (such as oxygen, moisture, and unwanted vapors released by food [like ammonia] during storage) and also releasing desirable compounds (such as antioxidants and antimicrobial agents) in order to keep quality and freshness of the food for longer duration (Berekaa 2015; Duncan 2011; Sozer and Kokini 2009). Applications of nanotechnology in meat packaging are summarized in Table 8.3.

Nanotechnology also helped to improve the mechanical, thermal, and permeation properties (temperature resistance, durability, gas permeation, moisture resistance) of packaging films of meat products. Various types of "nanocomposites" polymers incorporated with nanoparticles have been developed using nanotechnology with improved properties, which contain about 5% w/w nanoparticles (nanoclay, silver nanoparticles, gold nanoparticles, etc.). Nanoclay has

Table 8.3 Applications of Nanoparticle Materials in Meat Packaging

PACKAGING POLYMER	NANOMATERIAL	FUNCTIONALITY	REFERENCES
Polyamide 6 (PA6)	Nanoclay	Vacuum packaging of beef lions	Picouet et al. (2014)
Nanoparticle–Cellulose hybrid absorbent pad	Silver nanoparticles	Reducing bacterial growth	Fernández et al. (2010)
Cellulose	Silver nanoparticles	Minimally processed meat packaging	Lloret et al. (2012)
Low-density polyethylene (LDPE)	Ag and ZnO nanoparticles	Lower microbial count and increase in shelf life of chicken breast (restraining *E. coli*, *Pseudomonas aeruginosa*, and *Listeria monocytogenes* growth in chicken breast)	Panea et al. (2014)
Cellulose and collagen sausage casings	Silver nanoparticles	Improved antifungal and antibacterial properties	Fedotova et al. (2010)
Polyvinyl chloride (PVC)	Silver nanoparticles	Reducing foodborne pathogen growth and increasing the shelf life of minced meat	Mahdi et al. (2012)

inherent barrier properties for gases (O_2, CO_2) and controls the permeation of O_2 and CO_2 through the packaging films when added with polymer packaging films (Akbari et al. 2006; Ke and Yongping 2005). Various types of common polymers (polyimides and polyethylene terephthalate, nylons, ethylene-vinylacetate copolymer, polyamides, polyurethane, and polystyrene) had been tested for manufacturing nanoclay–polymer nanocomposite films for packaging of processed meat products, and the results showed these nanocomposite polymers had the potential of keeping meat product's quality and freshness for longer duration than regular polymer packaging films (Akbari et al. 2006). For producing this improved packing film, production polymers like polystyrene, polyamide, polyolefin, and nylons were reinforced with nanoscale-size fillers like clay, cellulose microfibers, silicates, and carbon nanotubes (De Azeredo 2009; Silvestre et al. 2011). Among these nanoscale fillers, clay is the most common filler used in food packaging industry due to its low cost and superior gas and moisture transmission properties (Silvestre et al. 2011). Reinforcing with nanoscale filler materials like nanoclay alters the path of O_2 molecules (circuitous path) through the packing film, which causes the change in barrier properties (gas transmission) of the films (Silvestre et al. 2011).

Applications of natural and organic biopolymers for packaging of meat and meat products are increasing in recent times due to the growing interest of consumers and industry (Rhim et al. 2013). Chitosan, carrageenan, and cellulose are the most commonly used biopolymers as packaging materials, and these biopolymers can also be reinforced with nanoscale fillers like nanoclay for improved packaging properties. Polyamide 6 (PA6) polymer film reinforced with nanoclay particles was used for vacuum packaging of beef lions, and this film showed improved oxygen transmission property (O_2 transmission rate reduced by 59%) than regular PA6 packaging films (Picouet et al. 2014). They also found that gas barrier and meat preservation properties could be achieved with thinner packaging films when reinforced with nanoclay particles.

In meat packaging, absorbent pads have been used to absorb moisture and exudate fluids from poultry, meat, and fish products, which helps to maintain freshness of the product as well as keep away unsanitary meat juices. The exuded fluids, absorbed by the pads, have potential to develop undesirable odor and bacterial growth. Silver nanoparticles were used with the cellulose-based absorption pad to reduce the possibility of *E. coli* and *S. aureus* growth. The results showed a decrease in the microbial count with increase in silver nanoparticle concentration when tested with pure pathogen cultuers (Fernández et al. 2009). Silver nanoparticle–cellulose hybrid material absorbent pad were also tested for modified atmospheric storage of beef for 11 days at refrigeration temperature (4 ± 0.2°C). The antimicrobial properties of the silver-cellulose absorbent pad kept the microbial load of all major bacteria (total aerobic bacteria, lactic acid bacteria, *Pseudomonas* spp., and *Enterobacteriacea*) 1 log CFU/g lower than that control beef samples (Fernández et al. 2010). Lloret et al. (2012) developed an absorbent pad with cellulose silver nanocomposites, packed minimally processed meat (poultry and beef) products, and stored at 4°C for 10 days. Assessment of the microbial count at an interval of 2 days for the entire storage period showed that the cellulose silver nanocomposite absorbent pad package had 90% less microbial load than the control samples.

Application of nanoparticles with antimicrobial capabilities such as silver, gold, and zinc nanoparticles directly on the meat product have some negative effects, such as seepage into the product and reduced microbial effect due to non-homogenous mixing with the product. But coating packaging polymer surface with antimicrobial nanomaterials and producing hybrid packaging films with polymer and antimicrobial nanoparticles help to increase overall antimicrobial effect as well as reduce the health hazard (Rodriguez et al. 2008; Sondi and Salopek-Sondi 2004; Véronique 2008). Silver and gold nanoparticles can be easily incorporated with packaging polymers to improve the antimicrobial properties of the packaging films (Duncan 2011). Silver nanoparticles exhibit greater antimicrobial effect against foodborne pathogens, cause damage to cellular membrane's structure, and increase toxicity due to the silver ions released from nanoparticles. These are the major reasons for the improved antimicrobial properties of silver nanoparticle-incorporated packing films (Duncan 2011; Lok et al. 2007; Morones et al. 2005; Ramachandraiah et al. 2015). Panea et al. (2014) blended nano Ag and ZnO particles in low-density polyethylene (LDPE) packaging films for chicken breast storage and tested antimicrobial activity against *E. coli*, *Pseudomonas aeruginosa*, and *L. monocytogenes* and migration of nanoparticles into meat from the packaging film. Meat packed with Ag+ZnO and LDPE material showed lower microbial count and oxygen depletion than the normal LDPE film package. These results also proved that adding Ag+ZnO NPs to LDPE packaging film increased the shelf life of chicken breast during storage as well as slowed down lipid oxidation. The seepage of Ag+ZnO into meat was also in the safe level permitted by European Union council regulation (Panea et al. 2014). Sausage casing made by biopolymers (cellulose and collagen) incorporated with silver nanoparticles had greater antifungal and antibacterial properties than the sausage casing made without silver nanoparticles (Fedotova et al. 2010).

Nanotechnology also helped to develop smart packaging technology for meat and meat products, in which the condition of the product or the surrounding environment was monitored continuously by the nanosensors affixed on the package (Yam et al. 2005). With these smart packaging nanosensors, temperature, moisture, and gas composition inside the package or the storage environment can be monitored. These nanosensors can also be used to detect the presence of foodborne pathogens and toxic substances in the product (Silvestre et al. 2011). The major advantage of these smart packaging nanosensors is the condition and integrity of the food, and the package can be monitored from the processing facility to consumer (Ramachandraiah et al. 2015). *E. coli* and *S. aureus* growth were inhibited in minced beef packed with a packaging film made by polyvinyl chloride (PVC) incorporated with silver nanoparticles, after seven days of storage at refrigeration temperature (4°C) (Mahdi et al. 2012). Their results proved that addition of silver nanoparticles with packaging polymer significantly reduced the foodborne pathogen growth, which increased the shelf life (minced meat might spoil in two days of storage with regular packaging) of minced meat. Incorporation of the electronic tongue with an array of nanosensors in the meat packaging helped to detect the odor and gases released by spoiled product, which could be used to identify the freshness of the meat and meat product in the package (Bowles and Lu 2014).

To check the oxygen concentration inside and outside of a modified atmospheric package (with no oxygen) containing beef, oxygen indicator nanosensors with titanium oxide and methyl bromide ink were placed inside and outside of the packaging film and then irradiated

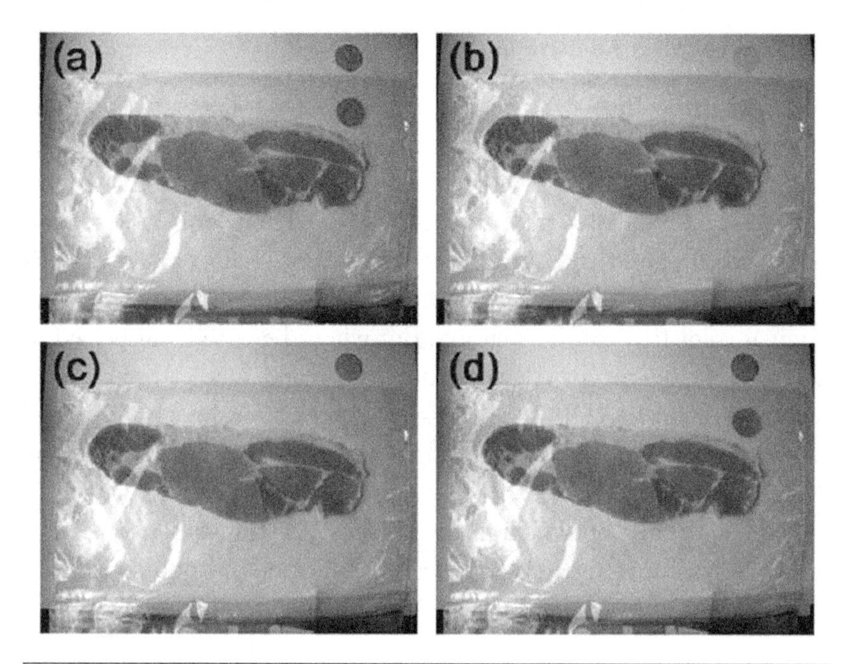

Figure 8.5 **(See color insert Figure 4.1.)** Nano O_2 sensors with UV-activated TiO_2 NPs and methylene blue indicator dye, one placed inside, and another one placed outside. (a) When the package is freshly sealed, (b) immediately after activation with UVA light irradiation, (c) few minutes after activation, and (d) when the package is opened. (Reproduced from Mills, A., *Chem. Soc. Rev.*, 34, 1003–1011, 2005.)

with the UV light. The sensors remained colorless until the absence of oxygen and turned into blue even for a small quantity of the oxygen presence (Figure 8.5) (Mills 2005). More detailed discussion about this sensor is given in Chapter 5. This sensor could be useful to check the integrity of the package as well as the quality of the meat product in the package from the packing plant to the consumer's hand.

A DNA sensor (with quartz crystal microbalance [QCM]) was used with MNPs to detect *E. coli* O157:H7, and this sensor successfully detected *E. coli* as little as 2.67×10^2 CFU/mL in the samples (Mao et al. 2006). MNPs conjugated with streptavidin, and these MNPs functioned as mass enhancers to amplify the QCM frequency due to the presence of *E. coli* DNA in the sample. Chen et al. (2008) developed a biosensor (circulating-flow piezoelectric sensor) with gold nanoparticles to detect *E. coli* O157:H7 and obtained a detection limit of 1.20×10^2 CFU/mL. In this biosensor also, gold nanoparticles served as mass enhancing and sequence verifying agents (Chen et al. 2008).

Detection of Meat Adulteration

A nanosensor with AuNPs was successfully tested for detecting adulteration of pork meat in chicken and beef meatballs (Ali et al. 2011, 2012, 2014; Inbaraj and Chen 2016; Sonawane et al. 2014). AuNPs (20 nm size) were coated with citrate and dispersed well in deionized water and remained dispersed well after three minutes of incubation of AuNPs with single-stranded DNA (chicken or beef DNA) and retained the original color (red) of the solution. But when the AuNPs dispersed in deionized water incubated with double-stranded DNA (swine DNA with beef or chicken DNA), AuNPs aggregated after three minutes and caused a color change of the solution from red to purple grey (Ali et al. 2012). Using this visible color change, pork adulteration during beef or chicken meatball production can be detected. Transmission electron microscopy (TEM) images (Figure 8.6) and the absorption spectrum

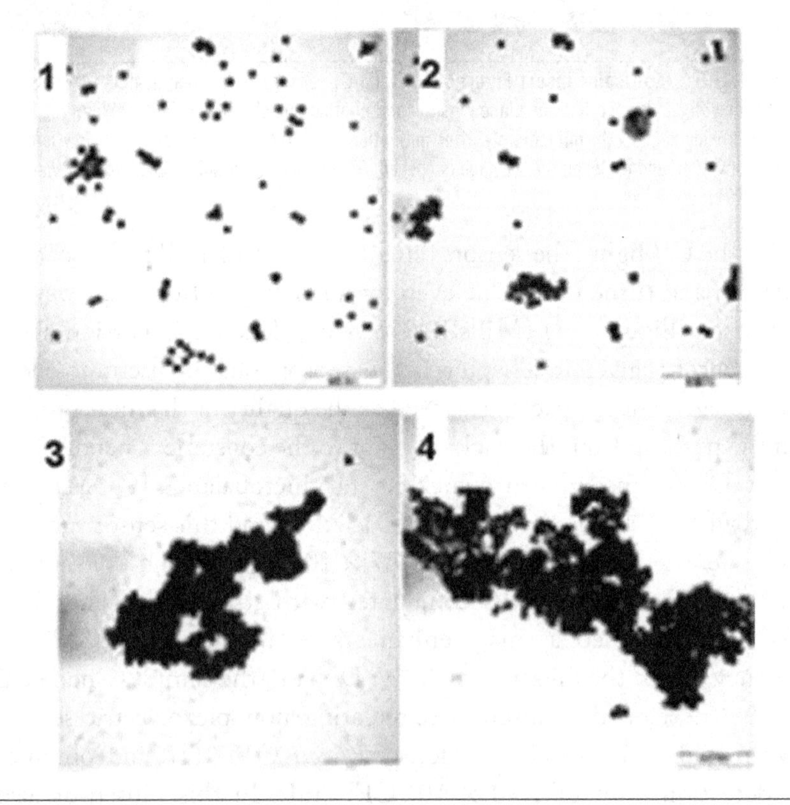

Figure 8.6 TEM images of AuNPs (1) dispersed in deionized water, (2) after 3 min incubation with single-stranded DNA, (3 and 4) after 3 min incubation with double-stranded DNA. (Reproduced from Ali, M.E. et al., *J. Nanomater.*, 1–7, 2012.)

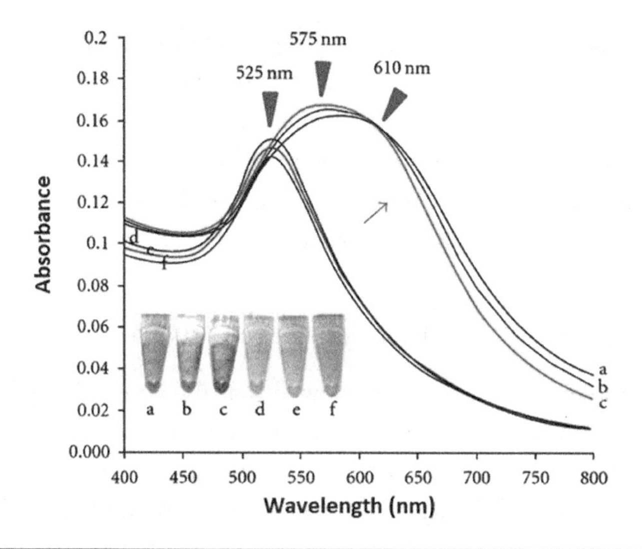

Figure 8.7 **(See color insert.)** Absorbance spectra and vials of swine DNA in mixed meatball prepared from (a) pure pork, (b) mixtures of pork–beef, (c) pork–chicken, (d) chicken–beef, (e) pure beef, and (f) pure chicken with AuNPs and deionized solution. (Reproduced from Ali, M.E. et al., *J. Nanomater.*, 1–7, 2012.)

(Figure 8.7) collected during their experiments clearly showed the aggregation phenomenon and color change when incubating double-stranded DNA with AuNP solution for three minutes (Ali et al. 2012). Studies by Ali et al. (2014, 2012, 2011) showed that size of the AuNPs played a major role in detection limit of pork adulteration in chicken and beef meatballs (detection limits were 6 μg/mL, 4 μg/mL, and 0.236 g/mL for AuNPs of 40, 20, and 3 nm, respectively). Ali et al. (2014) also proved that 20 nm size AuNPs produced more distinguished color change for pork adulteration in other meatballs than the 40 nm size AuNPs.

The freshness of the fish can be evaluated by measuring the xanthine and hypoxanthine levels (Mao and Yamamoto 2000; Mulchandani et al. 1989). A biosensor was developed to detect freshness of the fish using xanthine levels (Thandavan et al. 2013). Their electrochemical sensor had a nanointerface with Fe_3O_4 nanoparticles for quantitative measurement of xanthine; this sensor detected xanthine levels from 0.4 to 2.4 nM. Thandavan et al. (2013) also tested this sensor to detect the xanthine level from the fish extract and detected xanthine in two seconds. Cubukcu et al. (2007) modified a carbon paste electrode sensor with AuNPs to detect xanthine and hypoxanthine and tested with fresh and 15 days stored (at room temperature) tuna fish samples. Canned tuna fish was made into a paste for the test, and 5 mL of

0.5 M hyperchloric acid ($HClO_4$) was added to denature proteins. They found that the modified electrode produced an electrical signal (change in circuit current) for the fish sample stored for 15 days at room temperature due to the presence of hypoxanthine in the sample, and fresh samples did not generate any signal. They also found that the electrode responded well at the optimum pH level of 7.5.

Issues with Use of Nanotechnology in the Meat Industry

Recent studies and surveys also showed that the acceptability level was high among the consumers when the nanomaterial was present in packaging than when it was present directly in the meat product (Siegrist et al. 2008). But the major concern of using nanoparticles in meat packaging among the consumers is leaching of nanoparticles into meat product from the packaging films, and some studies found traces of nanoparticles (silver, titanium, gold) in the bloodstream and in organs of consumers when these nanoparticles were used for packaging of food products (Gurr et al. 2005; Kim et al. 2008; Rhim et al. 2013). Avella et al. (2005) found traces of nanoclay particles in food migrated from packaging material when using nanoclay incorporated biodegradable starch packaging films. Panea et al. (2014) also found Ag and ZnO particles in the chicken breast while using Ag/ZnO incorporated packaging polymers, but the level of Ag/ZnO in food product was within the minimum limit allowable by European Union regulation. The storage temperature and the storage duration were the two major factors limiting the level of nanoparticle migration to the food product from the packing film (Huang et al. 2011). Studies found that the acceptance level among consumers for using nanotechnology in food processing and packaging were high in North America, Korea, and had less positive acceptance in Europe (Chen et al. 2013; Cushen et al. 2012; Gaskell et al. 2005).

Summary

Application of nanotechnology in the meat industry is increasing day by day due to enormous development in the areas of detection and control of foodborne pathogens during processing and packaging using nanoscience and nanotechnology. Nanotechnology has a major impact in meat packaging by improving barrier properties and improving antimicrobial effects. Application of nanotechnology in rapid detection of foodborne pathogens will help the meat industry to decrease total production cost by reducing analysis time and will also increase production capacity. Most of the sensors developed using nanotechnology detect major foodborne pathogens in an accuarate manner quickly when compared with the regular conventional detection techniques and tools. Hybrid sensing techniques, which combine nanotechnology and advanced pathogen detection tools like PCR, ELISA, and fluorescence, yielded higher accuracy and showed great potential in using real-time production environment. Many of these nanosensors and detection techniques with nanomaterials were tested only with pure pathogen cultures and in vitro samples. Since food is a complex matrix, more research needs to be carried out to upgrade the sensors for use in the meat industry for rapid detection of foodborne pathogens in real time.

Apart from packaging and pathogen detection, nanotechnology has great potential applications in improving nutritional properties. Development of processing techniques and equipment to apply nanoparticles (nanopowders) along with the other ingredients will help to improve functional properties of meat products since nanoemulsification has the issue of instability due to oxidation. The success of applying nanotechnology in the meat industry does not only depend on the improvement in food safety but also the consumer's acceptance level through scientific studies. Researching the long-term effects of nanoparticle leaching into meat products from packaging films on the product and consumers health and upgradation of nanosensors for real-time detection of foodborne pathogens will help to improve food safety regulations while using nanotechnology in the meat industry as well as help to improve the consumer acceptance of nanotechnology in the meat industry. Improved food safety and consumer acceptance will yield an enormous boost in the application of nanotechnology in the meat industry.

References

Abdou E.S., Osheba A., Sorour M. (2012) Effect of chitosan and chitosan-nanoparticles as active coating on microbiological characteristics of fish fingers. *International Journal of Applied Research* 2(7):158–167.

Akbari Z., Ghomashchi T., Aroujalian A. (2006) Potential of nanotechnology for food packaging industry. *Proceedings of Nano and Micro Technologies in the Food and HealthFood Industries*, Amsterdam, the Netherlands, October 25–26.

Alfadul S., Elneshwy A. (2010) Use of nanotechnology in food processing, packaging and safety—Review. *African Journal of Food, Agriculture, Nutrition and Development* 10:2720–2739.

Ali M.E., Hashim U., Mustafa S., Man Y.B.C., Adam T., Humayun Q. (2014) Nanobiosensor for the detection and quantification of pork adulteration in meatball formulation. *Journal of Experimental Nanoscience* 9:152–160.

Ali M.E., Hashim U., Mustafa S., Man Y.B.C., Islam K.N. (2012) Gold nanoparticlesensor for the visual detection of pork adulteration in meatball formulation. *Journal of Nanomaterials* 1–7. doi:10.1155/2012/103607.

Ali M.E., Hashim U., Mustafa S., Man Y.B.C., Yusop M.H.M., Bari M.F., Islam K.N., Hasan M.F. (2011) Nanoparticle sensor for label free detection of swine DNA in mixed biological samples. *Nanotechnology* 22:195503.

Arshak K., Adley C., Moore E., Cunniffe C., Campion M., Harris J. (2007) Characterisation of polymer nanocomposite sensors for quantification of bacterial cultures. *Sensors and Actuators B-Chemical* 126:226–231.

Avella M., De Vlieger J.J., Errico M.E., Fischer S., Vacca P., Volpe M.G. (2005) Biodegradable starch/clay nanocomposite films for food packaging applications. *Food Chemistry* 93(3):467–474.

Berekaa M.M. (2015) Nanotechnology in food industry: Advances in food processing, packaging and food safety. *International Journal of Current Microbiology and Applied Science* 4:345–357.

Betala P.A., Appugounder S., Chakraborty S., Songprawat P., Buttner W.J., Perez-Luna V.H. (2008) Rapid colorimetric detection of proteins and bacteria using silver reduction/precipitation catalyzed by gold nanoparticles. *Sensing and Instrumentation for Food Quality and Safety* 2:34–42.

Bošković M., Baltić M.Ž., Ivanović J., Đurić J., Lončina J., Dokmanović M., Marković R. (2013) Use of essential oils in order to prevent foodborne illnesses caused by pathogens in meat. *Tehnologija Mesa* 54:14–20.

Bowles M., Lu J.J. (2014) Removing the blinders: A literature review on the potential of nanoscale technologies for the management of supply chains. *Technological Forecasting and Social Change* 82:190–198. doi:10.1016/j.techfore.2013.10.017.

Brody A.L., Bugusu B., Han J.H., Sand C.K., Mchugh T.H. (2008) Innovative food packaging solutions. *Journal of Food Science* 73:R107–R116.

Chaudhry Q., Scotter M., Blackburn J., Ross B., Boxall A., Castle L., Aitken R., Watkins R. (2008) Applications and implications of nanotechnologies for the food sector. *Food Additives and Contaminants Part a-Chemistry Analysis Control Exposure & Risk Assessment* 25:241–258.

Chen M.F., Lin Y.P., Cheng T.J. (2013) Public attitudes toward nanotechnology applications in Taiwan. *Technovation* 33:88–96.

Chen S.H., Wu V.C., Chuang Y.C., Lin C.S. (2008) Using oligonucleotide-functionalized Au nanoparticles to rapidly detect foodborne pathogens on a piezoelectric biosensor. *Journal of Microbiological Methods* 73:7–17.

Cubukcu M., Timur S., Anik U. (2007) Examination of performance of glassy carbon paste electrode modified with gold nanoparticle and xanthine oxidase for xanthine and hypoxanthine detection. *Talanta* 74:434–439. doi:10.1016/j.talanta.2007.07.039.

Cushen M., Kerry J., Morris M., Cruz-Romero M., Cummins E. (2012) Nanotechnologies in the food industry—Recent developments, risks and regulation. *Trends in Food Science & Technology* 24:30–46.

De Azeredo H.M. (2009) Nanocomposites for food packaging applications. *Food Research International* 42:1240–1253.

Du W.L., Niu S.S., Xu Y.L., Xu Z.R., Fan C.L. (2009) Antibacterial activity of chitosan tripolyphosphate nanoparticles loaded with various metal ions. *Carbohydrate Polymers* 75:385–389.

Duncan T.V. (2011) Applications of nanotechnology in food packaging and food safety: Barrier materials, antimicrobials and sensors. *Journal of Colloid and Interface Science* 363:1–24.

Duran N., Marcato P.D. (2013) Nanobiotechnology perspectives. Role of nanotechnology in the food industry: A review. *International Journal of Food Science and Technology* 48:1127–1134. doi:10.1111/ijfs.12027.

Fedotova A., Snezhko A., Sdobnikova O., Samoilova L., Smurova T., Revina A., Khailova E. (2010) Packaging materials manufactured from natural polymers modified with silver nanoparticles. *International Polymer Science and Technology* 37:T59–T64.

Fernández A., Picouet P., Lloret E. (2010) Cellulose-silver nanoparticle hybrid materials to control spoilage-related microflora in absorbent pads located in trays of fresh-cut melon. *International Journal of Food Microbiology* 142:222–228.

Fernández A., Soriano E., López-Carballo G., Picouet P., Lloret E., Gavara R., Hernández-Muñoz P. (2009) Preservation of aseptic conditions in absorbent pads by using silver nanotechnology. *Food Research International* 42:1105–1112.

Fu J., Park B., Siragusa G., Jones L., Tripp R., Zhao Y., Cho Y.J. (2008) An Au/Si hetero-nanorod-based biosensor for Salmonella detection. *Nanotechnology* 19:155502.

Gaskell G., Eyck T.T., Jackson J., Veltri G. (2005) Imagining nanotechnology: Cultural support for technological innovation in Europe and the United States. *Public Understanding of Science* 14:81–90.

GoodFood. (2008) *Food Safety and Quality Monitoring with Microsystems.* European Council, Brussels, Belgium.

Gurr J.R., Wang A.S., Chen C.H., Jan K.Y. (2005) Ultrafine titanium dioxide particles in the absence of photoactivation can induce oxidative damage to human bronchial epithelial cells. *Toxicology* 213:66–73.

Huang Y., Chen S., Bing X., Gao C., Wang T., Yuan B. (2011) Nanosilver migrated into food-simulating solutions from commercially available food fresh containers. *Packaging Technology and Science* 24(5):291–297.

Inbaraj B.S., Chen B. (2016) Nanomaterial-based sensors for detection of foodborne bacterial pathogens and toxins as well as pork adulteration in meat products. *Journal of Food and Drug Analysis* 24:15–28.

Joe M.M., Chauhan P.S., Bradeeba K., Shagol C., Sivakumaar P.K., Sa T. (2012) Influence of sunflower oil based nanoemulsion (AUSN-4) on the shelf life and quality of Indo-Pacific king mackerel (*Scomberomorus guttatus*) steaks stored at 20°C. *Food Control* 23:564–570.

Kaittanis C., Santra S., Perez J.M. (2010) Emerging nanotechnology-based strategies for the identification of microbial pathogenesis. *Advanced Drug Delivery Reviews* 62:408–423.

Ke Z., Yongping B. (2005) Improve the gas barrier property of PET film with montmorillonite by in situ interlayer polymerization. *Materials Letters* 59:3348–3351.

Kim Y.S., Kim J.S., Cho H.S., Rha D.S., Kim J.M., Park J.D., Choi B.S., Lim R., Chang H.K., Chung Y.H. (2008) Twenty-eight-day oral toxicity, genotoxicity, and gender-related tissue distribution of silver nanoparticles in Sprague-Dawley rats. *Inhalation Toxicology* 20:575–583.

Lloret E., Picouet P., Fernández A. (2012) Matrix effects on the antimicrobial capacity of silver based nanocomposite absorbing materials. *LWT-Food Science and Technology* 49(2):333–338.

Lok C.N., Ho C.M., Chen R., He Q.Y., Yu W.Y., Sun H., Tam P.K.H., Chiu J.F., Che C.M. (2007) Silver nanoparticles: Partial oxidation and antibacterial activities. *Journal of Biological Inorganic Chemistry* 12:527–534.

Luo P.G., Stutzenberger F.J. (2008) Nanotechnology in the detection and control of microorganisms. *Advances in Applied Microbiology* 63:145–181.

Mahdi S., Vadood R., Nourdahr R. (2012) Study on the antimicrobial effect of nanosilver tray packaging of minced beef at refrigerator temperature. *Global Veterinaria* 9:284–289.

Mao L.Q., Yamamoto K. (2000) Amperometric on-line sensor for continuous measurement of hypoxanthine based on osmium-polyvinylpyridine gel polymer and xanthine oxidase bienzyme modified glassy carbon electrode. *Analytica Chimica Acta* 415:143–150.

Mao X.L., Yang L.J., Su X.L., Li Y.B. (2006) A nanoparticle amplification based quartz crystal microbalance DNA sensor for detection of *Escherichia coli* O157: H7. *Biosensors & Bioelectronics* 21:1178–1185.

Mills A. (2005) Oxygen indicators and intelligent inks for packaging food. *Chemical Society Reviews* 34:1003–1011. doi:10.1039/b503997p.

Morones J.R., Elechiguerra J.L., Camacho A., Holt K., Kouri J.B., Ramírez J.T., Yacaman M.J. (2005) The bactericidal effect of silver nanoparticles. *Nanotechnology* 16:2346.

Mulchandani A., Luong J.H.T., Male K.B. (1989) Development and application of a biosensor for hypoxanthine in fish extract. *Analytica Chimica Acta* 221:215–222. doi:10.1016/S0003-2670(00)81958-3.

Olmedilla-Alonso B., Jimenez-Colmenero F., Sanchez-Muniz F.J. (2013) Development and assessment of healthy properties of meat and meat products designed as functional foods. *Meat Science* 95:919–930.

Panea B., Ripoll G., González J., Fernández-Cuello Á., Albertí P. (2014) Effect of nanocomposite packaging containing different proportions of ZnO and Ag on chicken breast meat quality. *Journal of Food Engineering* 123:104–112.

Picouet P., Fernandez A., Realini C., Lloret E. (2014) Influence of PA6 nanocomposite films on the stability of vacuum-aged beef loins during storage in modified atmospheres. *Meat Science* 96:574–580.

Ramachandraiah K., Han S.G., Chin K.B. (2015) Nanotechnology in meat processing and packaging: Potential applications—A review. *Asian-Australasian Journal of Animal Sciences* 28:290–302. doi:10.5713/ajas.14.0607.

Ravichandran M., Hettiarachchy N.S., Ganesh V., Ricke S.C., Singh S. (2011) Enhancement of antimicrobial activities of naturally occurring phenolic compounds by nanoscale delivery against *Listeria monocytogenes, Escherichia coli* O157: H7 and *Salmonella typhimurium* in broth and chicken meat system. *Journal of Food Safety* 31:462–471.

Rhim J.W., Park H.M., Ha C.S. (2013) Bio-nanocomposites for food packaging applications. *Progress in Polymer Science* 38:1629–1652.

Rodriguez A., Nerin C., Batlle R. (2008) New cinnamon-based active paper packaging against Rhizopusstolonifer food spoilage. *Journal of Agricultural and Food Chemistry* 56:6364–6369.

Salminen H., Herrmann K., Weiss J. (2013) Oil-in-water emulsions as a delivery system for n-3 fatty acids in meat products. *Meat Science* 93:659–667.

Shefer A., Shefer S.D. (2003) Biodegradable bioadhesive controlled release system of nano-particles for oral care products, U.S. Patent No. 6,565,873. May 20, 2003.

Shefer A., Shefer S.D. (2004) Biodegradable bioadhesive controlled release system of nano-particles for oral care products, U.S. Patent No. 6,790,460. September 14, 2004.

Siegrist M., Stampfli N., Kastenholz H., Keller C. (2008) Perceived risks and perceived benefits of different nanotechnology foods and nanotechnology food packaging. *Appetite* 51:283–290.

Silvestre C., Duraccio D., Cimmino S. (2011) Food packaging based on polymer nanomaterials. *Progress in Polymer Science* 36:1766–1782.

Singh P.K., Jairath G., Ahlawat S.S. (2016) Nanotechnology: A future tool to improve quality and safety in meat industry. *Journal of Food Science and Technology* 53:1739–1749. doi:10.1007/s13197-015-2090-y.

Sonawane S.K., Arya S.S., LeBlanc J.G., Jha N. (2014) Use of nanomaterials in the detection of food contaminants. *European Journal of Food Research & Review* 4:301.

Sondi I., Salopek-Sondi B. (2004) Silver nanoparticles as antimicrobial agent: A case study on *E. coli* as a model for Gram-negative bacteria. *Journal of Colloid and Interface Science* 275:177–182.

Sozer N., Kokini J.L. (2009) Nanotechnology and its applications in the food sector. *Trends in Biotechnology* 27:82–89.

Thandavan K., Gandhi S., Sethuraman S., Rayappan J.B.B., Krishnan U.M. (2013) Development of electrochemical biosensor with nano-interface for xanthine sensing—A novel approach for fish freshness estimation. *Food Chemistry* 139:963–969.

Valdés M.G., González A.C.V., Calzón J.A.G., Díaz-García M.E. (2009) Analytical nanotechnology for food analysis. *Microchimica Acta* 166:1–19.

Varshney M., Li Y. (2007) Interdigitated array microelectrode based impedance biosensor coupled with magnetic nanoparticle—Antibody conjugates for detection of *Escherichia coli* O157:H7 in food samples. *Biosensors and Bioelectronics* 22:2408–2414. doi:10.1016/j.bios.2006.08.030.

Véronique C. (2008) Bioactive packaging technologies for extended shelf life of meat-based products. *Meat Science* 78:90–103.

Villamizar R.A., Maroto A., Rius F.X., Inza I., Figueras M.J. (2008) Fast detection of *Salmonella infantis* with carbon nanotube field effect transistors. *Biosensors and Bioelectronics* 24:279–283.

Weiss J., Takhistov P., McClements D.J. (2006) Functional materials in food nanotechnology. *Journal of Food Science* 71:R107–R116.

Wu T., Wu C., Fu S., Wang L., Yuan C., Chen S., Hu Y. (2017) Integration of lysozyme into chitosan nanoparticles for improving antibacterial activity. *Carbohydrate Polymer* 155:192–200.

Xue X., Pan J., Xie H., Wang J., Zhang S. (2009) Fluorescence detection of total count of *Escherichia coli* and *Staphylococcus aureus* on water-soluble CdSe quantum dots coupled with bacteria. *Talanta* 77:1808–1813.

Yam K.L., Takhistov P.T., Miltz J. (2005) Intelligent packaging: Concepts and applications. *Journal of Food Science* 70(1):R1–R10.

Yamada K., Kim C.T., Kim J.H., Chung J.H., Lee H.G., Jun S. (2014) Single-walled carbon nanotube-based junction biosensor for detection of *Escherichia coli*. *PLoS One* 9(9):e105767.

Yang L., Li Y. (2006) Simultaneous detection of *Escherichia coli* O157: H7 and *Salmonella typhimurium* using quantum dots as fluorescence labels. *Analyst* 131:394–401.

Young J.F., Therkildsen M., Ekstrand B., Che B.N., Larsen M.K., Oksbjerg N., Stagsted J. (2013) Novel aspects of health promoting compounds in meat. *Meat Science* 95:904–911. doi:10.1016/j.meatsci.2013.04.036.

Yusop S.M., O'Sullivan M.G., Preuß M., Weber H., Kerry J.F., Kerry J.P. (2012) Assessment of nanoparticle paprika oleoresin on marinating performance and sensory acceptance of poultry meat. *LWT-Food Science and Technology* 46:349–355.

Zhao X., Ao Q., Du F., Zhu J., Liu J. (2010) Surface characterization of ginger powder examined by X-ray photoelectron spectroscopy and scanning electron microscopy. *Colloids and Surfaces B: Biointerfaces* 79:494–500.

Zhao X., Yang Z., Gai G., Yang Y. (2009) Effect of superfine grinding on properties of ginger powder. *Journal of Food Engineering* 91:217–222.

9

Applications in Wastewater Treatment

Introduction

Water covers nearly 70% of the earth's surface, and water is one of the major ingredients in the food and beverages industry. In Canada, the food and beverages industry is the fourth largest user of water, which uses around 400 million cubic meters of water per year (total manufacturing industrial use is around 4000 million cubic meters/year) (StatCanada 2011). More than 85% of the water is discharged as a wastewater after industrial use (3226.8 million cubic meters/year in Canada); 76.6% of this wastewater is discharged directly to surface freshwater, and 13% is discharged to tidewater bodies (StatCanada 2011). Based on a survey by Statistics Canada, nearly 34% of this wastewater is discharged into water bodies without any treatment, and most of the wastewater is treated only by primary and secondary treatments, which remove only part of pollutants in the wastewater. Only 12% of the industrial wastewater is treated by advanced wastewater treatments, which removes almost all the pollutants from the wastewater before releasing into water bodies. In the food and beverages industry, almost 50% (209.6 million cubic meters/year) of the wastewater is not treated before being released into water bodies or on land; approximately 98, 23.6, and 15.2 million cubic meters of wastewater is treated using primary (or mechanical), secondary (or biological), and advanced wastewater treatment processes before release into the environment (StatCanada 2011). The situation is similar in other parts of the world, and in developing countries, the condition is far worse than this. The release of industrial wastewater without proper treatment is causing water, land, and air pollution, which might cause health hazards to humans as well as to the environment. It is estimated that around 90% of the health diseases are linked to the polluted water in the developing countries, and most of the pollutants (toxic chemicals and pathogenic microbial cultures) found in the drinking water have a link to the untreated wastewater from municipal, agricultural, and manufacturing operations (Bora and Dutta 2014). Thus, proper treatment of manufacturing industrial wastewater is one of the key elements to ensure safe drinking water supplies to the world.

In the food and beverages sector, meat (beef, chicken, pork, and other products) processing (25,728 L of water per kg) and chocolate processing industries (17,196 L of water per kg) use a high amount of water (Wechsler 2015). The food and beverages industry uses water for various purposes, as an ingredient, for cleaning raw materials, cleaning food processing equipment and processing facilities, and cooling and heating operations. Therefore, the water requirement for the food and beverages industry is large. Among this huge amount of water use, only a small amount is used as an ingredient, and the rest of the water is discharged after use for processing operations at the food and beverages manufacturing facilities. The wastewater from food and beverages processing units contains a lot of organic matter (fat, carbohydrates, and proteins), nutrients, and some toxic compounds and heavy metals. Treatment of wastewater from food and beverages manufacturing facilities with proper methods will be beneficial to the environment (by removing hazardous pollutants). It will also provide economic benefits to the industry since the suitable wastewater treatment will help to recover useful nutrients and secondary by-products from the wastewater, as well as the water can be reused for secondary processing operations like cooling and heating operations. For example, the wastewater from dairy processing plants consists of large amounts of fat, protein, and lactose, and if a proper wastewater treatment process is applied, the protein can be recovered and used as caseins, other organic matters like lactose can be used for biomass and biofuel production, and the treated water can be used for cooling and generating steam for the heating and pasteurization process as well as for cleaning nonfood contact surfaces (Andrade et al. 2015).

In general, the food and beverages industries use conventional wastewater treatment like membrane separation as a primary treatment process to remove suspended solids and organic materials like fat and use biological treatments such as anaerobic and aerobic fermentation for the secondary process to remove other pollutants from the wastewater (Andrade et al. 2015). But these conventional wastewater treatment methods need high initial investment, high maintenance cost, and high energy cost for operating the wastewater treatment unit, as well as require large areas (Helmer and Hespanhol 2002). When considering economic benefits, the cost-benefit ratio is quite low for a comprehensive wastewater treatment facility to remove all the pollutants from the food and beverages manufacturing facilities. In developing countries, most of the food industries are small- or medium-scale units, which cannot afford the high cost of the wastewater treatment. Thus, the development of new cost-effective wastewater technologies with low energy and area requirement might help the food and beverages manufacturing industry to treat their wastewater in a proper way, which will be beneficial to the environment as well as to their economies. The wastewater treatment processes based on nanoscience and nanotechnology will be the best option for the food and beverages industry to remove most of the contaminants from the wastewater discharged from the processing plants, and also recover most of the components that might have nutritional as well as economic benefits (Hornyak et al. 2008). The unique capabilities of the nanoparticles and nanomaterials such as higher specific surface area for contact with the effluents, wide variety of functionality, and high reactivity with organic and inorganic pollutants from the food industry and the size (nanolevel) make the technologies based on nanoscience and nanomaterials applicable for treating wastewater from food processing facilities as well as for purifying the raw water used for various applications in the food and beverages industry (Cloete et al. 2010). Especially, the beverages industry extensively uses the nanotechnology-based processes to treat (to soften the hard water from the ground water or from open water sources) and to purify (remove unwanted heavy metals, pathogenic microbes) the raw water, as water is their major ingredient. Applications of nanoscience and nanotechnology for purifying water are well researched, and these nanotechnology-based water purification units purify the water effectively with low initial and operating cost (Baker 2004; Baruah et al. 2012a, 2012b; Hu et al. 2010; Ren et al. 2010). Use of nanoscience and nanotechnologies for treating wastewater from the food industry is gaining the attention of researchers and industry personnel because removal of organic and inorganic pollutants can be done at a lower cost than conventional treatment processes. The recent advancements in the applications of nanoscience and nanotechnology for treating wastewater from the food and beverages industry are discussed in detail in this chapter. Some of the technologies discussed below are also used for water purification.

Nanotechnology for the Wastewater Treatment Process

The advancements in nanoscience and nanotechnology help to improve processes for purification of water as well as for treatment of wastewater produced from the food and beverages industry. Broadly, the tested and applied advanced wastewater treatment technologies based on nanoscience and nanotechnology are: (a) nanofiltration, (b) photocatalysis, and (c) nanosorbents.

These three techniques are widely applied to purify the ground and drinking water and to treat the wastewater from agricultural operations, manufacturing industries, and the urban centers (Bora and Dutta 2014).

Nanofiltration

Membrane filtration technique is the most commonly used method for purification of drinking and domestic water as well as for wastewater treatment. In this technique, the solid organic and inorganic matters from the liquid phase are separated by a filter or membrane. This technique is commonly referred to as conventional filtration. Advanced membrane filtration techniques, such as microfiltration, ultrafiltration, nanofiltration, and reverse osmosis filtration, have been applied in water purification and wastewater treatment processes (Aroon et al. 2010; Cuartas-Uribe et al. 2010; Lau et al. 2012; Salehi 2014). The major differences between these filtration techniques are the pore size of the membranes and the pressure applied across the membrane for separation of the solid and liquid phase. Nanofiltration is the process of separating elements (normally solid and liquid phase) using the membranes (or filters) with nanoscale-level pores. The properties of nanofiltration, the pore size, and the pressure fall between the properties of ultrafiltration and reverse osmosis. The pore size of the nanofiltering membranes is less than that of micro and ultrafiltration techniques, and the pressure applied for separation (normally between 0.7 and 3 MPa [7 and 30 atm]) is less than the reverse osmosis process (normally between 2 and 101 MPa [20 and 100 atm]) (Baker 2004; Hong et al. 2006; Salehi et al. 2011). Nanofiltration membranes usually have a pore size of 1–5 nm, and most of the organic and inorganic solutes in the wastewater can be filtered out using these nanofiltration membranes. Nanofiltration is a membrane separation process driven by the pressure difference, and this technique is getting extensive attention among the researchers and industry personnel due to its advantages of high permeation rate and charge-based separation. Application of nanofiltration technology for the treatment of wastewater from dairy, brewery, winery, and other food processing facilities is becoming popular due to its low cost and high separation efficiency. Most of the organic and inorganic pollutants, as well as microbial components from the food and beverage processing facilities' wastewater, can be removed by this technique (Lipnizki 2010; Madaeni et al. 2011; Salehi 2014; Thanuttamavong et al. 2001, 2002; Vandanjon et al. 2002). Nanoclays, carbon nanotubes, metal oxide (zinc oxide [ZnO], titanium dioxide [TiO_2], zirconium dioxide [ZrO_2]), and silica dioxide (SiO_2) nanoparticles are the most common types of nanoproducts used for the fabrication of nanofilter (NF) membranes (Bora and Dutta 2014). Nanofilter membranes are generally made of synthetic polymeric materials, which are less expensive and have good flexibility. But the resistance against some organic and inorganic chemicals is less, and also synthetic NFs have high membrane fouling, which reduces the life of the filters. The conventional inorganic ceramic NFs demonstrate higher thermal resistance as well as resistance against most common organic and inorganic chemical compounds in the food and beverages industry wastewater than synthetic NFs; however, these have lower flexibility and higher cost of manufacturing. These drawbacks can be resolved by incorporating nanoclay and less expensive nanomaterials with conventional filter fabrication materials using the advanced nanotechnology methods (Van der Bruggen et al. 2008).

The NF membranes with the CNTs had a pore diameter of 1–10 nm, and the studies showed these CNT-based NF membranes had higher permeation flux for water and higher rejection rate for chemical and biological contaminants in the water and wastewater from industries (Falk et al. 2010; Noy et al. 2007; Wan et al. 2009). The major advantages of CNT-based NF membranes are that their chemical and thermal resistance is higher like the ceramic NF membranes, and they have higher flexibility like the synthetic polymer NF membranes (Wan et al. 2009). An NF membrane with 1 nm pore diameter fabricated with the freestanding CNTs developed by Karan et al. (2012) rejected almost all of the organic dyes from the simulated industrial wastewater, and the rejection rate for organic dyes was threefold higher when compared to the commercially available membrane filters.

Membrane filtration using nanotechnology and photocatalytic processes with nanomaterials are helping the food and beverages industry to treat their wastewater at low cost with high efficiency. Molybdenum trioxide (MoO_3)-based nanowires, and bismuth ($Bi_2Mo_3O_{12}/MoO_3$)-based nano-heterostructure were used to fabricate a filter for membrane filtration and photocatalytic probe for treating simulated wastewater (Liu et al. 2014). For α-MoO_3 nanowire fabrication, they used hydrothermal technique briefly described here. A solution with deionized water (34.0 mL), concentrated nitric acid (5.5 mL), and aqueous H_2O_2 solution (11.0 mL) was prepared and a commercial α-MoO_3 powder (1.44 g) was mixed with the above-prepared solution and stirred at room temperature for 3 days. About 35.0 mL of the dissolved yellow solution obtained after stirring was placed in a stainless-steel autoclave with Teflon lining and heated at 180°C for 1 day. The α-MoO_3 nanowires were obtained after cooling this solution to room temperature. The α-MoO_3 nanowire filter membrane for wastewater treatment was fabricated from these α-MoO_3 nanowires using a vacuum filtration (at 0.1 MPa pressure) method (Figure 9.1). For $Bi_2Mo_3O_{12}/MoO_3$ nano-heterostructure fabrication, first Bi (No_3).5 H_2O (5 mmol) was mixed with deionized water (160 mL) and stirred for 30 min. Then this suspension was mixed to α-MoO_3 nanowire solution prepared based on the previously discussed method using an ultrasonic homogenizer and then stirred in an oil bath set at a temperature of 80°C. After stirring for 6 h, the solution was filtered and then dried at 80°C to get a solid powder. The $Bi_2Mo_3O_{12}/MoO_3$ nano-heterostructure was obtained by calcining this solid powder at 500°C for 2 h (Figure 9.2). Performance of these wastewater treatment techniques was evaluated by purifying the simulated wastewater containing 50 μM rhodamine B (RhB) tracer dye and comparing results against filtration using commercial cellulose filter membrane. The scanning electron microscopy images of α-MoO_3 nanowires showed that these nanowires had a diameter of 100 nm, and the filtration performance test results showed that the

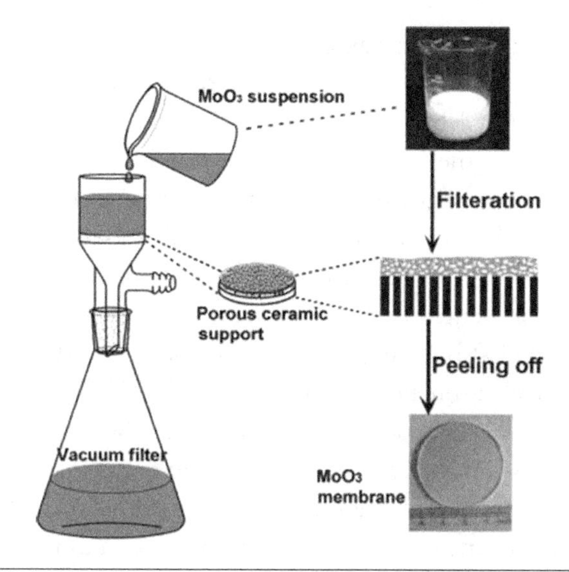

Figure 9.1 Schematic of the fabrication process of MoO_3–nanowire membrane by the vacuum filtration method. (Reproduced from Liu, T. et al., *Chem. Eng. J.*, 244, 382–390, 2014.)

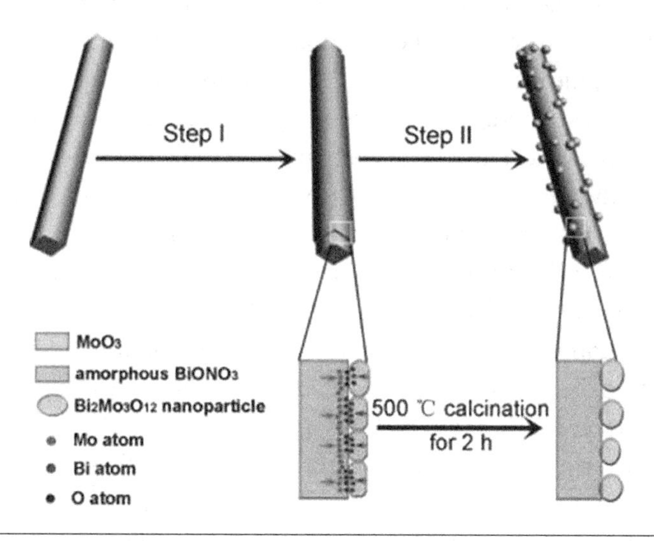

Figure 9.2 (See color insert.) The illustration the formation process for the $Bi_2Mo_3O_{12}/MoO_3$ nano-heterostructures. (Reproduced from Liu, T. et al., *Chem. Eng. J.*, 244, 382–390, 2014.)

commercial cellulose membrane removed only 4.5% of RhB from the wastewater, and the α-MoO_3 nanowire membrane removed all of the RhB molecules from the wastewater. The α-MoO_3 nanowire had an adsorption capacity of 38.50 µmol/g, and the membrane had a filtration capacity of 1.0 L/g for purifying simulated water with organic dyes

(50 μM). The absorbed organic RhB dye molecules can be removed from the α-MoO_3 nanowire membrane by calcination at a temperature of 350°C for 2 h, and the membrane can be reused for wastewater filtration in the food and beverages industry. The $Bi_2Mo_3O_{12}$/MoO_3 nano-heterostructure photocatalytic semiconductor degraded around 90% of the organic dye in 60 min of irradiation using visible light. When using the pure $Bi_2Mo_3O_{12}$ nanoparticles for simulated wastewater treatment, only 60% of the organic dye was degraded after 120 min of irradiation using visible light. Liu et al. (2014) attributed the increased electron-hole separation of $Bi_2Mo_3O_{12}$/MoO_3 nano-heterostructure with an aqueous solution, and the higher adsorption property of α-MoO_3 nanowires was the main reason for the improved filtration performance of photocatalytic treatment with $Bi_2Mo_3O_{12}$/MoO_3 nano-heterostructure and α-MoO_3 nanowires filtration membranes.

The dairy industry uses a high amount of water for sanitation and processing of various dairy products. The wastewater from the dairy industry contains protein, fats, and lactose from the milk, which can cause eutrophication of water bodies if not treated. This nutrient-rich wastewater from the dairy processing industry is treated with conventional biological and physicochemical wastewater treatments, but these involve high capital cost, as well as valuable nutrients in the wastewater may be removed along with the sludge. The membrane filtration technique is the cheap and best way of treating dairy industry wastewater as well as for recovery of nutrients. Several types of nanofiltration membranes (synthetic polymer, ceramic) were tested for treating dairy industry wastewater but resulted in the high amount of membrane fouling due to a high concentration of protein in the dairy processing industry wastewater. Chen et al. (2016) developed an integrated wastewater treatment unit for recycling of dairy industry wastewater and recovery of nutrients (caseins, acetate, butyrate), and hydrogen by combining isoelectric precipitation, nanofiltration, and anaerobic fermentation techniques (Figure 9.3). They simulated dairy industry wastewater by mixing skim milk powder with the deionized water (2381 mg in 1 L) using a centrifuge at 10,000 rpm. Then the artificial wastewater entered into the isoelectric precipitation chamber after adjusting the pH (between 3 and 7) by adding 1 M hydrochloric acid, where the protein components (casein) precipitated by adding

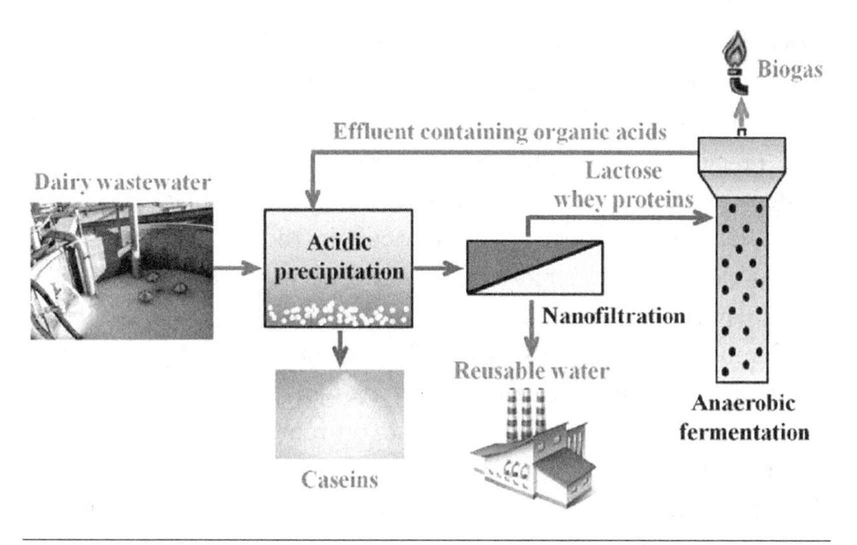

Figure 9.3 Schematic diagram of dairy wastewater treatment using an integrated wastewater treatment unit with acid (isoelectric) precipitation + Nanofiltration + Anaerobic fermentation process. (Reproduced from Chen, Z. et al., *Chem. Eng. J.*, 283, 476–485, 2016.)

different acids (sulfuric acid, hydrochloric acid, butyric acid, citric acid, and acetic acid). Then the supernatant liquid was sent to a nanofiltration chamber, which had a commercial nanofiltration membrane NF 270 (Dow-Filmtec, Edina, MN) and a volume of 12.5 mL and filtered at a pressure difference of 3000 kPa. For each replication, a new membrane was used and the NF membranes were pretreated by dipping in ethanol solution (50% v/v concentration) for 5 s followed by soaking in deionized water for 12 h. The NF retentate was sent to a continuous flow homemade bioreactor (up flow anaerobic sludge blanket [UASB] type with a volume of 7.5 L) and run for a hydraulic retention time of 12 h at 37°C ± 1°C (Chen et al. 2016). The results showed that organic acids like citric acid and butyric acid can also be used for casein precipitation, but the acidic precipitation by hydrochloric acid produced low turbidity supernatant (0.2 nephelometric turbidity units [NTUs]), which will be helpful to reduce membrane fouling during nanofiltration. The nanofiltration with the NF270 membrane removed most of the solids (96.7%) in the simulated wastewater, and also NF270 had greater antifouling performance (irreversible fouling [IF]) after acid precipitation (casein removal increased IF rate from 5.9% to 86.1%), and the permeation rate of NF270 was high after isoelectric precipitation of protein components. The anaerobic fermentation of retentate of nanofiltration produced more hydrogen (58.8% in the biogas produced from retentate) and more acetate and butyrate acids (26.0% and 51.9%, respectively) (Chen et al. 2016). This acid precipitation-nanofiltration-anaerobic fermentation integrated wastewater treatment can be implemented in the dairy industry. This could help recover most of the nutrients in the dairy wastewater as well as clear the water more effectively for reuse in cooling, heating, and steam production for dairy industry operations.

Membrane bioreactors (MBRs) have been tested for treating wastewater from pharmaceuticals, textile, food, and beverage processing facilities and showed good results on removing pollutants from the industrial wastewater. Since the dairy industry wastewater consists of organic substances from raw material, cleaning, and sanitizing agents, and secondary by products of the processes, it is hard to remove all the pollutants with conventional biological wastewater treatment processes. Andrade et al. (2014) tested MBR alone, and MBR+NF (NF as a second stage filtration) systems for the dairy industry wastewater treatment and measured the chemical oxygen demand (COD) and total solid contents to check the efficiency of the systems (Figure 9.4). The polytherimide microfiltration membrane of the

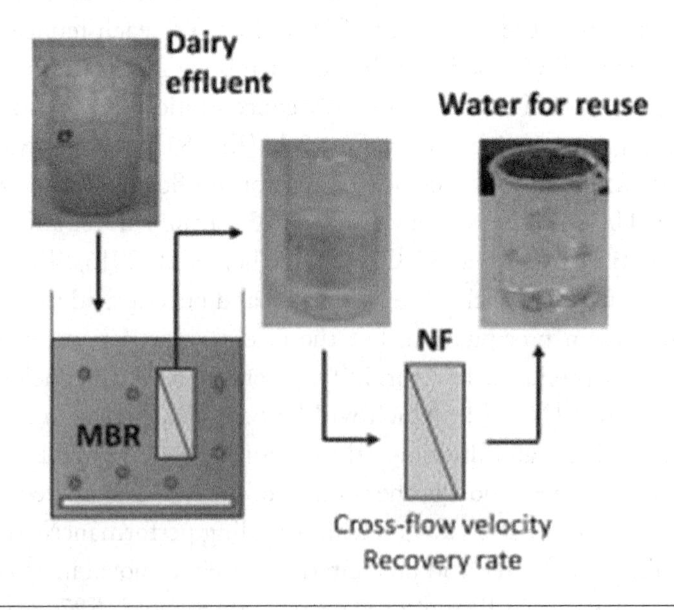

Figure 9.4 **(See color insert.)** Dairy wastewater treatment. (Reproduced from Andrade, L. et al., *Sep. Purif. Technol.*, 126, 21–29, 2014.)

MBR unit designed and developed by PAM Selective Membranes, Rio De Janerio, Brazil, had a pore size of 0.5 μm and a permeate flux of 18.2 L/(h.m^2). The MBR system alone effectively reduced COD (by 98%, and 86%, 89% of total nitrogen and phosphorus, respectively) from the wastewater collected from a dairy processing plant. But the total solids content was high, which prevented the reuse of permeate (water) for secondary dairy processing operations (cooling and heating) and for sanitation purposes. When the nanofiltering unit with NF 90 membrane (Dow-Filmtec, Edina, MN) was added as a second stage cleaning, more than 93% of the total solids were removed from the wastewater, and overall COD reduction was 99.9%. The quality tests of permeate water showed that permeate water can be reused for steam generation, cooling, and cleaning of nonfood contact surfaces of the processing facility and transportation trucks (Andrade et al. 2014). They also found a feed flow rate of 7.8 m/s to the nanofiltering unit yielded lower membrane fouling and better permeate flux. Andrade et al. (2015) tested the filtration efficiency and economic benefits of this integrated wastewater treatment unit with a membrane bioreactor (MBR) and nanofiltration techniques for the dairy industry wastewater for a longer run from actual dairy processing facilities. Six samples of 150 L wastewater from a commercial dairy factory in Minas Gerais, Brazil, were collected and treated with a laboratory-scale cleaning unit with an MBR (designed by PAM Selective Membranes, Rio De Janerio, Brazil) and a nanofiltration system. The MBR had a polyetherimide microfiltration module with a pore size of 0.5 μm, area 0.044 m^2, and hydraulic permeability 177 L/h.m^2.bar (17.7 L/h.m^2.MPa) (Figure 9.5a). The sludge from the same dairy industry where the wastewater was collected was inoculated in the biological tank for 40 days prior to the test, and then the wastewater samples were run through the MBR at continuous run mode at a hydraulic retention time of 6 h and permeate flow rate of 0.70 L/h. During this period, the MBR permeate was automatically backwashed once in every 15 min for 15 s at a flow rate of 2.0 L/h. Then the permeate from the MBR was treated with the nanofiltration unit consisting of commercial NF 90 membrane (Dow-Filmtec, Edina, MN), which had a permeability rate for water of 2.3 L/h.m^2.bar (230 3 L/h.m^2.kPa) and placed inside an 8.9 cm diameter steel cell NF membrane chamber (Figure 9.5b). The NF filtration was performed at a pressure difference of 1000 kPa,

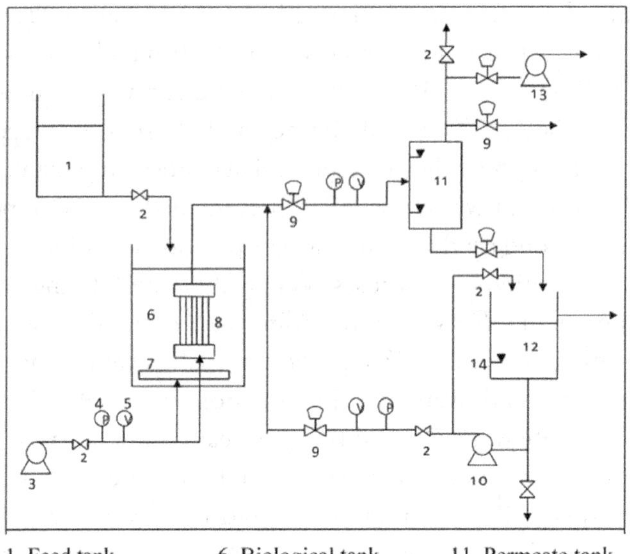

1. Feed tank
2. Needle valve
3. Air compressor
4. Manometer
5. Flow indicator
6. Biological tank
7. Air diffuser
8. Membrane module
9. Solenoid valve
10. Diaphragm pump
11. Permeate tank
12. Permeate and backwash tank
13. Vacuum pump
14. Level sensor

(a)

Thermometer
Manometer
Valve
Wastewater
Rotameter
Permeate
Membrane cell
Supply Tank
Pump

(b)

Figure 9.5 Schematic of microbial bioreactor (MBR) system (a) and nanofiltration (NF) unit (b) used for dairy industry wastewater treatment. (Reproduced from Andrade, L. et al., *Braz. J. Chem. Eng.*, 32, 735–747, 2015.)

and the permeate flux through the NF membrane was measured. The COD, color, and the total solids in the wastewater collected from the dairy industry permeates of MBR and NF systems, and the retentate of the NF unit were measured to check the efficacy of the integrated wastewater treatment process. Andrade et al. (2015) also performed a cost-benefit study of this unit to check the economic feasibility to use it in large-scale dairy processing facilities. They also analyzed pH; presence of metals like copper, magnesium, and zinc; and the alkalinity of the NF permeate (clean water) to evaluate the reusability of the treated water. The results showed the CODs of the raw wastewater, MBR permeates, and NF permeates were 3274, 34, and 4 mg/L, respectively; the total solid (TS) contents were 3366, 1783, and 233 mg/L; and the color values were 2173, 35, and 15 Hu, respectively. Their results showed clearly that the MBR process removed most of the organic matter content in the dairy wastewater and part of the total solids. The NF process removed the remaining dissolved solids in the effluent. The NF process also removed the metals, sodium, and improved the color of the wastewater. The physicochemical tests of NF permeate also proved that this NF filtered water met the quality regulations of water used for heating, cooling, and steam generation, as well as the general sanitation process (cleaning floors and nonfood contact areas). The economic analysis showed this integrated technique might cost 6.82–9.99 Brazil Real/m^3, and the implementation of this unit in a dairy processing facility might pay back the investment for this wastewater treatment unit in 1–3 years based on the cost of the local treated water.

Parham et al. (2013) fabricated a ceramic nanofiltration membrane with a CNT and tested it for removal of microorganisms and heavy metals from the simulated wastewater. A commercial porous ceramic with pore size between 300 and 500 μm made up of Al_2O_3 (81%), SiO_2 (14%), K_2O (2.5%), Na_2O (2.5%), and other minute oxides like TiO_2, MgO, Fe_2O_3, and Cr_2O_3 (2.5%) was used as a growing medium for CNT, which used Camphor ($C_{10}H_{16}O$) as a source of carbon and nickel (II) nitrate hexahydrate crystals ($Ni[NO_3]_2 \cdot 6H_2O$) as catalyst precursory agent. The porous ceramic was dipped into an acetone solution and placed inside a tube. Then it was heated inside a furnace at 700°C for 30 min, and then camphor acetone solution (40% concentration) was injected slowly (0.8 mL/h) into the quartz tube for 90 min, and then the CNT/ceramic filter was obtained after bringing the furnace to room temperature (Figure 9.6). This CNT filter was functionalized by an oxidation process using air at 400°C for 2 h. The simulated wastewater was artificially prepared by mixing yeast and heavy metals into the deionized water. The results showed

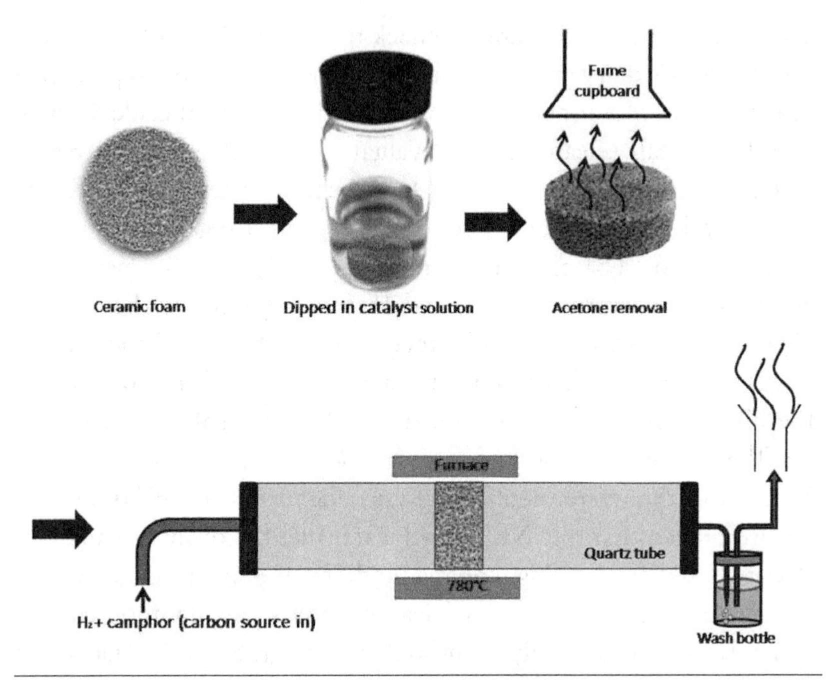

Figure 9.6 CNT/ceramic filter fabrication process. (Reproduced from Parham, H. et al., *Carbon*, 54, 215–223, 2013.)

the CNT/ceramic filters with the lengths of 50 and 70 mm removed nearly 98% of the yeast from the simulated wastewater when the flow rate was 20 mL/h. The filtration efficiency for yeast was less when the flow rate was higher (40 and 60 mL/h), as well as the length of the filter was less (20 mm). This CNT/ceramic filter with 20 mm length itself removed 99% of heavy metals (Fe^{2+}, Zn^{2+}, Cu^{2+}, and Mn^{2+}) when the flow rate of feed was between 2 and 30 mL/h. The cost of CNTs is higher than the regular synthetic nanofiltering membranes used for wastewater treatment but Parham et al. (2013) suggested this CNT/ceramic filter had higher thermal stability, and it can be reusable for a long duration. The metal ions that settled on the filter can be removed by cleaning the filter using (HNO_3) acidic solution, and the microorganisms can be removed by adding acetone, which could help to maintain the filtration efficiency (Parham et al. 2013).

Luo et al. (2011) developed a two-stage membrane filtration technique for dairy wastewater treatment. In this technique, ultrafiltration (UF) was used as the first stage of cleaning to filter out proteins and lipids from the dairy industry wastewater. The permeate of

UF was fed into a nanofiltration unit where the lactose was separated as a retentate and used for biogas production, and the permeate of the NF was water, which could be reused for industrial secondary applications. They tested three types of commercially available UF membranes (UP005P [Microdyn-Nadir, Wiesbaden, Germany], Ultracel PLGC [Millipore, Billerica, MA], and UH030P [Microdyn-Nadir, Wiesbaden, Germany]) for separating protein and lipids from the simulated dairy industry wastewater (skim milk mixed with the deionized water through centrifugation) and five types of commercial NF membranes (NF270 [Dow-Filmtec, Edina, MN], NF90 [Dow-Filmtec, Edina, MN], Nanomax50 [Millipore, Billerica, MA], Desal-5 DL [GE-Osmonics, Trevose, PA], and Desal-5 DK [GE-Osmonics, Trevose, PA]) for removing lactose and purifying water. They used the protein and lipids separated from the wastewater to grow algae, which was used for biodiesel and biofuel production, and the lactose obtained from NF was used for biogas production through anaerobic digestion. All the membranes used in their study were cleaned using 50% ethanol to remove all of the residues from manufacturing and immersed in deionized water for 24 h prior to the experiment. A transmembrane pressure (TMP) of 500 kPa was maintained for the UF stage, and 3000 kPa TMP was used for nanofiltration stage. The high retention rate for the protein and low lactose retention rate was used as selection parameters for the UF membrane. The membranes Ultracel PLGC and UP005P retained all of the protein from the simulated dairy wastewater, and the Ultracel PLGC and UH030P allowed most of the lactose to pass through the membranes. Also, the fouling index of Ultracel PLGC was very low (2.4%). Their results showed the Ultracel PLGC membrane was suitable for first stage filtration of dairy wastewater. Luo et al. (2012a) suggested the 25% diluted protein- and lipid-rich retentate of the UF system could be mixed with *Chlorella* Sp., algae, which had the capability of producing 59 mg L^{-1} day^{-1} biomass. The permeate of UF filtration using Ultracel PLGC membrane was used as a feed for second-stage nanofiltration, and the high lactose retention rate, high through flux, and the low inorganic salt content in the permeate were used as selection parameters for the NF membrane. Among all the five NF membranes, NF270 had the highest lactose retention rate (97.8%), but the salt concentration in permeate was 37.8%. The permeate from the

NF90 membrane had lower inorganic salt content, but the permeate flux was lower (10.5 L m^{-2} h^{-1} bar^{-1} (1050 L m^{-2} h^{-1} kPa^{-1}) than that of NF270 membrane (13.4 L m^{-2} h^{-1} bar^{-1} (1340 L m^{-2} h^{-1} kPa^{-1}). Their results showed the NF270 membrane had the lowest operating cost and the highest efficiency among the five tested NF membranes. A comparative study (Luo et al. 2012a) of single-stage nanofiltration with NF270 membrane, and a two-stage UF (with Ultracel PLGC membrane) + NF (with NF270 membrane) filtration showed that the IF index of a single-stage NF was 59.10%, whereas the IF value of the two-stage filtration was 2.64%. The protein and lipids in the raw dairy industry wastewater were the reason for higher amounts of membrane fouling during nanofiltration, and the removal of these protein and lipids using UF as a pretreatment to NF reduced the membrane fouling. In addition, the separated protein and lipids could be used for biodiesel and biofuel production, which might be economically and environmentally beneficial for the dairy processing industry.

A nanofiltration rotating disk membrane (NF-RDM) was developed with the commercial NF270 membrane to treat the dairy industry wastewater (Luo et al. 2010). This NF-RDM disk had a diameter of 7.76 cm and had vanes with a width of 2 mm and height of 6 mm. The disk was fixed with the shaft, and it rotated at a speed of 2500 rpm, which generated a large shear force (Figure 9.7). Luo et al. (2010) artificially made simulated wastewater by mixing skim milk with deionized water, and the membrane was pretreated with deionized water

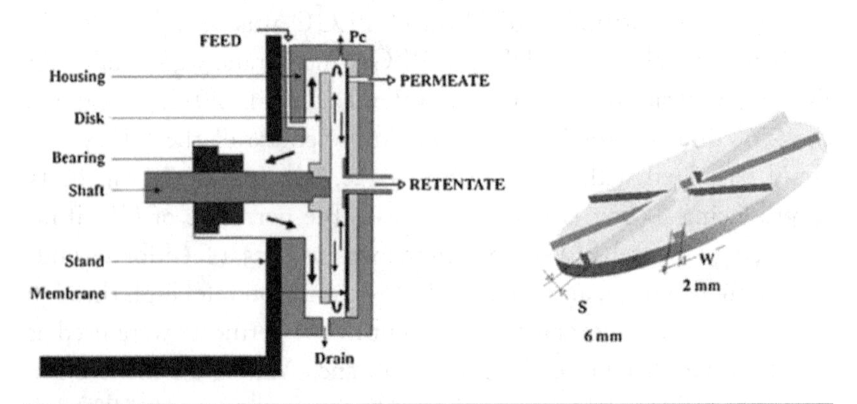

Figure 9.7 Nanofiltration rotating disk membrane (NF-RDM) module for dairy industry wastewater treatment. (Reproduced from Luo, J. et al., *Chem. Eng. J.*, 163, 307–316, 2010.)

for a minute at a transmembrane pressure of 4 MPa. This NF-RDM produced a higher amount of permeate flux when compared with the regular nanofiltering membranes at the same flow rate and same rejection rate for simulated dairy industry wastewater. The fouling rate was also low at higher shear rates and higher transmembrane pressure, but the irreversible fouling increased when the unit was operated below the critical flux rate. Luo et al. (2010) also tested this NF-RDM treatment unit for real dairy industry wastewater and found the permeate flux and permeate quality was similar to that of when it was tested in the laboratory with simulated dairy industry wastewater. They also found the energy required for treating 1 m^3 of dairy industry wastewater with this NF-RDM was less than the regular filtration techniques, and the membrane fouling was completely removed (>99%) in this NF-RDM membrane since the operation at a higher shear rate reduced the concentration polarization within a short time.

Luo et al. (2012b) tested the fouling behavior of this NF-RDM unit under various pressures and shear rates and found that high shear rate ($>10^5$ s^{-1}) and high transmembrane pressure (4 MPa) produced a stable permeate flux of 420 L m^{-1} h^{-1} for 4 h. After this stable flux period, they noticed some membrane fouling due to the lactose and calcium in the dairy wastewater. The use of alkaline cleansing agent P3-ultrasil 10 (Ecolab, St. Paul, MN) reduced the fouling and kept the permeate flux rate at a high level beyond the stable flux period (4 h). Thus, there is no need of any pretreatment of dairy industry wastewater, which also reduces the capital cost as well as the operational cost of the dairy industry's wastewater treatment plants. This chemical cleansing agent P3-ultrasil 10 not only prevented the formation of inorganic and organic components from caking but also had the ability to clean up the pores of the nanofiltration membrane. Unfortunately, it also caused swelling of the NF membrane, which might result in failure of filtration unit during long operating time. Even though alkaline cleaning reduces the irreversible fouling rate of the NF-RDM membrane, it is necessary to find a suitable cleaning material to clean up the membrane pores, which does not harm the nanofiltering membrane material.

Luo and Ding (2011) tested the effect of the dairy industry wastewater's pH on the efficacy of the NF-RDM for treating simulated dairy industry wastewater as well as real dairy industry wastewater and found the pH of the initial feed had a greater influence on the filtration ability of NF-RDM membranes. Luo et al. (2011) tested the NF-RDM membrane discussed above (Luo et al. 2010) with the simulated dairy industry wastewater with the pH range of 6.30 to 10.10 and two batches of real dairy industry wastewater at pH of 8.72 and 9.56. Their results showed that at acidic pH range (<7), the NF-RDM filtration system had a higher permeate flux but removed less soluble salt in the dairy wastewater, and at alkaline pH (>7) the issue of membrane fouling was reduced but the concentration polarization reduced the permeate flux too. Luo and Ding (2011) suggested the NF-RDM filtration unit performed well when the pH of dairy industry wastewater was between 7 and 8. Their tests with the real dairy industry wastewater showed that NF-RDM decreased the turbidity from 101.00 to 0.56 and the chemical oxygen demand from 297 to <15 mgO_2/L when the feed pH was 8.72. But the treatment of real dairy industry wastewater had lower permeate flux since the pH range of feed was in the alkaline range.

Luo et al. (2012a) tested the NF-RDM system at various trans-membrane pressures and shear rates to determine a threshold permeate flux to apply this new shear-based nanofiltration system in the dairy industry wastewater treatment plants. The simulated dairy wastewater was artificially made by mixing ultra-high temperature (UHT) skim milk with deionized water in a ratio of 1:2 and 1:3 to test the effect of solid matter concentration on permeate flux. Luo et al. (2012a) found the threshold permeate flux of this NF-RDM using simulated wastewater (1:2 diluted skim milk) was 6.1×10^{-5} ms^{-1} at a high shear rate (1.81×10^5 s^{-1}), and this permeate flux rate remained constant for 6 h of testing. Luo et al. (2012a) also noticed the membrane fouling was below 20% at or below this permeate flux, and when the flux increased above this threshold flux, reversible membrane fouling increased even at the higher shear rate. Their result proved that the threshold permeate flux of this shear-enhanced filtration with NF-RDM unit was significantly higher than critical fluxes of regular cross-flow nanofiltration and ultrafiltration membranes, and this higher and more stable permeate flux for longer operating time showed this shear-enhanced NF-RDM system could be used practically in large-scale dairy processing facilities to treat the wastewater effectively and economically.

Dairy processing plants use a lot of detergents in clean-in-place (CIP) units, and these contribute to around 40% of pollutants from the dairy processing industry. Recovery of detergents from the CIP will help to reduce the environmental pollution by reducing the amount of detergents in the wastewater, and it can be economically beneficial to the industry since the recovered detergents can be reused for CIP. Fernández et al. (2010) developed a nanofiltration system for treating wastewater from a CIP unit in a dairy processing plant operated with commercial single-phase detergent (Deptal EVP, Hypred®, Dinard, France) containing sodium and potassium hydroxide. A commercial polysulfone spiral-wound nanofiltering membrane (model: KOCH MPS-34, Koch Membrane Systems, Wilmington, MA) with an active filtration area of 1.4 m^2 and molecular weight cut-off (MWCO) value of 200 Da was used for treating the wastewater from a commercial yogurt manufacturing facility's CIP unit, and the long run test was carried out for 1800 h at an effluent temperature and feed rate of 70°C and 1650 L/h, respectively. Their results showed that the nanofiltration system recovered more than 75% of single-phase detergent from the wastewater and had a permeate flux of 45 L/h throughout the experimental period, and the selected NF membrane performed well even at high pH range (>12). Nanofiltration also reduced the chemical oxygen demand from >6 g/L for the feed to <2 g/L for the permeate, as well as hardness level from >10 mg Ca^{2+}/L for the feed to <4 mg Ca^{2+}/L for the permeate. The single-phase detergent recovered by their nanofiltration system was used for cleaning the UHT pilot plant's yogurt-filling machine. The cleaning efficiency of the detergent was measured by bioluminescence analysis technique (using spectrophotometer) and cell count (plating method). Their results showed that the surface of the filling machine cleaned using recycled single-phase detergent had low cell count (10 CFU/mL) and low relative light units (<100 RLU). These results proved that nanofiltration systems can be used in large-scale industrial facilities to treat the wastewater from the dairy processing industry, and the recovered cleaning agents can also be reused for cleaning and sanitation purposes, which can increase the economic benefits of the wastewater treatment unit.

A dual-layer hollow-fiber nanofiltering membrane unit was tested for removing heavy metals like cadmium (Cd^{2+}), chromium ($Cr_2O_7^{2-}$), and lead from the wastewater (Zhu et al. 2014). They fabricated hollow-fiber membrane using a dry-jet wet-phase inversion process by co-extruding polybenzimidazole (PBI) and polyether sulfone (PES) dopes simultaneously through a triple-orifice spinner. After extrusion, the hollow-fiber membranes were immersed in water for three days to complete the wet-phase inversion process and remove all the solvents from the filter membranes. Then the membranes were immersed in a glycerol aqueous (50% weight basis) solution for two days and finally air dried at room temperature. The scanning electron microscopy images showed that the pore radius of the dual-layer hollow-fiber nanofiltering membranes was 0.4 nm, and the MWCO of this dual-layer membrane was 338 Da (Figure 9.8). To test the heavy metal rejection of this developed membrane, they prepared simulated wastewater by mixing 200 mg of salt ($MgCl_2$, $Pb[NO_3]_2$, $CdCl_2$ and $Na_2Cr_2O_7$) to 1 L water. Their filtration results showed that the dual-layer PBI/PES hollow-fiber nanofiltering membrane removed 95% of Cd^{2+} and 93% of Pb^2 from the wastewater. The $Cr_2O_7^{2-}$ removal was low when the pH of wastewater was in the acidic range, and the rejection rate was as high as 98.9% when the pH of the solution was 12. Zhu et al. (2014) attributed the hollow-fiber NF membrane was negatively charged at a high level when the pH of the wastewater solution was high, which might result in the high rejection rate of $Cr_2O_7^{2-}$ during filtration. They also attributed the very narrow-size pore distribution and the amphoteric charge property of the dual-layer hollow-fiber NF membrane resulted in the higher removal rate of heavy metal ions from the wastewater. Their results showed the potential of this simple

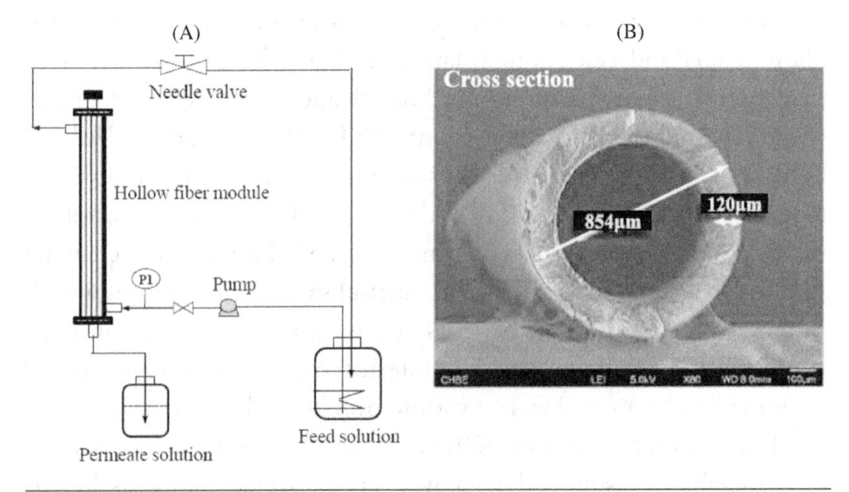

Figure 9.8 (A) Schematic view of wastewater treatment with dual-layer hollow-fiber membrane, and (B) SEM image of cross-sectional view of the hollow-fiber NF membrane. (Reproduced from Zhu, W.P. et al., *J. Memb. Sci.*, 456, 117–127, 2014.)

dual-layer nanofiltering membrane to remove heavy metals from the wastewater of food and beverages processing facilities, which might reduce the environmental pollution from this wastewater.

Photocatalysis

Photocatalysis is degradation of pollutants in the wastewater from the industries (food and beverage, textile, pharmaceutical, agriculture), and it is defined as "change in the rate of a chemical reaction or its initiation under the action of ultraviolet (UV), visible, or infrared radiation in the presence of a substance—the photocatalyst—that absorbs the light and is involved in the chemical transformation of the reaction partners" (McNaught and Wilkinson 1997). In recent times, photocatalysis-based wastewater treatment systems have been tested for purification of water by degradation of most of the pollutants present in the industrial wastewater using a light-activated nanostructured photocatalytic medium (commonly metal oxide semiconductors) (Bora and Dutta 2014). The photocatalytic metal oxide semiconductors absorb the light energy and generate an electron-hole (e-h) pair when the absorbed light energy is higher than its band gap. This photo-generated e-h pair produces highly reacting ions (oxidizing or reducing atoms) like superoxides (O_2^-) and hydroxyl ions ($OH^{\cdot}$) in water when the semiconductor comes in contact with the water. Then these highly reactive components react with the organic and inorganic pollutants in the wastewater and degrade them through secondary reactions. In some cases, photo-generated electrons or the hole directly move from the photocatalytic semiconductor surface to the pollutant molecules and degrade the polluting component in the wastewater (Bora and Dutta 2014).

Photocatalysis is a complex surface reaction phenomenon, and it is explained in the following five simple steps by Pirkanniemi and Sillanpää (2002) and Bora and Dutta (2014) and is schematically shown in Figure 9.9:

1. Diffusion of reactants to the surface of the catalyst
2. Adsorption of the reactants on the surface of the catalyst
3. Reaction at the surface of the catalyst
4. Desorption of the products from the surface of the catalyst
5. Diffusion of the products from the surface of the catalyst

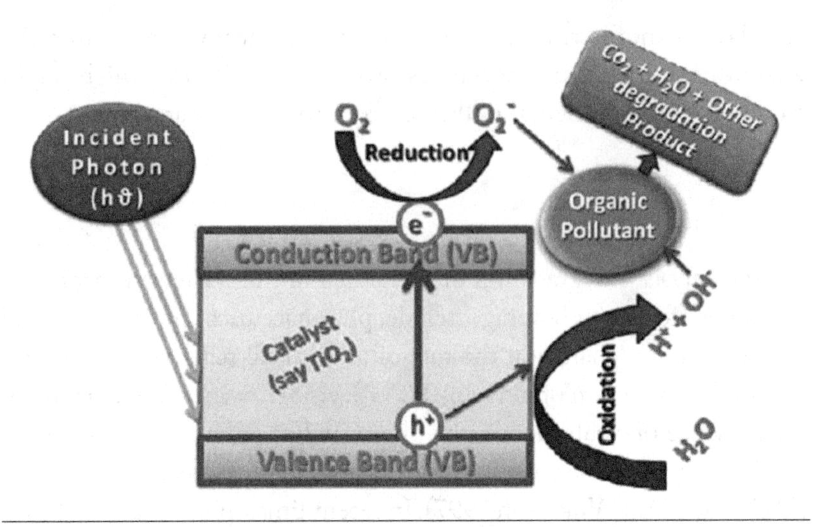

Figure 9.9 Graphical representation of nanotechnology based photocatalysis process. (Reproduced from Singh, P. et al., *Nano Res.*, 2, 1–10, 2016.)

The generation of electron-hole (e-h) pair when the light is absorbed by the photocatalysis semiconductor defines the effectiveness of the photocatalytic process. Since the lifespan of the e-h pair is very short, effective utilization of this free electron and hole for degradation of organic and inorganic pollutants through secondary reactions is important. The nanostructured semiconductor materials are preferred over bulk semiconductors since they have more surface-to-volume ratio, which increases the number of electrons and holes present at the surface of the semiconductor. The most commonly used nanostructured semiconductors for photocatalysis process are ZnO, zinc sulfide (ZnS), cadmium sulfide (CdS), ferric oxide (Fe_2O_3), and TiO_2. Most of these nanostructured semiconductors absorb UV light, but the capital cost is high for the application of a UV light source. Utilization of natural solar radiation and the visible light band of the solar radiation for the photocatalysis process is the current interest among the researchers since the solar radiation contains nearly 46% visible light, which is cost effective when compared with UV-light-based nanostructured photocatalytic semiconductor systems (Bora and Dutta 2014).

The photocatalysis processes using nanostructured ZnO and TiO_2 semiconductors were used to degrade the organic contaminants in the wastewater from industries (textile, pharmaceutical, food and beverages), which degraded the environmentally hazardous organic compounds into carbon dioxide and water (Baruah et al. 2010; Chan et al. 2011; Danwittayakul et al. 2013). The ZnO and TiO_2 semiconductors were also tested for degradation of inorganic pollutants like ammonia, nitrates, and cyanide from the wastewater through a photocatalytic reaction. The ammonia in the wastewater was degraded by the oxidation process initiated through photocatalysis using CdS-Titanate nanotubes (Chen et al. 2011). The nanostructured photocatalytic semiconductors were also tested to remove/degrade heavy metals in the industrial wastewater, which have a significant negative impact on the environment as well as public health if directly released to a body of water or on land. The nanosized TiO_2 nanoparticles have been successfully tested to remove cadmium (Cd) from the wastewater (Skubal et al. 2002), and more than 90% of the Cd from the wastewater was recovered on the TiO_2 nanoparticles. The wastewater from the food and beverages processing facilities also contains harmful bacteria like *Escherichia coli* and *Staphylococcus aureus*, and the ZnO and TiO_2-based nanostructured semiconductors demonstrated great potential to inhibit these bacteria in the wastewater through the photocatalytic process (Baruah et al. 2012a, 2012b; Jaisai et al. 2012; Sapkota et al. 2011). Application of these nanostructured photocatalytic semiconductors in wastewater treatment has been tested for various industries, and some of the techniques applied in the food and beverages industry are discussed in this section.

The polyurethane (PU)-based membrane was developed with silver titanium dioxide ($AgTiO_2$) nanoparticles and $AgTiO_2$ nanofibers to treat the dairy industry wastewater (effluent) using the photocatalysis process (Kanjwal et al. 2015). The PU-based membranes were also used alone for dairy effluent (DE) treatment, but the photocatalytic process separated nanocatalytic elements, which created another secondary pollutant from the wastewater. Kanjwal et al. (2015) prepared membranes with PU, tetrahydrofuran (THF), titanium (IV) isopropoxide, silver nitrate, and N, N-dimethylformamide (DMF) using a sol-gel preparation technique. Next, 4 g of titanium (IV) isopropoxide was mixed with 14% DMF, acetic acid was added drop by drop until the solution turned transparent. Then 0.2 g of silver nitrate was dissolved in 0.8 g of DMF, and this solution was mixed into the above-prepared titanium (IV) isopropoxide and DMF solution. This mixture was stirred for 10 min and then dried for 2 h at 80°C. The $AgTiO_2$ nanoparticles with a diameter of 200 nm were obtained by grinding this dried gel and calcinated for 1 h at 600°C. The $AgTiO_2$ nanofibers were obtained with an average diameter of 180 nm. The $AgTiO_2$ nanoparticles and nanofibers (10% concentration) were mixed with PU dissolved in THF and DMF using electrospinning at room temperature for 1 h. The dairy wastewater from a commercial dairy plant (Purmil Ltd, Cheollabukdo, South Korea) was treated with the photocatalytic membranes of pure PU, PU+$AgTiO_2$ nanoparticles, and PU+$AgTiO_2$ NFs at visible light. Only 10% DE was degraded after 2 h when the PU-alone membrane was used, but the PU+$AgTiO_2$ NFs photocatalytic membrane degraded more than 70% of DE within 5 min, and most of the DE (95%) was degraded after 2 h of treatment. The PU membrane incorporated with $AgTiO_2$ nanoparticles degraded 75% of DE after 2 h. However, this photocatalytic treatment of dairy industry wastewater did not change the pH of the effluent. Kanjwal et al. (2015) attributed the incorporation of $AgTiO_2$ nanoparticles, and $AgTiO_2$ NFs with the polyurethane improved the formation of photo-induced electrons and electron holes when irradiating with visible light, which enhanced the degradation of polluting compounds present in the dairy processing facility wastewater. Kanjwal et al. (2016) developed hybrid nanofibers incorporating TiO_2 and $AgTiO_2$ with silicone coating for photocatalytic degradation of dairy wastewater with high water flux, and they found these hybrid nanofibers degraded more than 60% of the DE, with the water flux and hydrogen production of 12,229 L/m^2 h and 3399 mL/g h, respectively.

The meat industry uses a great amount of water, mainly for sanitizing and cleaning the processing area and machines and the cooling and heating equipment, as well as cleaning carcasses, and the wastewater from the meat processing industry contains microbial compounds like *Bacillus subtilis* and *E. coli*, biodegradable organic wastes (proteins and fats), and chemical residues like bovine albumin, phenyl. All these elements are hazardous to the environment, humans, and animals who consume the water if these are directly released to water or land. Mostafa and Darwich (2014) developed a nanocomposite with copper and silver nanoparticles and chitosan to treat the wastewater from the meat industry. They developed hybrid nanocomposites with polyaniline-grafted amino-modified chitosan immobilized with copper and silver nanoparticles. These developed nanocomposites with the aromatic chitosan-modified amino acid, which grafted with polyaniline, the contained copper and silver nanoparticles generated active-oxygen species by the interaction between the metal nanoparticles, and H_2O_2 completely degraded bovine albumin and tryptophan in the meat industry wastewater. These hybrid nanocomposites also inhibited all (100%) of the foodborne pathogens (*B. subtilis* and *E. coli*) present in the wastewater from the meat processing industry (Mostafa and Darwich 2014).

Nanosorbents

Sorption is a process in which the substance called sorbent absorbs another substance called sorbate through a physical or chemical reaction. In wastewater treatment, sorbents are commonly used as a separation or purification media to absorb the organic and inorganic contaminants from the water. First, the contaminants transport from the wastewater to the sorbent surface and absorbed at the surface of the sorbent media, and finally move inside the sorbent by which the contaminants are removed from the polluted wastewater (Bora and Dutta 2014). Nanoparticle-based sorbents have an advantage of a higher surface-to-volume ratio than the normal bulk sorbents, which helps to improve the efficacy of the wastewater treatment process. The nanosize pores of the nanosorbents also improve the sorption capacity of the sorbents, and most of the nanosorbents are reusable after removal of absorbed contaminants through heating or rinsing of the sorbent media. Carbon-based nanomaterials (activated carbon nanoparticles, CNTs), magnetic nanoparticles, metal oxide-based nanomaterials (TiO_2, alumina, tungsten oxide, iron oxide), and silica are the most commonly used nanosorbents for water and wastewater treatment (Çeçen and Aktaş 2011; Hu et al. 2010; Kuo et al. 2008; Li et al. 2007; Luo et al. 2011, Zhang et al. 2010, 2013). The organic dyes from the textile industry wastewater were successfully removed by using tungsten oxide nanosorbents, and the ferrous-oxide-based nanosorbents demonstrated greater absorption potential for heavy metal ions present in the industry wastewater (Hu et al. 2010; Iram et al. 2010; Jeon and Yong 2010; Nassar 2010). The magnetic nanoparticles and ferrous-oxide-based nanosorbents are easily separated from the water (or aqueous solution) after absorbing pollutants in the wastewater using a permanent magnet.

The olive milling industry produces a large amount of wastewater from the milling process, as well as cooling and heating operations. This wastewater is highly phytotoxic due to the residues of production operations, and has higher bacterial toxicity due to the microbial load in the raw materials as well as processing steps. Removing the large quantities of these organic and suspended solids in the olive mill wastewater using conventional techniques like membrane filtration, centrifugation, oxidation, and natural evaporation could not reduce the toxicity level to an acceptable limit (Paraskeva and Diamadopoulos 2006). Researchers found the combination of UF, nanofiltration, and reverse osmosis operations nearly reduced 80% of the toxic substances in the olive mill wastewater, but the cost of operation was high (Niaounakis and Halvadakis 2006; Paraskeva et al. 2007). Integrating anaerobic fermentation as pretreatment of olive milling wastewater with suitable pretreatment techniques like membrane filtration or sedimentation to remove suspended solids had the ability to improve the efficacy of the wastewater treatment.

The wine industry produces a large amount of wastewater from the preprocessing of raw materials (washing of fruits, cleaning of all raw materials), processing operations (crushing, pressing, bottling, and packaging), and processing facility cleaning and sanitation operations. This wastewater contains large amounts of phenols and other toxic substances, which pollute the air, land, and water if released from the wineries without any treatment. The conventional biological treatments (aerobic and anaerobic) of winery wastewater treatment do not reduce toxic levels to regulatory limits (Monge and Moreno-Arribas 2016). Rytwo et al. (2013) developed a technique for the olive mill and winery wastewater treatment using nanocomposites of polymer and nanoclay. They used Poly-DADMAC polymer and sepiolite clay mineral to make the hybrid nanocomposite material by mixing 50 g of sepiolite with a 50 mL of poly-DADMAC polymer solution (1 g of polymer/100 mL warm water) using an ultrasonicator for 2 h.

Rytwo et al. (2013) also made chitosan–nanoclay composite, as well as a polymer–Volclay composite using the similar technique and tested their efficacy for clearing wastewaters obtained from commercial olive milling facility (Kadoori Ltd. Olive Mill, Lower Galilee, Israel) and a commercial winery (Galil Mountain Winery, Yiron, Israel). The sedimentation rate and profiles of all the treatments were collected using a dispersion analyzer (Model: LUMiSizer 6110, L.U.M. GmbH, Germany) by analyzing the light intensity through the solution. To increase the cost benefit and effectiveness of the wastewater treatment, Rytwo et al. (2013) also sequentially added nanocomposite material with wastewater (adding 20 mL of nanocomposite to 1000 mL effluent, centrifuge for 5 min, remove clear water after centrifugation, add 900 mL of raw effluents with the remaining sludge after certification, then add 4 mL of nanocomposite and centrifuge again for 5 min). The dispersion analysis results of olive mill wastewater showed that the nanocomposite with polymer and sepiolite with a concentration of 970 mg polymer/g of sepiolite gave a light intensity over 60% after 5 min centrifugation at a relative centrifugal force (RCF) of 32.8 g (whereas the light intensities of distilled water and raw olive mill wastewater were 93% and 10%, respectively). The nanocomposite with chitosan and sepiolite had more than 60% light intensity after 2 min of centrifugation when the chitosan concentration was 490 and 650 mg/g clay. Their results also showed that the nanocomposites with lower polymer concentration in clay had lower clarification rate, and the high polymer (or chitosan) with clay also had lower light intensity. When treating the winery wastewater with the polymer–sepiolite (concentration 50 mg polymer/g of clay), chitosan–sepiolite (60 g chitosan/g of clay), and polymer–Volclay (130 g polymer/g of Volclay), light intensity was more than 70% after 3 min of centrifugation at a RCF of 5.2 g (light intensity of raw winery wastewater was 22%). The turbidity removal from the wastewater was used to test the efficacy of sequential nanocomposite addition technique. The turbidity levels of raw olive mill wastewater and winery wastewaters were 1570 and 754 NTUs, respectively, and the sequential addition of polymer–sepiolite nanocomposite removed more than 90% of turbidity level after 6 cycles. Rytwo et al. (2013) found around 350 g of polymer–sepiolite nanocomposite was used to clarify 1 m^3 of effluent from the olive milling and wine-making facilities, and they estimated this sequential addition of nanocomposite for sedimentation of organic contaminants and soluble solids in the olive milling and winery wastewater cost 1.1 and 2.6 Euros per m^3 of effluents, respectively.

Conclusion

The wastewater treatment technologies commonly used in food and beverages manufacturing facilities have the ability to control the physical, chemical, and biological contaminants in the wastewater to a certain level but cannot eliminate all of the contaminants. Normally, water treated using the conventional mechanical and biological wastewater treatment methods is not reusable since the contaminant level in the treated water is higher than the legal limit. These techniques are also not cost effective due to the high initial investment, continuing operating cost and low recovery of valuable nutrients and other materials from the retentate. Nanotechnology-based wastewater treatment methods show higher efficiency for purifying wastewater for secondary industrial applications, as well as the higher amount of nutritional and economically beneficial materials recovery from the retentate from lab scale and from large industry scale. The nanotechnology-based nanofiltration technique using synthetic and nanocomposite-based membranes with nanoscale pores helped to recover most of the compounds from the food industry wastewater and at the same time gave lower membrane fouling when compared with microfiltration or UF methods. The photocatalysis using nanomaterials with the help of natural solar radiation can reduce the cost of operation, and the researchers found the high reactivity and sensitivity of the photocatalysis nanomaterials resulted in higher efficiency of the wastewater treatment systems. The higher surface-volume ratio and higher absorption ability of nanosorbents improve the removal of hazardous pollutants from the wastewater, and these nanosorbent materials can be used again for wastewater treatment after a simple and minimal thermal or physical process. Some of the nanotechnology-based wastewater treatment techniques have been incorporated along with conventional wastewater treatment for either primary or secondary processing steps to completely remove the pollutants from the wastewater and purify the water to be suitable for reuse in the manufacturing facility operations. The effectiveness of the nanotechnology-based wastewater treatment technologies was high in removing the organic and inorganic contaminants, toxic substances, heavy metal particles, and the pathogenic microbial cultures found in the food and beverages wastewater. Newer technologies based on nanoscience and nanotechnology are evolving every day due to the continuous research on improvement in process and cost effectiveness. The studies also found the treated water from nanotechnology-based wastewater treatment processes can be reused in the secondary processing operations (like cleaning and sanitation of nonfood contact areas of processing units and the facility, cooling process, steam generation for heating), which will reduce the water requirement for the food processing industry, resulting in improved cost effectiveness. The consumer acceptance and socioeconomic image of the food and beverages industry can be improved by applying nanotechnology-based wastewater treatment methods, because the treated water can be used for secondary processes, and most of the contaminants, which cause damage to the land, water, and air, can be removed. The recent and continuous advancements in the nanoscience and nanotechnology-based wastewater treatment techniques will make these units an essential part of the food and beverages industry in the near future.

References

Andrade L., Mendes F., Espindola J., Amaral M. (2014) Nanofiltration as tertiary treatment for the reuse of dairy wastewater treated by membrane bioreactor. *Separation and Purification Technology* 126:21–29.

Andrade L., Mendes F., Espindola J., Amaral M. (2015) Reuse of dairy wastewater treated by membrane bioreactor and nanofiltration: Technical and economic feasibility. *Brazilian Journal of Chemical Engineering* 32:735–747.

Aroon M., Ismail A., Matsuura T., Montazer-Rahmati M. (2010) Performance studies of mixed matrix membranes for gas separation: A review. *Separation and Purification Technology* 75:229–242.

Baker R.W. (2004) *Membrane Technology and Applications.* John Wiley & Sons, London, UK. pp. 96–103.

Baruah S., Jaisai M., Dutta J. (2012a) Development of a visible light active photocatalytic portable water purification unit using ZnO nanorods. *Catalysis Science and Technology* 2:918–921.

Baruah S., Jaisai M., Imani R., Nazhad M.M., Dutta J. (2010) Photocatalytic paper using zinc oxide nanorods. *Science and Technology of Advanced Materials* 11:055002.

Baruah S., Pal S.K., Dutta J. (2012b) Nanostructured zinc oxide for water treatment. *Nanoscience and Nanotechnology-Asia* 2:90–102.

Bora T., Dutta J. (2014) Applications of nanotechnology in wastewater treatment—A review. *Journal of Nanoscience and Nanotechnology* 14:613–626.

Çeçen F., Aktaş Ö. (2011) Water and wastewater treatment: Historical perspective of activated carbon adsorption and its integration with biological processes. In: *Activated Carbon for Water and Wastewater Treatment: Integration of Adsorption and Biological Treatment*. John Wiley & Sons, New York. pp. 1–11.

Chan S.H.S., Wu T.Y., Juan J.C., Teh C.Y. (2011) Recent developments of metal oxide semiconductors as photocatalysts in advanced oxidation processes (AOPs) for treatment of dye waste-water. *Journal of Chemical Technology and Biotechnology* 86:1130–1158.

Chen Y.C., Lo S.L., Ou H.H., Chen C.H. (2011) Photocatalytic oxidation of ammonia by cadmium sulfide/titanate nanotubes synthesised by microwave hydrothermal method. *Water Science and Technology* 63:550–557.

Chen Z., Luo J., Chen X., Hang X., Shen F., Wan Y. (2016) Fully recycling dairy wastewater by an integrated isoelectric precipitation–nanofiltration–anaerobic fermentation process. *Chemical Engineering Journal* 283:476–485.

Cloete T.E., De Kwaadsteniet M., Botes M. (2010) *Nanotechnology in water treatment applications*. Caister Academic Press, Nortfort, UK: 1–195.

Cuartas-Uribe B., Vincent-Vela M., Alvarez-Blanco S., Alcaina-Miranda M., Soriano-Costa E. (2010) Application of nanofiltration models for the prediction of lactose retention using three modes of operation. *Journal of Food Engineering* 99:373–376.

Danwittayakul S., Jaisai M., Koottatep T., Dutta J. (2013) Enhancement of photocatalytic degradation of methyl orange by supported zinc oxide nanorods/zinc stannate (ZnO/ZTO) on porous substrates. *Industrial & Engineering Chemistry Research* 52:13629–13636.

Falk K., Sedlmeier F., Joly L., Netz R.R., Bocquet L. (2010) Molecular origin of fast water transport in carbon nanotube membranes: Superlubricity versus curvature dependent friction. *Nano Letters* 10:4067–4073.

Fernández P., Riera F.A., Álvarez R., Álvarez S. (2010) Nanofiltration regeneration of contaminated single-phase detergents used in the dairy industry. *Journal of Food Engineering* 97:319–328.

Helmer R., Hespanhol I. (2002) *Technology Selection, Water Pollution Control: A Guide to the Use of Water Quality Management Principles*. CRC Press, Taylor & Francis Group, London, UK. pp. 45–87.

Hong S.U., Miller M.D., Bruening M.L. (2006) Removal of dyes, sugars, and amino acids from NaCl solutions using multilayer polyelectrolyte nanofiltration membranes. *Industrial & Engineering Chemistry Research* 45:6284–6288.

Hornyak G.L., Dutta J., Tibbals H.F., Rao A. (2008) *Introduction to Nanoscience*. CRC Press, Taylor & Francis Group, New York. pp. 59–104.

Hu H., Wang Z., Pan L. (2010) Synthesis of monodisperse Fe_3O_4@ silica core–shell microspheres and their application for removal of heavy metal ions from water. *Journal of Alloys and Compounds* 492:656–661.

Iram M., Guo C., Guan Y., Ishfaq A., Liu H. (2010) Adsorption and magnetic removal of neutral red dye from aqueous solution using Fe_3O_4 hollow nanospheres. *Journal of Hazardous Materials* 181:1039–1050.

Jaisai M., Baruah S., Dutta J. (2012) Paper modified with ZnO nanorods—Antimicrobial studies. *Beilstein Journal of Nanotechnology* 3:684–691

Jeon S., Yong K. (2010) Morphology-controlled synthesis of highly adsorptive tungsten oxide nanostructures and their application to water treatment. *Journal of Materials Chemistry* 20:10146–10151.

Kanjwal M.A., Alm M., Thomsen P., Barakat N.A., Chronakis I.S. (2016) Hybrid matrices of TiO_2 and TiO_2-Ag nanofibers with silicone for high water flux photocatalytic degradation of dairy effluent. *Journal of Industrial and Engineering Chemistry* 33:142–149.

Kanjwal M.A., Barakat N.A., Chronakis I.S. (2015) Photocatalytic degradation of dairy effluent using $AgTiO_2$ nanostructures/polyurethane nanofiber membrane. *Ceramics International* 41:9615–9621.

Karan S., Samitsu S., Peng X., Kurashima K., Ichinose I. (2012) Ultrafast viscous permeation of organic solvents through diamond-like carbon nanosheets. *Science* 335:444–447.

Kuo C.Y., Wu C.H., Wu J.Y. (2008) Adsorption of direct dyes from aqueous solutions by carbon nanotubes: Determination of equilibrium, kinetics and thermodynamics parameters. *Journal of Colloid and Interface Science* 327:308–315.

Lau W., Ismail A., Misdan N., Kassim M. (2012) A recent progress in thin film composite membrane: A review. *Desalination* 287:190–199.

Li Y., Zhao Y., Hu W., Ahmad I., Zhu Y., Peng X., Luan Z. (2007) Carbon nanotubes—The promising adsorbent in wastewater treatment. In: *Journal of Physics: Conference Series*. IOP Publishing, Bristol, UK. pp. 698–702.

Lipnizki F. (2010) *Membrane Technology, Volume 3: Membranes for Food Applications*. Wiley-Vch Verlag GmbH & Co. KGaA, Weinheim, Germany.

Liu T., Li B., Hao Y., Yao Z. (2014) MoO_3-Nanowire membrane and $Bi_2Mo_3O_{12}/MoO_3$ nano-heterostructural photocatalyst for wastewater treatment. *Chemical Engineering Journal* 244:382–390. doi:10.1016/j.cej.2014.01.070.

Luo J., Ding L. (2011) Influence of pH on treatment of dairy wastewater by nanofiltration using shear-enhanced filtration system. *Desalination* 278:150–156.

Luo J., Ding L., Wan Y., Jaffrin M.Y. (2012a) Threshold flux for shear-enhanced nanofiltration: Experimental observation in dairy wastewater treatment. *Journal of Membrane Science* 409:276–284.

Luo J., Ding L., Wan Y., Paullier P., Jaffrin M.Y. (2010) Application of NF-RDM (nanofiltration rotating disk membrane) module under extreme hydraulic conditions for the treatment of dairy wastewater. *Chemical Engineering Journal* 163:307–316.

Luo J., Ding L., Wan Y., Paullier P., Jaffrin M.Y. (2012b) Fouling behavior of dairy wastewater treatment by nanofiltration under shear-enhanced extreme hydraulic conditions. *Separation and Purification Technology* 88:79–86.

Madaeni S., Yasemi M., Delpisheh A. (2011) Milk sterilization using membranes. *Journal of Food Process Engineering* 34:1071–1085.

McNaught A.D., McNaught A.D. (1997) *Compendium of Chemical Terminology*. Blackwell Science, Oxford, UK.

Monge M., Moreno-Arribas M.V. (2016) Applications of nanotechnology in wine production and quality and safety control. In *Wine Safety, Consumer Preference, and Human Health*, Springer, Cham, Switzerland: 51–69.

Mostafa T.B., Darwish A.S. (2014) An approach toward construction of tuned chitosan/polyaniline/metal hybrid nanocomposites for treatment of meat industry wastewater. *Chemical Engineering Journal* 243:326–339. doi:10.1016/j.cej.2014.01.006.

Nassar N.N. (2010) Rapid removal and recovery of Pb (II) from wastewater by magnetic nanoadsorbents. *Journal of Hazardous Materials* 184:538–546.

Niaounakis M., Halvadakis C.P. (2006) Physical Processes. In *Olive Processing Waste Management: Literature Review and Patent Survey*, 2nd edn. Elsevier, Amsterdam, the Netherlands: 107–113.

Noy A., Park H.G., Fornasiero F., Holt J.K., Grigoropoulos C.P., Bakajin O. (2007) Nanofluidics in carbon nanotubes. *Nano Today* 2:22–29.

Paraskeva C., Papadakis V., Tsarouchi E., Kanellopoulou D., Koutsoukos P. (2007) Membrane processing for olive mill wastewater fractionation. *Desalination* 213:218–229.

Paraskeva P., Diamadopoulos E. (2006) Technologies for olive mill wastewater (OMW) treatment: A review. *Journal of Chemical Technology and Biotechnology* 81:1475–1485.

Parham H., Bates S., Xia Y., Zhu Y. (2013) A highly efficient and versatile carbon nanotube/ceramic composite filter. *Carbon* 54:215–223. doi:10.1016/j.carbon.2012.11.032.

Pirkanniemi K., Sillanpää M. (2002) Heterogeneous water phase catalysis as an environmental application: A review. *Chemosphere* 48:1047–1060.

Ren X., Zhao C., Du S., Wang T., Luan Z., Wang J., Hou D. (2010) Fabrication of asymmetric poly (m-phenylene isophthalamide) nanofiltration membrane for chromium (VI) removal. *Journal of Environmental Sciences* 22:1335–1341.

Rytwo G., Lavi R., Rytwo Y., Monchase H., Dultz S., König T.N. (2013) Clarification of olive mill and winery wastewater by means of clay–polymer nanocomposites. *Science of the Total Environment* 442:134–142.

Salehi F. (2014) Current and future applications for nanofiltration technology in the food processing. *Food and Bioproducts Processing* 92:161–177.

Salehi F., Razavi S.M., Elahi M. (2011) Purifying anion exchange resin regeneration effluent using polyamide nanofiltration membrane. *Desalination* 278(1–3):31–35.

Singh P., Abdullah M.M., Ikram S. (2016) Role of nanomaterials and their applications as photo-catalyst and sensors: A review. *Nano Research & Applications* 2:1–10.

Skubal L., Meshkov N., Rajh T., Thurnauer M. (2002) Cadmium removal from water using thiolactic acid-modified titanium dioxide nanoparticles. *Journal of Photochemistry and Photobiology A: Chemistry* 148:393–397.

StatCanada. (2011) *Industrial Water Use 2011.* Statistics Canada, Ottawa, ON. Available from http://www.statcan.gc.ca/pub/16-401-x/16-401-x2014001-eng.pdf.

Thanuttamavong M., Oh J., Yamamoto K., Urase T. (2001) Comparison between rejection characteristics of natural organic matter and inorganic salts in ultra-low-pressure nanofiltration for drinking water production. *Water Science and Technology: Water Supply* 1:77–90.

Thanuttamavong M., Yamamoto K., Oh J.I., Choo K.H., Choi S.J. (2002) Rejection characteristics of organic and inorganic pollutants by ultra-low-pressure nanofiltration of surface water for drinking water treatment. *Desalination* 145:257–264.

Van der Bruggen B., Mänttäri M., Nyström M. (2008) Drawbacks of applying nanofiltration and how to avoid them: A review. *Separation and Purification Technology* 63:251–263.

Vandanjon L., Cros S., Jaouen P., Quéméneur F., Bourseau P. (2002) Recovery by nanofiltration and reverse osmosis of marine flavours from seafood cooking waters. *Desalination* 144:379–385.

Wan R., Lu H., Li J., Bao J., Hu J., Fang H. (2009) Concerted orientation induced unidirectional water transport through nanochannels. *Physical Chemistry Chemical Physics* 11:9898–9902.

Wechsler L. (2015) Water usage in food processing. *OMEP Newsletter*, Oregon Manufacturing Extension Partnership, Portland, OR. Available from https://www.omep.org/water-usage-in-food-processing/.

Zhang S., Niu H., Hu Z., Cai Y., Shi Y. (2010) Preparation of carbon-coated Fe_3O_4 nanoparticles and their application for solid-phase extraction of polycyclic aromatic hydrocarbons from environmental water samples. *Journal of Chromatography A* 1217:4757–4764.

Zhang S., Xu W., Zeng M., Li J., Li J., Xu J., Wang X. (2013) Superior adsorption capacity of hierarchical iron oxide@ magnesium silicate magnetic nanorods for fast removal of organic pollutants from aqueous solution. *Journal of Materials Chemistry A* 1:11691–11697.

Zhu W.P., Sun S.P., Gao J., Fu F.J., Chung T.S. (2014) Dual-layer polybenzimidazole/polyethersulfone (PBI/PES) nanofiltration (NF) hollow-fiber membranes for heavy metals removal from wastewater. *Journal of Membrane Science* 456:117–127.

10

Safety Considerations, Consumer Acceptance, and Regulatory Framework

Introduction

New technologies are introduced in all manufacturing sectors each and every day due to the growing research and development in many fields of science and engineering. Nanotechnology is one of the major emerging technologies in recent times, and the applications of nanoscience and nanotechnology in all manufacturing sectors are growing faster day by day. Nanotechnology-based food products and processes have been developed in order to: improve the physicochemical, organoleptic, and nutritional properties of food products, increase the nutritional availability, increase the shelf life of food products, and improve the quality of food products. Nanotechnology-based materials and processes help the food and beverages industry to improve the functionalities of the food and also to reduce the use of chemical substances since the higher surface-to-mass ratio nature of nanomaterials provides similar kinds of functionalities as the large amount of regular-size materials (Amenta et al. 2015; Chaudhry et al. 2010). The food and beverages industry is facing a major obstacle in winning the consumers' acceptance of the nanotechnology-based food and beverage products since the application of nanotechnology in food and beverages is directly related to the consumers' health due to direct consumption of these products. The higher surface-area-to-mass ratio of nanomaterials is also playing a negative role in public perception about nanotechnology-based food products because the higher ratio may increase the absorbance of the nanomaterials through the organ tissues, which may cause cell damage, and consumer may receive overdose of the nutrients due to the increased bioavailability (Amenta et al. 2015; Couch et al. 2016; Martirosyan and Schneider 2014; NIOSH 2009).

The recent developments in communication technologies and social media have increased the consumer awareness about the food they are taking, and the food and beverages industry is compelled to satisfy their concerns. A survey among the European Union (EU) general public by the German Federal Institute of Risk Assessment showed the more acceptance towards the application of nanotechnology in non-food products and more reluctance to accept the use of nanotechnology-based products and processes in food or agricultural operations (Chaudhry et al. 2010; Coles and Frewer 2013; Kumari and Yadav 2014). Chaudhry et al. (2010) compared this negative perspective about the application of nanotechnology in food to the negative view shown by the European public about the introduction of genetically modified crops and products and the food irradiation process in the past. They suggested proper demonstration of benefits of nanotechnology-based products to consumers and addressing the safety concerns and health risks might help to win the public acceptance.

Around the world, most of the countries are working towards the regulatory laws and guidelines for the proper use of nanotechnology-based materials and processes in food and beverages industry, but there is no specific single harmonized law available about regulating the application of nanotechnology in all the aspects of food processing operations. The European Union Commission developed some new regulations and altered some other existing regulatory guidelines for safety evaluation, risk assessment, and authorization of nanotechnology-based food products (Amenta et al. 2015; Coles and Frewer 2013). To date, the EU is the only region that mandates to display the food ingredient with the term "nano" in the ingredients list on a food label if any of the ingredients is used in nanosize in the manufacturing process. The USA and Canada have a voluntary preapproval consulting process with the food industries about the use of nanotechnology-based products and processes in food (Amenta et al. 2015). Australia and New Zealand did a safety assessment of commonly used nano-based products like titanium dioxide (TiO_2), silver (Ag), silica, and published a toxicology datasheet of those products to improve consumer awareness (Amenta et al. 2015; Drew and Hagen 2016). Most of the

Asian, African, and South American countries are regulating the use of nanotechnology in the food and beverages industry using available food safety regulations and working towards nanotechnology-specific regulations (Amenta et al. 2015; Gupta et al. 2012b). The safety concerns about nanotechnology, perspectives of consumers, government and other regulators, industry and other stakeholders of the food and beverages sector, and the current regulations regarding nanotechnology-based applications with respect to food and beverages industry are discussed in this chapter.

Safety Concerns of Nanotechnology-Based Products and Processes in Food and Beverages

Titanium dioxide is one of the food additives in use for a long time to impart color in food, and it is approved as a food additive in most parts of the world. The Food and Agriculture Organization (FAO) and World Health Organization's (WHO) Codex Alimentarius and the USA allow up to 1% w/w concentration of TiO_2 in food without mentioning it in the food label's ingredients list (FDA 2015; Larsen et al. 2008). Australia and the EU had also approved TiO_2 as a food-coloring agent (E171) for decades (European Commission 2004). The use of food ingredients and additives in nanoparticle (NP) form is increasing in the food and beverages industry in recent times due to advantages like an increase in nutritional bioavailability and improved textural and sensory properties of food while using nanoform materials. The regulations for use of TiO_2 do not have any restrictions about the particle size of TiO_2, and most of the foods containing TiO_2 have TiO_2 in nanoform (Jovanović 2015). A study by Weir et al. (2012) found nearly 36% of the TiO_2 in food in nanoscale, and Chen et al. (2013) found that chewing gum from six different manufacturers contained TiO_2 in nanoscale level in the range of 2.4–7.5 mg/piece of chewing gum. That the surface-to-volume ratio of the NP might improve the absorption of the NPs into the human body is one of the major concerns among consumers and researchers, and toxicology of nanosize TiO_2 particles in food has not been studied extensively. Jones et al. (2015) tested the absorption of TiO_2 NPs in the human gastrointestinal (GI) tract using in vivo and in vitro studies and compared this with the absorption of microsize TiO_2 particles in the GI tract. The commercially available TiO_2 in 3 different particle sizes (15 nm, <100 nm, and <5 μm) were used for this study, and in a human in vivo study, TiO_2 absorption was analyzed by checking the TiO_2 level in urine and blood samples collected before and after giving TiO_2 oral doses at a level of 5 mg/kg of body weight from 9 volunteers. The urine samples were collected from each volunteer 24 h before giving TiO_2 dose and 72 h after TiO_2 dose. Blood samples were collected just before the TiO_2 dose, and 2, 4, 24, and 48 h after the TiO_2 dose. For the in vitro study, the TiO_2 samples (all three sizes) were mixed with artificially made (mix of pepsin with saline solution) simulated gastric fluid (SGF) at a rate of 5 mg/mL and incubated at 37°C. The SGF was analyzed after 1 h of incubation, and a control sample without incubation was also analyzed for comparison. The penetration ability of TiO_2 NPs across the human gut barrier was analyzed with the help of a simulated gut cell wall lining made of colorectal adenocarcinoma (Caco-2) cell line. The Caco-2 cell layers were incubated at 37°C with the solution containing TiO_2 particles (all 3 sizes, 50 and 100 μg/mL) for 3 h, and the transepithelial resistance value was measured before and after TiO_2 application. The particle analysis of incubated TiO_2 particles with SGF showed all three sizes (15 nm, <100 nm, and <5 μm) of TiO_2 agglomerated within an hour of ingestion, which showed the TiO_2 particles agglomerate inside the stomach within a short time after intake into the human body irrespective of the particle size of TiO_2. The penetration across the gut barrier analysis results showed only <0.05% of the applied TiO_2 dose was observed in the basal chamber of Caco-2 cell layer. The in vivo analysis with human volunteers showed no significant differences in TiO_2 level in serum and blood samples after giving TiO_2 oral dose with all 3 sizes of TiO_2 particles, but the TiO_2 level in urine samples from 100 nm and 5 μm size TiO_2 doses were significantly higher than the 15 nm size TiO_2 dose (Jones et al. 2015). They found the pre-dose TiO_2 level of 15 nm treatment was also lower, which might be the reason for lower TiO_2 levels in post-dose analysis and attributed that the seasonal variation might be the reason for this difference in TiO_2 levels between 15 nm size TiO_2 dose and the other two treatments (100 nm and 5 μm). The results of the persistence and absorption analysis study showed that the TiO_2 particles agglomerate quickly in the stomach after intake, and these did not get absorbed in the human body or penetrated into the gut barrier after entering the human body (Jones et al. 2015). But the TiO_2 used for this study was not a food-grade one, and the rate and extent of agglomeration of TiO_2 particles might be affected by various parameters of the human digestion system, which were not considered in this study. Long-time effects and longer duration of TiO_2 exposure were not studied.

The biodistribution and the uptake of nano and microsize TiO_2 particles were tested using in vitro study carried out using simulated gut epithelium, and in vivo study carried out using rats (MacNicoll et al. 2015). The TiO_2 at 6 different particle sizes (15 nm, 25 nm, 40 nm, 50 nm, 120 nm, and 5 μm) were used for oral doses (5 mg/kg of body weight) for 8-weeks-old rats, and the urine, blood, and feces samples were collected before and after (2, 24, 48, 72, and 96 h) administering oral doses for the in vivo study. The analysis of urine and blood samples from the rats fed with TiO_2 oral dose showed no significant change in TiO_2 levels in urine and blood before and up to 96 h after giving a TiO_2 dose. All of the treatments showed a higher amount of TiO_2 in feces samples collected from the rats, and the excretion rate of TiO_2 in feces varied between rats. The GI tract collected from the rats after the test had traces of TiO_2 on the GI tract, but there was no absorption of TiO_2 found in the GI tract cells. The in vitro analysis showed there was no translocation of TiO_2 across the Caco-2 gut epithelium layers, but the increase in TiO_2 concentration decreased the Caco-2 cell viability irrespective of TiO_2 particle size (Figure 10.1) (MacNicoll et al. 2015). The TiO_2 traces were even found in feces samples of the rats collected four days after oral intake, which showed the slower phase of TiO_2 removal from the biological surfaces. MacNicoll et al. (2015) attributed some TiO_2 might enter into intracellular surfaces, and more studies were needed to check the amount of intracellular TiO_2 concentration in order to examine the overall effect of TiO_2 in the GI tract.

Sang et al. (2013) measured the effect of nano-TiO_2 intake for a longer duration on the health of mice using an in vivo study. The nano-TiO_2 food-grade particles with a particle size of 5–6 nm were administered to mice at a rate of 10 mg/kg body weight of mice per day for 15, 30, 45, 60, 75, and 90 days, and then mice were anesthetized using the ether solution. The spleen was removed from the mice and kept in the freezer until analysis. The TiO_2 content in the spleen

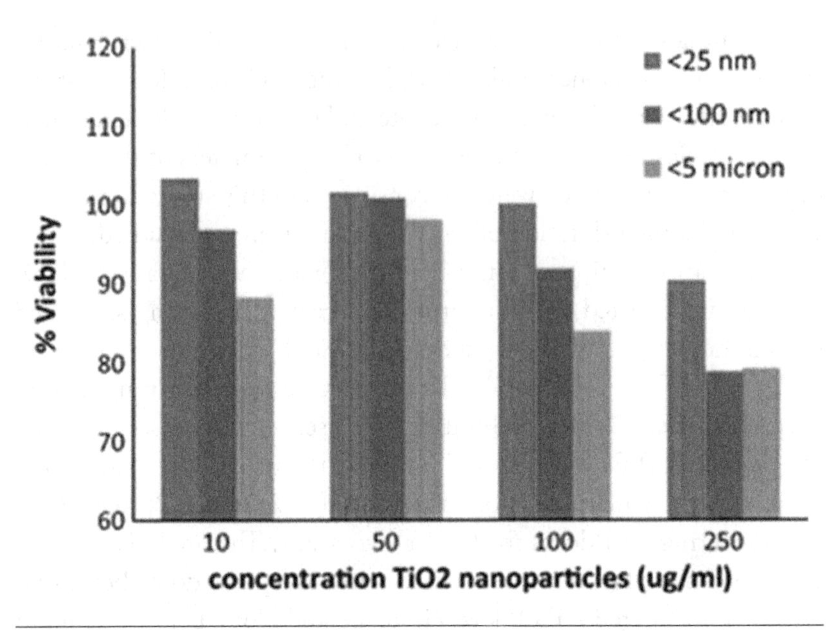

Figure 10.1 The viability of Caco-2 cells after 24 h exposure at different concentrations of TiO_2 nanoparticles. (Reproduced from MacNicoll, A. et al., *J. Nanopart. Res.*, 17, 1–20, 2015.)

was measured using inductively coupled plasma-mass spectrometry (ICP-MS) (Model: Thermo Elemental X7; Thermo Electron Co., Beverly, MA, USA), and the histopathological changes in the spleen of mice were examined using an optical microscope (Model: U-III Multipoint Sensor System, Nikon, Japan). The body weight analysis of mice showed the mice fed with nano-TiO_2 for more than 45 days had significantly lower body weight than the control samples. The spleen weight and the TiO_2 level in the spleen were increased with the nano-TiO_2 application time, and the spleen of the control samples did not have any traces of TiO_2 (Sang et al. 2013). The reactive oxygen species generation rate of the mice spleen also increased with the increase in TiO_2 exposure time. The histopathological analysis of the spleen showed severe changes, like macrophage infiltration, cell necrosis, lymphocyte proliferation, and fatty degeneration for the mice administered with nano-TiO_2 for longer duration, and the severity of the damages increased with an increase in TiO_2 exposure time (Figure 10.2). Sang et al. (2013) found the cyclooxygenase-2 enzyme (involved with inflammation of cells or organs) level in the mice spleen increased (12.64%–64.06%) with an increase in nano-TiO_2 exposure duration. These results showed the longer the nano-TiO_2 exposure,

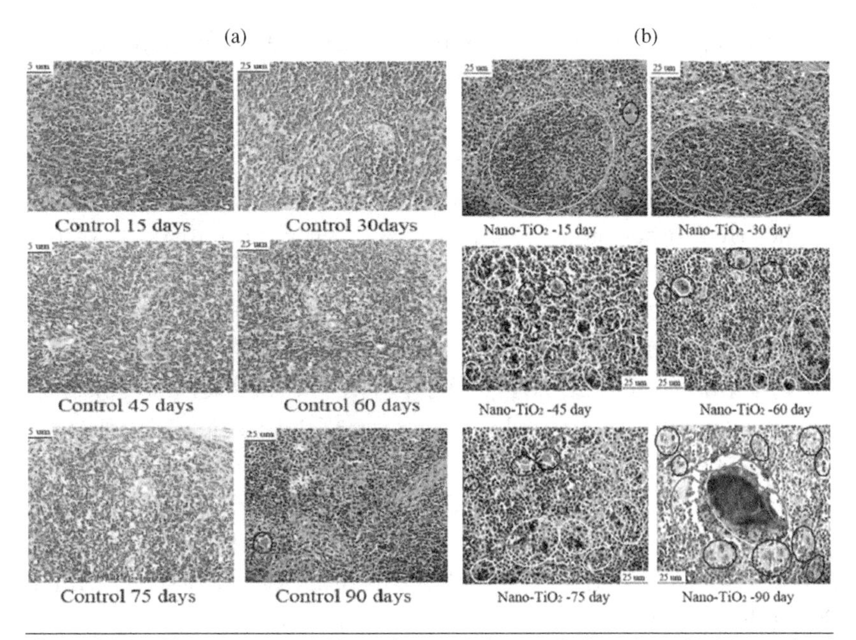

Figure 10.2 **(See color insert.)** Histopathological changes of mice spleen with (a) control and (b) nano-TiO_2 exposure for different durations (Black circle: macrophage infiltration, green circle lymphocyte proliferation; yellow circle: nano-TiO_2 aggregation; red circle: fat degeneration and cell necrosis). (Reproduced from Sang, X. et al., *J. Agricult. Food Chem.*, 61, 5590–5599, 2013.)

the more severe the histopathological changes due to the aggregation of TiO_2 in the spleen. Sang et al. (2013) attributed the higher surface-to-volume ratio of TiO_2 NPs in the food additive might cause the higher absorption of TiO_2 and higher penetration into cells, which might lead to severe histopathological changes under prolonged exposure to NPs, which may not be the case with the use of larger particle-size TiO_2 food additives.

The in vivo studies with the mice showed the distribution of titanium (Ti) particles in various organs of the mice (kidney, lung, liver, spleen, and heart) when the nano-TiO_2 particles were administered through intraperitoneal injection (IP) or intravenous (IV) injection in a single dose or daily administration for two weeks (Geraets et al. 2014; Li et al. 2010; Shinohara et al. 2014; Umbreit et al. 2012; Xie et al. 2011; Yamashita et al. 2011). Shinohara et al. (2014) mixed 2 g of nano-TiO_2 particles (average size of 23 nm) with 50 mL of disodium phosphate solution and administered it to mice at a rate of 0.93 mg/kg body weight in a single dose through IV injection. The analysis of blood samples collected after 5 and 15 min of IV injection from the mice showed the Ti levels of 420 and 19 ng/mL (2.80% and 0.13% of the given dose), which showed the rapid decrease of Ti levels in the blood right after administration, but the Ti level remained around 4 ng/mL after 4 h of administration. The Ti level in liver tissues was higher than other organ tissues after 6 h of IV administration (liver had 94% of administered dose). The Ti found in the spleen, lung, blood, kidney, and heart after 6 h of IV administration were 2.00%, 0.17%, 0.03%, 0.02%, and 0.01% of the given nano-TiO_2 dose, respectively (Shinohara et al. 2014). The Ti level in the liver and spleen of the mice stayed at a higher level even after 30 days of administration, and the Ti levels in spleen and blood from dose-administered mice samples were significantly ($p < 0.01$) higher than the control samples for the whole experiment duration (Figure 10.3). The result from this study also showed that the Ti level in other organ tissues (kidney, heart, lung, and blood) decreased over time in the nano-TiO_2 administered mice.

Another study by Cho et al. (2013) showed no significant increase in the Ti level in organ tissues (spleen, liver, brain, and kidney) when the nano-TiO_2 particles, with an average particle size of 26 nm, were administered daily through gavage to rats continually for 13 weeks with dosage rates of 260.4, 520.8, and 1041.5 $mgkg^{-1}day^{-1}$. The analysis of the gastric juice collected from the rats showed that the TiO_2 remained in the acidic gastric juice in stable condition throughout the administration time (Cho et al. 2013).

Most of the studies with animals and human samples showed the TiO_2 particles tend to agglomerate in the GI tract when it is taken as a food additive in NP sizes, and the TiO_2 does not dissolve in acidic gastric juices quickly (Cho et al. 2013; Xie et al. 2011; Yamashita et al. 2011). The absorption of TiO_2 in the GI tract was low, and the penetration ability of TiO_2 into the human and animal organ cells was also low when TiO_2 was used in NPs (Jones et al. 2015; MacNicoll et al. 2015). But some studies showed that traces of TiO_2 remained in the human and animal organs when nano-TiO_2 was administered for a longer duration or at higher doses, and the cell damage increased

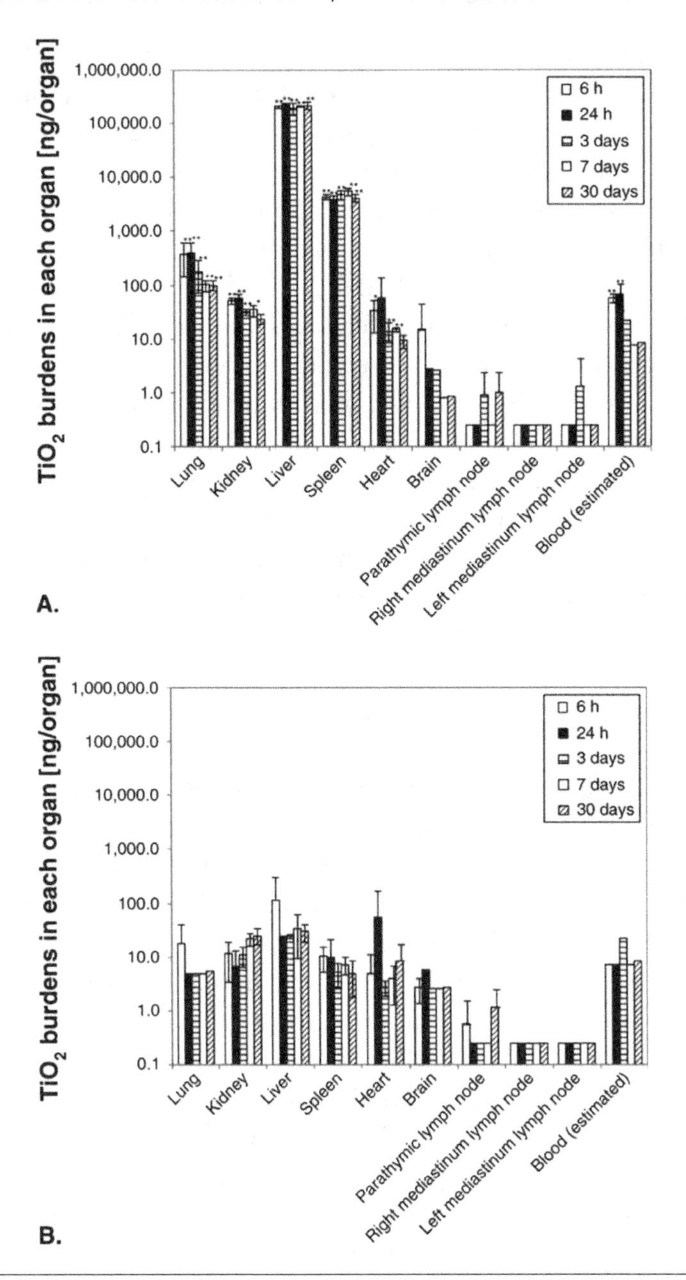

Figure 10.3 Detected TiO_2 levels at various organ tissues of the (A) administration group and (B) control group mice over time after intravenous (IV) administration of nano-TiO_2. (Reproduced from Shinohara, N. et al., *Nanotoxicology*, 8, 132–141, 2014.)

with increase in nano-TiO_2 exposure duration (Sang et al. 2013; Shinohara et al. 2014). So the data available so far are inconclusive, and not many new studies are being carried out to test the toxicity of the nano-TiO_2 by oral administration (or dietary exposure) in recent times. The studies about the toxicity of dietary exposure of TiO_2 were conducted more than 20 years ago, and those researchers did not consider the particle size of TiO_2 and the grade (whether TiO_2 is food grade or not) when they conducted toxicity research. Thus, additional scientific research is needed to test the toxicity of food-grade TiO_2 when it is used in NP sizes as a food additive in order to determine the short-term and long-term effects on the human health and reduce the consumer's concerns about health hazards.

Amorphous silica (SiO_2) has been used as a food additive for a long time, and the European Food Safety Authority (EFSA) found the dietary intake of SiO_2 up to a level of 0.3–0.8 mg/kg body weight per day had no ill effect on human health. Similar to TiO_2, use of nanoengineered SiO_2 NPs in food preparation raises consumer's concern about absorption into human organs and penetration into organ cells due to the higher surface-to-volume nature of the NPs. Several studies have been conducted in recent years to test the clinical and hematological changes in the organs due to the use of nanosize amorphous SiO_2 (Buesen et al. 2014; Fruijtier-Polloth 2012; Kim et al. 2014a; van der Zande et al. 2014; Wolterbeek et al. 2015; Yoshida et al. 2014). Most of the studies showed the use of amorphous SiO_2 in nanosize did not have any toxicity when administered orally or gavage doses to mice (Buesen et al. 2014; Kim et al. 2014a; Wolterbeek et al. 2015; Yoshida et al. 2014; Yun et al. 2015), but some studies showed little to moderate clinical toxicity when tested with mice in vivo studies (Fruijtier-Polloth 2012; van der Zande et al. 2014). Four different types of SiO_2—negative charge (SiO_2 phosphate), positive charge (SiO_2 amino), neutral charge (SiO_2 polyethylene glycol), and raw SiO_2 without any surface modification—with average particle size of 40–50 nm were administered to rats through gavage dosing for 28 days at a rate of 1000 mg/kg body weight per day, and then hematological and clinical examinations were performed to check the effect of nano-SiO_2 (Buesen et al. 2014). The results showed the all four types of SiO_2 NPs got agglomerated in the GI tract into microsize particles, and the neutral-charged SiO_2 polyethylene glycol NPs formed a larger-size agglomerate than the other three types of the SiO_2 NPs. They found no adverse hematological and clinical effects in the rats after the gavage dose for 28 days with all four types SiO_2 NPs. The SiO_2 NPs without any surface modification showed very little agglomeration and absorbed the acute-phase proteins when the haptoglobin and α2-macroglobulin were measured from the tested samples. The metabolomics changes were also measured, and these changes stayed below the threshold level, which causes ill effects, in all the four types of SiO_2 NPs oral dosage applications (Buesen et al. 2014).

Kim et al. (2014a) tested the adverse effect of negatively and positively charged SiO_2 NPs of two different particle sizes of 20 nm and 100 nm in rats using clinical or histopathological changes during 90 days of gavage administration of SiO_2 NPs at 3 different daily dosage levels: high (2,000 mg/kg body weight), medium (1,000 mg/kg body weight), low doses (500 mg/kg body weight). The ninety-day tests showed no death of rats or any adverse clinical effects for all three dosage levels and both the SiO_2 NP applications. There were no significant differences in aspartate aminotransferase, creatine kinase, and the lymphocyte levels between blood samples collected from high-dose-administered rats and control rats (Kim et al. 2014a).

Shumakova et al. (2014) conducted a two-phase study to check the adverse effect of intake of the SiO_2 NPs using rats. In the first phase, rats were administered with a gavage dose of SiO_2 NPs with particle sizes of 5–30 nm at a rate of 0.1, 1.0, 10.0, 100.0 mg/kg body weight per day for 30 days, and in second phase, the same concentrations of SiO_2 NPs were administered via diet for 62 days. The results proved there were no adverse clinical effects on any of those dosage levels. Similar kinds of results were obtained by Yun et al. (2015) when the 12 nm particle size SiO_2 NPs mixed with distilled water were fed to rats for 91 days at 3 different dosage levels (245, 490, and 980 mg/kg body weight per day). There were some concerns among the general public about the adverse effect of synthetic amorphous SiO_2 NPs on reproductive performance and the developmental issues in the next-generation animals and humans due to the aggregation and settlement of SiO_2 NPs in organ tissues. Wolterbeek et al. (2015) tested the toxicity of synthetic amorphous SiO_2 NPs on the reproduction system of Wistar rats using a two-generation experiment. Male (116) and female (116) rats were orally administered with NM200 synthetic SiO_2 NPs (mixed with ultrapure water) at four different dosage levels (0 [control], 100, 300, and 1000 mg/kg body weight per day) for 2 generations, and the body weight and food consumption of animals were measured to check the abnormality and clinical signs. The reproductive performance of rats was analyzed using the mating, fertility, live births, viability, and lactation indices, and for the developmental performance of the second generation, the body weight of pubs was measured on 1, 4, 7, 10, 14, 17, and 21 postnatal days. The results showed no adverse effect on body weight or food consumption due to the use of synthetic amorphous SiO_2 NPs. The mating index of rats orally administered with amorphous SiO_2 NPs were between 96% and 100%, and there was no significant difference between control and treated animals. The female fertility and fecundity indices were 96%–100%, and 89%–98%, for first and second generation, respectively, and there were no significant differences observed in testicular sperm count and daily sperm production between control and high dose (1000 mg/kg body weight per day) treatment male rats in both the generations. Necropsy tests also showed no significant changes in organ weights after oral administration of synthetic amorphous SiO_2 NPs for two generations even at a high dose (1000 mg/kg body weight per day). Therefore, Wolterbeek et al. (2015) suggested the dosage level of 1000 mg/kg body weight could be used as No Observed Adverse Effect Level for synthetic amorphous SiO_2 NPs while developing guidelines for usage of synthetic amorphous SiO_2 NPs in human and animal foods.

A study was conducted by van der Zande et al. (2014) to check the biodistribution and the adverse effects of using food-grade synthetic amorphous SiO_2 NPs using rats. The commercially available food-grade hydrophilic synthetic amorphous silica (SAS; 7 nm particle size) and another nanostructured SiO_2 (NM-202; particle size 10–25 nm) were tested with a dosage levels of 100, 1000, or 2500 mg/kg body weight per day of SAS, and 100, 500, or 1000 mg/kg bodyweight per day of NM-202 along with control for 28 days. The ICP-MS analysis of the rats after 28 days of oral administration of SAS and NM-202 doses showed a higher amount of silica in most of the organs (liver, spleen, and kidney) when the application dosage was low or medium (Figure 10.4). Daily body weight and tissue weight measurement of

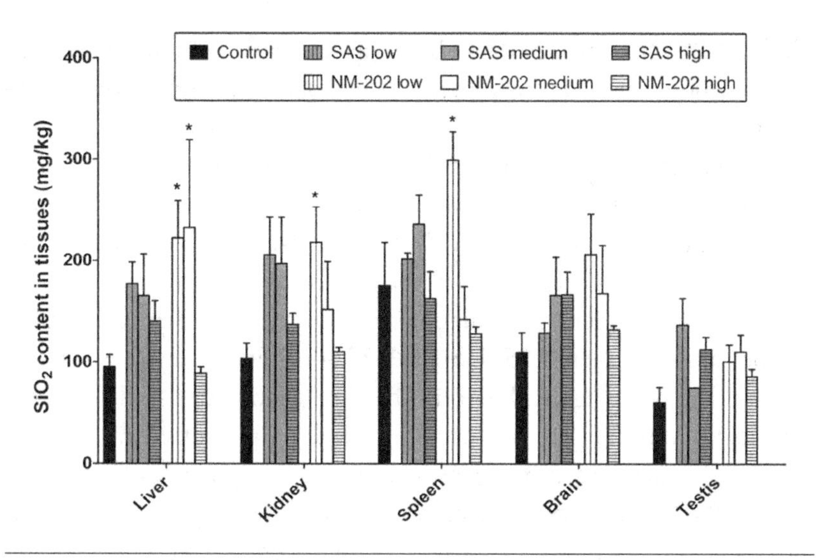

Figure 10.4 Silica level in various organ tissues of rats administered (oral) with different dosage levels of SAS and NM-202 for 28 days. (Reproduced from van der Zande, M. et al., *Part. Fibre Toxicol.*, 11, 1–8, 2014.)

control and treated rats showed no treatment-related effects, and there were no immunotoxic effects observed due to SAS or NM-202 dosage after 28 days. The quantitative histopathological analysis showed no significant changes in organ tissues after 28 days of SAS or NM-202 exposure, but when the higher doses were orally administered for 84 days, fibrosis induction in the liver were observed in NM-202-treated rats, and a fibrosis induction-related gene expression was also observed in the 84 days high-dose NM-202-administered rats (van der Zande et al. 2014). No accumulations of silica in the liver were detected in the NM-202 dose administered rats, whereas silica accumulations in the liver were observed after 84 days in the high-dose SAS-administered rats. They stated the effects on high-dose NM-202 administered rats were moderate, linked this mild adverse effect to the mild transcriptome changes in the liver, and suggested more studies by exposing humans to silica dosage levels to further investigate the toxicity of silica on human organ tissues.

Silver in nanoform has been used in recent times in various manufacturing industries due to its antimicrobial properties, and the silver nanoparticles (AgNPs) are most commonly marketed and applied NPs throughout the world in food and beverage processing and medical equipment manufacturing industries (Gaillet and Rouanet 2015;

US EPA 2010). In the food and beverages processing industry, AgNPs are commonly used as an ingredient or coating agent of food packaging materials, and also used for antibacterial coating of food processing equipment (Bouwmeester et al. 2009; Chaudhry et al. 2008). So direct contact of AgNPs with food is very imminent in the food and beverages industry. The AgNPs were also used as a supplementary diet and food additive in the past decade due to the enormous growth of nanoscience and nanotechnology (Chaudhry et al. 2008; Gaillet and Rouanet 2015; US EPA 2014). Silver is legally approved as a food coloring agent (E174) in the European Union and Australia (in alcoholic beverages and candy) as well as permitted to use as a microbial control agent for good manufacturing practices (Drew and Hagen 2016). The US Environmental Protection Agency (US EPA) suggested a daily oral dose limit of 5 μg/kg body weight per day for no adverse effect on human health (US EPA IRIS 1996), and a survey by Hadrup and Lam (2014) showed daily Ag intake of <0.4 μg/day in Italy and 27 μg/day in UK. But there are no clear data available on a daily intake of Ag in nanoform.

The cytotoxicity and genotoxicity of application of commercial AgNPs on mice health were tested using the AgNPs with the particle size of 10 nm dissolved in 1% citric acid, and no genotoxicity was observed (Kim et al. 2013). The acute oral toxicity and dermal toxicity of rats, eye and dermal irritation of rats, eye and acute dermal corrosion tests on rabbits, skin sensitization tests on guinea pigs, bacterial reverse mutation using the Ames test, and in vitro chromosome aberration tests were carried out using Organization for Economic Cooperation and Development test protocols and the Good Laboratory Practice guidelines. Kim et al. (2013) found the use of AgNPs caused patchy erythema in only one guinea pig out of 20 tested animals and showed some cytotoxicity effects in chromosome aberration tests. They did not observe any genotoxicity, abnormal clinical signs, and irritation or corrosion in eyes and skin in the tested animals up to a dosage level of 2000 mg/kg body weight (Kim et al. 2013).

The zinc oxide (ZnO) NPs are used in the food and beverages industry in nanocomposites for packaging and chemical-sensing agents. The toxic effect of ZnO NPs on rat health was tested with two different types of ZnO NPs (negatively charged commercial $ZnO^{AE100[-]}$, and positively charged commercial $ZnO^{AE100[+]}$) with an average particle size of 100 nm administered through gavage dosing in a 90-day study (Kim et al. 2014b). Three different daily dosage levels were administered (31.25 mg/kg body weight for low dose, 125 mg/kg body weight for medium dose, and 500 mg/kg body weight for high-dose treatments) for 90 days, and blood biochemical and hematological analyses were carried out on the test subjects. The rats fed with high doses of ZnO NPs showed significant hematological and blood chemical changes when compared with the control, and the histopathological examination of rats after 90 days of continuous high-dose ZnO NPs gavage administration showed significant damages in the stomach, eye, pancreas, and prostate gland tissues irrespective of particle charge (Kim at al. 2014b). The rats with medium doses of ZnO NPs showed some damage in the eyes (retinal atrophy lesion), but no other ill effects in the other organs. Low-dose administration (31.25 mg ZnO NPs/kg body weight) did not show any adverse effects after 90 days of the test, which proved the application dosage plays a major role in the damage of organs of a rat when ZnO is used as nanosized particles.

The distribution of Ag in various parts of the body after AgNP intake was tested by many researchers who found Ag particles in various organ tissues were similar to the large Ag particles intake (Lankveld et al. 2010; Loeschner et al. 2011; van der Zande et al. 2012). But in some studies, higher Ag concentrations in liver and kidney tissues were observed (Park et al. 2010; Yun et al. 2015). The Ag concentration in the liver, spleen, kidney, and lung was higher in the 91-day gavage AgNP (particle size of 11 nm) administered rats when compared with control animals, and the Ag concentration was higher in the blood and feces too (Yun et al. 2015). The Ag concentrations in the kidney, liver, lung, and brain were higher when AgNPs (22, 42, and 71 nm particle size) were administered via gavage to mice for 11 days, and the Ag concentrations in these organs were less in the animals administered with larger-size AgNPs (323 nm particle size) with the same application rate (1 mg/kg body weight per day). But the study by van der Zande et al. (2012) showed no significant differences in Ag concentrations in the blood between regular $AgNO_3$ (9 mg/kg body weight per day) and AgNP (9 mg/kg body weight per day) administered rats via gavage for 28 days. They observed the higher amount of Ag in the gastrointestinal tract tissues and statistically higher Ag concentrations in spleen and testis of the

AgNP-administered rats. They also found the Ag concentration in blood and feces of mice administered with AgNPs decreased more than 90% within a week after the Ag intake stopped, but the Ag concentration in the brain and testis remained at a high level even after 2 months of exposure. Austin et al. (2012) observed higher Ag concentrations in all of the tested organ tissues (liver, kidney, spleen, lung, and visceral yolk) of pregnant mice administered with AgNPs (particle size 50 nm) when compared with the control mice, but there were no adverse immunotoxicity effects after 28 days of AgNP injection on gestation days 7, 8, and 9 at a dose level of 60 μg Ag/mouse. Tavares et al. (2012) and Kovvuru et al. (2015) found DNA damage and DNA deletions (in mice embryos) after exposure to AgNPs (5–150 nm particle size) at various dose levels and durations implying genotoxicity with AgNPs, but Kim et al. (2013) did not observe any genotoxicity in Chinese hamsters due to the gavage administration of commercial AgNPs at 2000 mg/kg body weight dose for 14 days. Thus, the results of the studies conducted to test the effect of using AgNPs on toxicity and clinical abnormality are inconclusive, and there is no solid evidence to declare the use of AgNPs causes new risks other than the health risks of using regular Ag components in the food and beverages. But the studies about the application of AgNPs for controlling foodborne microorganisms showed AgNPs damages the cell wall of the foodborne pathogens, by which AgNPs inhibit the development of adverse microorganisms in food and beverages during processing, storage, and logistics. These results created a negative impression about application of AgNPs directly as a food additive, packing material with direct food contact, or coating of packaging film, which might lead to migration of AgNPs to food. The concern is that the AgNPs might damage cells of human organs similar to the cell damage they cause to microorganisms when they enter into human body via food and beverages. Thus, more research is needed to test the long-term exposure effects on humans to remove this negative image and create more awareness about the AgNP applications in food and beverages.

Consumer's Perceptions on Acceptance of Nanotechnology in Food and Beverages

Nanotechnology is one of the emerging technologies in the world at present, and it is facing challenges similar to other technologies like genetically modified organisms (GMOs) when it is applied in the agriculture and food processing industry. The success of a new technology always depends on the societal perspective and acceptance (Frewer et al. 2011). The socio-psychological factors play a major role in the social acceptance of an emerging technology and method (Gupta et al. 2012a). The perceptions about the applications of nanotechnology in food and beverages differ based on various factors, like demography, knowledge level, and awareness about nanotechnology and type of application itself (Gupta et al. 2012a, 2012b; Handford et al. 2014; Ho et al. 2011; Pidgeon et al. 2009; Priest and Greenhalgh 2012; Siegrist et al. 2008; Stampfli et al. 2010; Wesley et al. 2014). A study by Ho et al. (2011) found the general public perceives high risk and low benefit through the application of nanotechnology in food and other manufacturing sectors, and they also found the general public wanted to reduce the public funding for the nanotechnology-related research programs. Among all the manufacturing sectors, application of nanotechnology-based materials and processes in health- and food-related products was considered more controversial among the general public (Siegrist et al. 2007). Pidgeon et al. (2009) found the applications of nanotechnology in products that penetrate into the human body (by oral intake or some other means) were considered riskier than the other applications. Schnettler et al. (2014) conducted a study in southern Chile to check the public acceptance of nanotechnology in sunflower oil processing and packaging by using 400 people. They checked the factors influencing the decision of buying a product manufactured or packed using nanotechnology and found the brand name of a product was the major deciding factor followed by the price of the product. Schnettler et al. (2014) also found around 44% of the participants preferred sunflower oil processed or packed without any nanotechnology-based applications, 35% of the people showed positive approach towards nanotechnology in sunflower oil processing or packaging, and 21% of people showed greater interest in sunflower oil produced with less cholesterol using nanotechnology.

Stampfli et al. (2010) conducted a survey of 514 people from the German-speaking part of Switzerland to check their willingness to accept the 6 types of nanotechnology-based applications (UV-protection packaging films, individually modified foods [tomato juice, orange juice] using nanotechnology, high-nutritional value bread, supplements for cancer prevention, packaging films for increasing shelf life of food products, food containers with antibacterial properties) in food and food-packing operations. The survey results showed that the public was more acceptable of nanotechnology applications in food packaging than the food products modified with nanotechnology. People considered the nano-modified tomato or orange juice posed more risks and less benefit, so their willingness to buy those products was less. The perception about nanotechnology-based antibacterial food containers, shelf life increasing, and UV-protection packaging films was more positive, they thought these applications had more benefit, and their willingness to buy was also high (Stampfli et al. 2010). The results also showed that trust and attitude towards newer but well-informed technologies like gene technology played a major role in the perceiving of potential risks and benefits of the application of nanotechnology in food and food packaging. But a similar kind of study conducted in France by Vandermoere et al. (2011) showed that the French public had a negative perspective about the use of nanotechnology for food packaging and modifying individual food products like nanotechnology-based juices. A study by Greehy et al. (2011) to check the acceptance of application of nanotechnology in food found how the food industry and the research community address the potential health and other risks of nanotechnology-based food products was the major deciding factor for acceptance or rejection of nanotechnology-based products. These results show how the views about the application of nanotechnology in sensitive sectors like food and beverages differ among demography and cultures (Gupta et al. 2012b).

A study was conducted in northwestern Europe to check the views of the experts from various stakeholders of the society (industry, academia, government and regulatory agencies, consumers, and media) on the main factors affecting the societal response about the application of nanotechnology in various fields (water purification, nanofabric, food and beverage processing, food packaging, targeted delivery of drugs and nutrients from nanoencapsulated foods, nano sports goods, smart dust, nanochemical sensors, pesticides, and easy-to-clean surfaces) (Gupta et al. 2012a,b). The experts from various fields (totally 17 experts) were interviewed face-to-face for about 50 min each about the application of nanotechnology in 15 different areas, and the results were analyzed using principal component analysis and generalized procrustes analysis techniques. The principal component 1 (PC1) scores from the analysis showed that the experts believed the application of nanotechnology in water purification, targeted delivery of drugs, nanochemical sensors, nanofuel cells, nanoimplantation devices, and the soil and water treatments had higher social benefits, and they thought societal acceptance might be high on nanotechnology application in these areas due to their necessity and usefulness. The PC1 scores for application of nanotechnology in pesticides, smart dust for air purification, Radio-frequency identification (RFID) tags, cosmetics, and fabric were low, and the experts thought the application of nanotechnology in these areas were less beneficial, and the societal acceptance might be less due to their less usefulness. The analysis also showed experts had a neutral view on the application of nanotechnology in food packaging, surface-cleaning agents, and nanoencapsulation of food for targeted nutrient delivery (Figure 10.5). Gupta et al. (2012b) found the usefulness and need of the technology played major roles in deciding the societal acceptance of the applications in the various fields. Application of nanotechnology in the medical field (targeted drug delivery, nanoimplantation devices) was considered most acceptable by the experts, and the application in environmental safety and protection (water purification, soil and wastewater treatment, and chemical nanosensors) was considered more beneficial and might have more social acceptance. Gupta et al. (2012b) also found the "distance from the end user" (whether the nanomaterial came into contact with the user or the nanoparticle migrated into the human body or the food and beverages) was another major factor influencing the acceptance of the nanotechnology among the consumers, and in the food and beverages manufacturing domain, this "distance from the end user" is the important factor influencing the consumer's perspective and acceptability of nanotechnology. The analysis of the results also

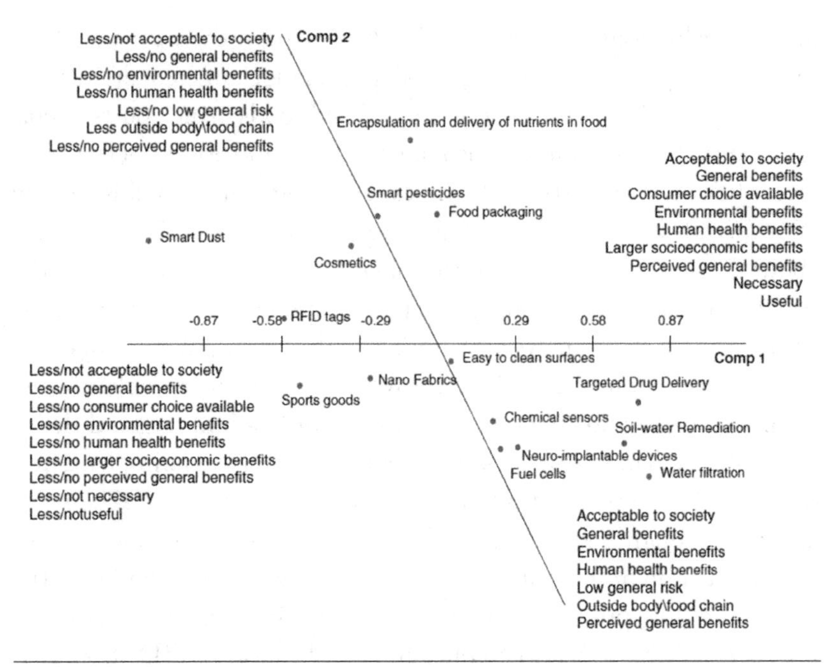

Figure 10.5 Expert opinion on acceptability of applications of nanotechnology in various fields based on PC1 and PC2 scores. (Reproduced from Gupta, N. et al., *J. Nanoparticle Res.*, 14, 857–871, 2012b.)

showed a majority of the experts (15 out of 17 experts) directly compared the application of nanotechnology in the agriculture and food sector with the genetically modified plants and foods, which showed how the newly developed technologies, which have links with food or agriculture, may get a negative social perception similar to the socially controversial techniques like GMOs. They also emphasized the importance of the research in the area of removing consumers' (or laymen's) concerns about migration of nanoparticles into food from the packages or the migration of nanomaterials into human body (Gupta et al. 2012a,b).

Capon et al. (2015) conducted a survey through computer-assisted telephone interviews with the Australian public (1355), academics (301), government regulatory organizations (19), and industry (21) people to check the perceptions of the risks involved with applying nanotechnology in various fields (food, agriculture, cosmetics, medicine, sports goods) and the trust of public and other stakeholders towards scientists working on nanotechnology. Analysis of this survey results using chi-squared analysis showed the Australian public had more concern about nanotechnology when it applied to food and agriculture, which they thought had more direct impact on their health (84.8% agreed nanotechnology in food had more health risk) than it applied in cosmetics or sunscreen manufacturing. The general public also showed more concern when the nanoengineered products were directly applied to food but thought risks were fewer when the nanotechnology was applied for indirect food processing operations (like using food packaging materials with nanoparticles). The academics also expressed more concern about the risk of applying nanotechnology in food (55.4%), followed by application in cosmetics and pesticides (39.5% and 39.0%, respectively). The personnel from government agencies ranked application of nanotechnology in cosmetics/sunscreen had more risks, and the business sector ranked nanotechnology-based pesticides had more risk. The trust analysis through this survey showed the general public expressed low trust with the scientists working on nanotechnology and the health professionals to safeguard them from any potential risks arising from materials developed using nanoscience and nanotechnology, but all the stakeholders (public, academia, business, government) showed more trust of scientists than the health department, government, journalists, and politicians (Capon et al. 2015). These results proved that the general public perception about risk of applying nanotechnology in food and beverages is high when compared with the other sectors, and they are expecting that the development of regulations and policies for applications of nanotechnology in the food and beverages sector with more caution can help to increase the public's trust and change the negative perceptions.

A survey of community attitudes towards nanotechnology among the Australian public by the Ipsos Social Research Institute, Richmond, Australia, showed the public awareness about nanotechnology was increased to 87% in 2012, which was 51% in 2005, but the awareness about application of nanotechnology in food and beverage processing operations was less when compared to other applications of nanotechnologies like nanotechnology-based solar panels, sunscreen, and cosmetics. Only 17% of the interviewed public told they heard about the nanotechnology-based food packaging materials for keeping the food fresh for a long time, as well as to monitor the quality and freshness of the packed food (Ipsos Social Research Institute 2012). Even though the awareness about nanotechnology applications in the food and beverages industry among the Australian public was low, the responses from the people who knew about nanotechnology were more positive towards the application of nanotechnology for food packaging (57%) and improvement of nutritional benefits of food product using nanotechnology (47%) than the negative comments. This survey also showed that the age group and gender also play a major role in deciding the benefits and risks associated with applying nanotechnology in food and beverages. Men (61%) and younger (age group: 16–30 years) people (73%) supported more the nanotechnology-based food packaging than the females (55%) and the older (65% of people in the age group of 31–50 years).

Regulations about Application of Nanotechnology in the Food and Beverages Industry

There is no single general international regulation about the use of nanoscience and nanotechnology in foods until now, and various countries are working on their own regulatory framework about the use of nanomaterials, nanoengineered particles, and nanotechnology-based processing operations in various fields like medicine, food and beverages manufacturing, cosmetics, and agricultural operations. Most of the regulations available at present in the European Union, Australia, and North America are covered by the provisions of other legislation (e.g., "horizontal legislation" in the European Union, which covers a whole range of environmental subject areas, such as agriculture, food, water, and is not so specific to a single area [Europa 2017]), which have broad coverage ranges and cover the application of nanotechnology too (Cushen et al. 2012).

In Europe, gap analysis studies were carried out by UK Food Standards Agency (FSA 2008) and Chaudhry et al. (2006) about the currently available food safety laws in relation to the application of nano-based materials and technologies in food and beverages, as well as the risks involved in those applications. Both studies found the lack in regulatory processes and recommended to implement approval framework for nano-based materials and processes prior to use in food and beverages manufacturing processes (Chaudhry et al. 2008). Even though there is no specific legislation related to use of nanomaterials as food additives, the EFSA's Framework Directive 96/77/EC covers it by considering the food or beverages manufactured using nanosize food additive as a "novel" food additive, or by making proper amendments to the regulations already approved for the same food additive used in macro size in food and beverages (Chaudhry et al. 2008). The European Union released four proposed regulations related to use of nanotechnology-based products and processes in food and beverages in July 2006, which cover the area of food flavoring agents (Regulation: COM/2006/0427 final), additives (Regulation: COM/2006/0428 final), enzymes (Regulation: COM/2006/0425 final), and the food products and processes approval procedure (Regulation: COM/2006/0423 final). These new regulations give more power to EFSA to evaluate and assess the safety and risk of new food additives, flavoring agents, and food enzymes, as well as to assess the use of nanosized materials for these purposes in the food and beverage industry. When the nanoscale materials are used as a food ingredient, the European legislation "Regulation (EC) 258/97" regulates the product through a mandatory requirement of pre-market approval by treating either the food as "novel food" or the ingredient as "novel food ingredient." In this legislation, "novel food" is defined as a food product or any one of the ingredients of the food product that was not consumed as a human food in the community before May 1997 (Chaudhry et al. 2008). The use of nanotechnology-based food ingredients falls in either one of the following two categories as defined by this Regulation (EC) 258/97 (Chaudhry et al. 2008):

1. "Foods and food ingredients with a new or intentionally modified primary molecular structure" or
2. "Foods and food ingredients to which has been applied a production process not currently used, where that process gives rise to significant changes in the composition or structure of the foods or food ingredients which affect their nutritional value, metabolism or level of undesirable substances."

The major drawback in this legislation is, the declaration of a food product as "novel food" solely depends on the manufacturer. If the company decided not to claim a product as novel since they used same food ingredients in macro size in the past, or they thought there was no significant differences in the nutritional value, metabolism, or level of undesirable substances between nanofoods and its macro-size counterpart, then EFSA's safety evaluation based on this Regulation (EC) 258/1997 is not required (Chaudhry et al. 2008). And this legislation also has a clear view if a food ingredient was used before May 1997, since the legislation does not have proper definition about the particle size of food ingredients. Nanotechnology-based packaging materials have been tested widely and pose great potential in the application of packing food and beverages with increased shelf life and better quality. The European Union's Regulation (EC) 1935/2004 covers the use of nanocomposites for food packaging materials, considering these as "Food Contact Materials (FCM)." This regulation not only covers the food packaging material (like the film, boxes, containers) but also covers the food processing equipment, shipping materials, and the cooking utensils (Chaudhry et al. 2008; European Commission 2004). Based on this legislation, the EU regulations directives 82/711/EEC and 85/572/EEC provide the testing protocols for checking the migration of packaging material components to food, by which the safety of the product can be ensured by checking whether this measured migration level stayed below the established safety limit of each component of the FCM. The food and beverages manufacturing industry is forced to comply to test the migration of all the FCM components and declare to EFSA for approval of the food packaging materials by this regulation. Thus, there is no need for separate regulation for nanotechnology-based food packaging materials, since (EC) 1935/2004 covers all of the components of packaging materials and their properties (Chaudhry et al. 2008; Cubadda 2013).

Based on the feedback from the Regulation (EC) 258/97, the EU developed new regulatory measures for testing and regulating food products containing nanoscale food additives and food ingredients. The Regulation (EC) 1333/2008 requires the food additive to go through new EFSA risk assessment if the particle size of the additive is altered by new processing operations, and the Regulation (EU) 257/2010 also allows EFSA to reevaluate the food additives approved before 2009 (Rossi et al. 2014). Based on this new regulation, food additives with nanosize particles, such as E171 and E551, have been reevaluated by EFSA. The EU also approved the plastic food packaging materials with titanium nitride NPs (particle size of 20 nm), carbon NPs (particle size of 10–300 nm), and the SAS (particle size 1–100 nm) for application in the food and beverage industry through the Regulation (EU) 10/2011. Regulations (EC) 450/2009 and (EC) 1333/2008 regulate the use of nanosized materials through a pre-approval from EFSA to use as an intelligent food packaging material and direct release food ingredient, respectively. The food labeling regulation of EU "Regulation (EU) 1169/2011" requires the food and beverage manufacturers to label all of the nanoengineered ingredients of the food product in their ingredients' lists with the clear term "nano" with the ingredient (Couch et al. 2016; Rossi et al. 2014).

In the US, the Food and Drug Administration (FDA) regulates the food, beverages, cosmetics, and drug industry under the Federal Food, Drug, and Cosmetic Act (FFDCA), and the FDA is working along with all of the research centers and universities working on nanotechnology in the US to frame regulatory guidelines for the industry as well as regulatory bodies to streamline the use of nanotechnology-based products and processes in food and beverages, cosmetics, medicine, and other industries. Until now, the FDA does not have general regulatory guidelines for testing or approving all of the food and beverages containing nanoengineered materials or manufactured using nanotechnology-based processes, but it takes a "case by case" approach using the presently available regulatory guidelines. Based on the health and safety risk of the nanosize materials in a food product, the FDA decides its regulatory process, but consideration of the particle size is not the main parameter for deciding the safety evaluation and risk assessment tests of a food or beverage product (Breggin et al. 2009). Based on FDA regulations, all of the food additives need pre-market approval, and other food ingredients and packaging materials that have "generally recognized as safe" status do not require any pre-market approval from the FDA (Amenta et al. 2015). The current FFDCA legislation does not provide any approval mechanism or safety assessment requirements for food and beverages containing nanomaterials or manufactured using nanotechnology. The FDA released guidelines about the safe use of nanotechnology in food, cosmetics, and animal feed: "Considering Whether a FDA-Regulated Product Involves the Application of Nanotechnology," "Assessing the Effects of Significant Manufacturing Process Changes, Including Emerging Technologies, on the Safety and Regulatory Status of

Food Ingredients and Food Contact Substances, Including Food Ingredients that are Color Additives," and "Use of Nanomaterials in Food for Animals" (US FDA 2014a, 2014b, 2014c). Based on this guidance, the working definition of a nanomaterial is formed as the "materials or end products may also exhibit similar properties or phenomena attributable to a dimension(s) outside the nanoscale range of approximately 1–100 nm" (US FDA 2014). FDA regulation does not consider all the products having nanomaterial as a hazardous product and use the available data about larger versions of the substance and toxicity of the substances as a screening tool. There is no mandatory requirement of pre-market and pre-manufacture safety evaluation of food ingredients, and FCM contains nanomaterials, but FDA recommends evaluation of safety and risk of nanoscale FCM and food ingredients prior to use (Amenda et al. 2015).

In Canada, Health Canada (which consists of the Public Health Agency of Canada and the Canadian Food Inspection Agency) regulates the food industry, and until now there is no regulation formed specifically for use of nanotechnology-based materials in food and beverages manufacturing. But it regulates the food and beverage industry and other manufacturing sectors on use of nanotechnology through the existing regulations and the "Regulatory Framework for Nanomaterials under Canadian Environmental Protection Act 1999" (Nanoportal 2017). The Canadian regulation and the working definition of nanomaterials is more similar to the USDA's regulation, and the use of nanomaterials in the agricultural and food products is also covered under the New Substances (NS) program of Environment Canada (Nanoportal 2017). This NS program requires an evaluation of substances with nanosize materials to analyze the toxicity profile of the product (Amenta et al. 2015). Health Canada and Environment Canada are working along with their USA counterpart agencies to develop common regulatory protocols and guidelines for the use of nanomaterials and nanotechnology-based processing operations in the food and beverage industry through the Canada–US Regulatory Cooperation Council (RCC) from February 2011 with the mission of "increase alignment in regulatory approaches for nanomaterials between Canada and the US to reduce risk to human health and the environment, promote the sharing of scientific and regulatory expertise, and foster innovation" (Nanoportal 2017). The RCC's nano-initiative work plan was started in 2012 with following five working elements:

1. "Principles: Identification of common principles for the regulation of nanomaterials to help ensure consistency for industry and consumers in both countries.
2. Priority-Setting: Identification of common criteria for determining characteristics of industrial nanomaterials of concern/no-concern.
3. Risk Assessment/Management: Sharing of best practices for assessing and managing the risks of industrial nanomaterials.
4. Commercial Information: Characterization of existing commercial activities and identifying gaps and priorities for future knowledge gathering for industrial nanomaterials.
5. Regulatory Cooperation in Areas of Emerging Technologies: Development of a model framework outlining key elements and approaches to regulating products and applications of emerging technologies with respect to potential impacts on the environment, human health, food and/or agriculture." (Nanoportal 2017).

The report of the RCC's nano-initiative was completed on February 2014, which needed substantial changes in US and Canadian regulations and recommends the food and beverages industry to apply for approval of new food additives, ingredients, and FCMs (including packaging materials) in both the territories at the same time. The industry stakeholders in USA and Canada proposed an alternate strategy of conducting a pre-notification consultation (PNC) with the regulatory agencies in both countries about the nanomaterial application at the RCC Nanotechnology Results Workshop. But in this process, the regulatory agencies in the both the jurisdictions totally depend on the industries for voluntary PNC submission and consultation and does not have any mandatory requirements for submission of PNCs for all the products containing nanomaterials.

In Australia and New Zealand, the Food Standards Australia New Zealand (FSANZ) regulates the food industry and made amendments in its FSANZ application handbook to regulate the use of nanomaterials and nanotechnology-based processes in food and beverages manufacturing as well as manage the potential health risks of application of nanotechnology-based materials and processes in food and beverages (Amenda et al. 2015). The National Industrial Chemical Notification and Assessment Scheme released hazard analysis datasheets about the TiO_2 NPs and AgNPs to inform and educate the general public and industry about the toxicity of nanomaterials (Bowles and Lu 2012). In Switzerland, the Federal Office of the Public Health regulates the use of nanomaterials in agricultural and food sector using the existing safety regulations and adopted mandatory pre-market approval process for use of nanomaterials as food additives (FAO/WHO 2013). Russia regulates the food industry through the "Sanitary Rules and Regulations," and application of nanotechnology-based materials and processes have been streamlined by Russian Corporation of Nanotechnologies and Nanotechnology Industry Development Program (Amenta et al. 2015). Similar to Switzerland, Russia also implemented safety levels for use of TiO_2 NPs, AgNPs, and carbon nanotubes through sanitary regulations (FAO/WHO 2013). Even though there are no specific laws regarding nanomaterials used in food and beverages, South Africa regulates the application of nanotechnology in food and beverages using the "Foodstuff, Cosmetics and Disinfectant Amendment Act 2007." Brazil is the leading Latin American country in research and development of nanotechnology-based products and still working on implementing specific regulations for nanotechnology-based products and processes in food and beverages (Amenta et al. 2015).

Similar to the other parts of the world, most of the Asian countries like Japan, China, India, and South Korea are working on the development of regulatory framework for the application of nanotechnology in food and beverages, and at present using the existing regulations. Korea released "Guidance on Safety Management of Nano-based Products" in 2011, and Malaysia regulates the use of nanotechnology-based products through "The Nanotechnology Safety Related Act," revised "The Food Act 1983," and a revised "The Food Regulations 1985" (Nano Malaysia 2017). Taiwan uses a certification system called "Nano Mark" to evaluate and certify the products containing nano-based materials. Iran formed a Nanotechnology Committee for development of guidelines for safety assessment and authorization of nanotechnology-based food, beverages, cosmetics, medicines, and pharmaceutical products. Iran also introduced an industrial standard certification process for nanotechnology-based products and maintains a database of standard procedures and nano-based materials through Iran Nanosafety, and Iran Nanoproducts websites (Nanosafety 2017; Nanoproduct 2017). Thailand developed regulations for nanotechnology-based products through the National Nanotechnology Center and introduced a certification process (NanoQ label) via the Nanotechnology Association of Thailand. But this NanoQ label was issued for paints, textile, and household plastic products only, and there is no data available on NanoQ-label-certified food and beverage products (Amenda et al. 2015). The review work by Bumbudsanpharoke and Ko (2015) explains in detail about the status of the current regulations and the legislation on the application of nanotechnology-based food packaging films and containers (food contact materials), and Table 10.1 outlines these current regulations briefly.

Conclusions

Application of nanotechnology-based materials and processes in the food and beverages industry is growing in a rapid manner, and the recent developments in the production and processing of nanoparticles and nanotechnology-based unit operations created enormous opportunities for developing new food and beverage products with improved properties and benefits. The major advantages of application of nanotechnology-based materials and processes in the food and beverages industry include improved or novel products with new texture/taste, increased bioavailability of micro and macro nutrients, increased shelf life and freshness, advanced intelligent and active food packaging, enhanced food traceability, and improved and economically beneficial solid and liquid waste management systems.

In the food and beverages industry, nanotechnology is mainly applied in the packaging sector since the packaging films incorporated with nanoparticles of clay and Ag provide improved mechanical,

Table 10.1 Currently Available Regulations and Their Present Status around the World about Application of Nanotechnology-Based Materials for Food Packaging

COUNTRY/REGION	STATUS OF REGULATIONS AND LEGISLATION
European Union countries	In 2011, the EFSA published a scientific opinion entitled "Guidance on the risk assessment of the application of nanoscience and nanotechnologies in the food and feed chain" (FAO/WHO 2013). Four nanomaterials have been studied: 1. Silicon dioxide (Food Packaging Related) 2. Titanium nitride (Food Packaging Related) 3. Silver hydrosol 4. Calcium carbonate There is no particular legislation in the EU concerning nanomaterials for food contact or food packaging. However, there is related existing legislation. Therefore, producers or manufacturers must comply with these existing regulations or legislation: • Commission Regulation (EU) No 202/2014: On materials and articles intended to come into contact with food and repealing (European Commission 2014). • Commission Regulation (EU) No 1282/2011: On plastic materials and articles intended to come into contact with food (European Commission 2011). • Commission Regulation (EC) No 975/2009: relating to plastic materials and articles intended to come into contact with foodstuffs (European Commission 2009b). • Commission Regulation (EC) No 450/2009: On active and intelligent materials and articles intended to come into contact with food (European Commission 2009a).
Russia	All industries who use nanomaterials must follow with the Regulation of the Chief State Health Officer of the Russian Federation dated July 23, 2007, N 54 (FAO/WHO 2013). Risk evaluations have been performed since 2012. Research is ongoing. The results will be used for new regulations (FAO/WHO 2013).
China	No specific regulation about nanomaterials for food packaging (FAO/WHO 2013). Applications for using nanominerals or food ingredients have been rejected by regulatory authorities, but the safety evaluation of nanotechnology in foods continues to be discussed (FAO/WHO 2013; Takeuchi et al. 2014).
Japan	Nanotechnology was specified as one of the priority research targets in the Third Science and Technology Basic Plan for 2006–2010 by the Japanese government (FAO/WHO 2013; Takeuchi et al. 2014). The Japanese Ministry of Health, Labor, and Welfare launched a 6-year program (2009–2014) called the "Research project on the potential hazards, etc. of nanomaterials" (FAO/WHO 2013).

(*Continued*)

Table 10.1 (*Continued*) Currently Available Regulations and Their Present Status around the World about Application of Nanotechnology-Based Materials for Food Packaging

COUNTRY/REGION	STATUS OF REGULATIONS AND LEGISLATION
South Korea	The Republic of Korea government released guidance on safety management of nano-based products in 2011. A new research plan for nanomaterial safety based on "The First Master Plan on Management of Nanomaterials Safety" is effective for 2012 (Hwang et al. 2012a, 2012b). • The Ministry of Food and Drug Safety (MFDS) undertakes a variety of research related to nanosafety in food as well as food packaging, and they are planning to propose new guidelines and safety regulations regarding nanofood packaging within a few years (Hwang et al. 2012a, 2012b).
Malaysia	No safety assessments or regulations specific to nanomaterials in the food and agriculture sectors were found on Malaysian government websites (FAO/WHO 2013; Takeuchi et al. 2014).
Indonesia	No safety assessments or regulations specific to nanomaterials in the food and agriculture sectors were found on Indonesian government websites (FAO/WHO 2013).
Oceania (Australia and New Zealand)	There is no regulation or organization in Australia focused on nanomaterial for food packaging (FAO/WHO 2013). FSANZ recently published an article describing its regulatory approach to nanoscale materials in the International Food Risk Analysis Journal (FAO/WHO 2013; Fletcher and Bartholomaeus 2011). The primary focus is not on the size of the material per se but on materials likely to exhibit physicochemical and/or biological novelty (FAO/WHO 2013). At present, the approval process for new substances applied for food packaging is generally unnecessary if there is an approval in the EU or US (Tager 2014).

Source: Bumbudsanpharoke, N., and Ko, S., *J. Food Sci.*, 80, 910–923, 2015.

thermal, gas, and moisture transmission properties, which help to keep the food products and beverages fresh and safe for a long time, and these nanocomposite food-packaging films and containers also have improved antimicrobial properties. Similar to the all the new technologies, nanotechnology also has pros and cons. The major safety concerns about applying nanotechnology in food and beverages are caused by the large surface-to-mass ratio nature of nanoparticles, which may increase the absorption of chemicals into human organs, and the migration of nanoparticles from the FCMs to food might also cause health risks. Studies conducted in Europe and other parts of the world regarding this migration of metal and other nanoparticles from FCM to food showed there was some migration from FCM to food, but the Ag and other nanoparticle contents in the food remains below the maximum allowable level in the food set by the regulatory agencies based on risk assessment tests. But there is lack of research related to health risks of long-term exposure to nanomaterials. The perspective about the application of nanotechnology in food and beverages vary based on the demography, age group, sex, occupation, culture, and the knowledge about the nanoscience and nanotechnology. Surveys among various stakeholders of the food and beverages industry in various parts of the world showed that, still, a major portion of people have some reluctance towards using nanoengineered products directly into the food products and having positive perceptions about the application of nanotechnology in food packaging. A clear demonstration of benefits and the risks associated with the use of nanotechnology to the common public in food products without compromising the developments in nanoscience and nanotechnology in the agriculture and food sector might help the scientific community and the industry to win the consumer's acceptance.

Even though the applications of nanotechnology in the food and beverages industry are increasing enormously, the regulations and laws regarding ensuring food safety and health risks are lagging throughout the world. Only a few countries like the European Union countries and Switzerland have regulations and legislation specifically for the application of nanotechnology-based products and processes in agriculture and food sector. Most of the other countries like USA, Canada, Australia, and New Zealand regulating nanotechnology application in food and beverages are using existing regulations or making amendments to the existing regulations, and also developing guidelines for the food and beverages industry for safe use of nanotechnology in food and beverages. In this global and open market era, harmonized and uniform regulations are needed to regulate the application of nanotechnology in the food and beverages industry. Knowledge, information sharing, and collaboration on forming regulations and guidelines among the countries are needed to frame safety guidelines and procedures for guaranteeing safe products with no risks to the consumers as well as the environment, at the same time without jeopardizing the development of novel food and beverage products with improved and new or improved sensory, textural, and nutritional properties.

References

Amenta V., Aschberger K., Arena M., Bouwmeester H., Moniz F.B., Brandhoff P., Gottardo S., Marvin H.J., Mech A., Pesudo L.Q. (2015) Regulatory aspects of nanotechnology in the agri/feed/food sector in EU and non-EU countries. *Regulatory Toxicology and Pharmacology* 73:463–476.

Austin C.A., Umbreit T.H., Brown K.M., Barber D.S., Dair B.J., Francke-Carroll S., Feswick A., Saint-Louis M.A., Hikawa H., Siebein K.N. Goering P.L. (2012) Distribution of silver nanoparticles in pregnant mice and developing embryos. *Nanotoxicology* 6(8):912–922.

Bouwmeester H., Dekkers S., Noordam M.Y., Hagens W.I., Bulder A.S., De Heer C., Ten Voorde S.E., Wijnhoven S.W., Marvin H.J., Sips A.J. (2009) Review of health safety aspects of nanotechnologies in food production. *Regulatory Toxicology and Pharmacology* 53:52–62.

Bowles M., Lu J. (2012) Review on nanotechnology in agricultural products logistics management. *Computing and Networking Technology (ICCNT), 2012 8th International Conference on*, IEEE. pp. 415–420.

Breggin L., Falkner R., Jaspers N., Pendergrass J., Porter R. (2009) Securing the promise of nanotechnologies: Towards transatlantic regulatory cooperation. Report, Chatham House, London, UK.

Buesen R., Landsiedel R., Sauer U.G., Wohlleben W., Groeters S., Strauss V., Kamp H., van Ravenzwaay B. (2014) Effects of SiO_2, ZrO_2, and $BaSO_4$ nanomaterials with or without surface functionalization upon 28-day oral exposure to rats. *Archives of Toxicology* 88:1881–1906. doi:10.1007/s00204-014-1337-0.

Bumbudsanpharoke N., Ko S. (2015) Nano-food packaging: An overview of market, migration research, and safety regulations. *Journal of Food Science* 80(5):910–923.

Capon A., Gillespie J., Rolfe M., Smith W. (2015) Perceptions of risk from nanotechnologies and trust in stakeholders: A cross sectional study of public, academic, government and business attitudes. *BMC Public Health* 15:424 (1–13).

Chaudhry Q., Blackburn J., Floyd P., George C., Nwaogu T., Boxall A., Aitken R. (2006) Final Report: A scoping study to identify gaps in environmental regulation for the products and applications of nanotechnologies. Defra, London, UK.

Chaudhry Q., Castle L., Watkins R. (2010) *Nanotechnologies in Food*. Royal Society of Chemistry, London, UK.

Chaudhry Q., Scotter M., Blackburn J., Ross B., Boxall A., Castle L., Aitken R., Watkins R. (2008) Applications and implications of nanotechnologies for the food sector. *Food Additives and Contaminants* 25:241–258.

Chen X.X., Cheng B., Yang Y.X., Cao A., Liu J.H., Du L.J., Liu Y., Zhao Y., Wang H. (2013) Characterization and preliminary toxicity assay of nano-titanium dioxide additive in sugar-coated chewing gum. *Small* 9:1765–1774.

Cho W.S., Kang B.C., Lee J.K., Jeong J., Che J.H., Seok S.H. (2013) Comparative absorption, distribution, and excretion of titanium dioxide and zinc oxide nanoparticles after repeated oral administration. *Particle and Fibre Toxicology* 10:1–9.

Coles D., Frewer L. (2013) Nanotechnology applied to European food production—A review of ethical and regulatory issues. *Trends in Food Science & Technology* 34:32–43.

Couch L.M., Wien M., Brown J.L., Davidson P. (2016) Food nanotechnology. *Agro Food Industry Hi Tech* 27:1.

Cubadda F. (2013) Applications and prospects of nanotechnologies in the food sector. In: Cubadda, F. et al. (Eds.) *Nanomaterials in the Food Sector: New Approaches for Safety Assessment*. Istituto Superiore di Sanità, Rome, Italy. p. 3.

Cushen M., Kerry J., Morris M., Cruz-Romero M., Cummins E. (2012) Nanotechnologies in the food industry—Recent developments, risks and regulation. *Trends in Food Science & Technology* 24:30–46.

Drew R., Hagen T. (2016) *Potential Health Risks Associated with Nanotechnologies in Existing Food Additives*. Food Standards Australia New Zealand, Melbourne, Australia.

Europa. (2017) *Nanotechnology*. European Commission, Brussels, Belgium.

European Commission. (2004) European Parliament and Council Directive 94/36/EC of June 30, 1994 on colours for use in foodstuffs. Official Journal of the European Communities. Available from http://ec.europa.eu/food/fs/sfp/addit_flavor/flav08_en.pdf.

European Commission. (2009a) Commission regulation (EC) No 450/2009 of May 29, 2009 on active and intelligent materials and articles intended to come into contact with food. *Official Journal of European Union* L135:3–11.

European Commission. (2009b) Commission regulation (EC) No 975/2009 of October 19, 2009 amending directive 2002/72/EC relating to plastic materials and articles intended to come into contact with foodstuffs. *Official Journal of the European Union* L274:3–8.

European Commission. (2011) Commission regulation (EU) No 1282/2011 of November 28, 2011 amending and correcting Commission regulation (EU) No 10/2011 on plastic materials and articles intended to come into contact with food. *Official Journal of the European Union* L328:22.

European Commission. (2014) Commission regulation (EU) No 202/2014 amending regulation (EU) No 10/2011 on plastic materials and articles intended to come into contact with food. *Official Journal of the European Union* L62:13–15.

FAO/WHO. (2013) *State of the Art on the Initiatives and Activities Relevant to Risk Assessment and Risk Management of Nanotechnologies in the Food and Agriculture Sectors*. Food and Agriculture Organization of the United Nations and World Health Organization, Rome, Italy.

FDA. (2015) Summary of color additives for use in the United States in foods, drugs, cosmetics, and medical devices. United States Food and Drug Administration, Silver Spring, MD.

Fletcher N., Bartholomaeus A. (2011) Regulation of nanotechnologies in food in Australia and New Zealand. *International Food Risk Analysis Journal* 1:33–40.

Frewer L.J., Bergmann K., Brennan M., Lion R., Meertens R., Rowe G., Siegrist M., Vereijken C. (2011) Consumer response to novel agri-food technologies: implications for predicting consumer acceptance of emerging food technologies. *Trends Food Science and Technology* 22(8):442–456.

Fruijtier-Pölloth C. (2012) The toxicological mode of action and the safety of synthetic amorphous silica—A nanostructured material. *Toxicology* 294:61–79.

FSA. (2008) *A Review of Potential Implications of Nanotechnologies for Regulations and Risk Assessment in Relation to Food.* Food Standards Agency, London, UK.

Gaillet S., Rouanet J.M. (2015) Silver nanoparticles: Their potential toxic effects after oral exposure and underlying mechanisms—A review. *Food and Chemical Toxicology* 77:58–63.

Geraets L., Oomen A.G., Krystek P., Jacobsen N.R., Wallin H., Laurentie M., Verharen H.W., Brandon E.F., De Jong W.H. (2014) Tissue distribution and elimination after oral and intravenous administration of different titanium dioxide nanoparticles in rats. *Particle and Fibre Toxicology* 11:30.

Greehy G., McCarthy M., Henchion M., Dillon E., McCarthy S. (2011) An exploration of Irish consumer acceptance of nanotechnology applications in food. *Proceedings in Food System Dynamics* 175–198.

Gupta N., Fischer A.R., Frewer L.J. (2012a) Socio-psychological determinants of public acceptance of technologies: A review. *Public Understanding of Science* 21:782–795.

Gupta N., Fischer A.R.H., van der Lans I.A., Frewer L.J. (2012b) Factors influencing societal response of nanotechnology: An expert stakeholder analysis. *Journal of Nanoparticle Research* 14:857–871. doi:10.1007/s11051-012-0857-x.

Hadrup N., Lam H.R. (2014) Oral toxicity of silver ions, silver nanoparticles and colloidal silver—A review. *Regulatory Toxicology and Pharmacology* 68:1–7.

Handford C.E., Dean M., Spence M., Henchion M., Elliott C.T., Campbell K. (2014) *Nanotechnology in the Agri-Food Industry on the Island of Ireland: Applications, Opportunities and Challenges.* Safefood, Cork, Ireland.

Ho S.S., Scheufele D.A., Corley E.A. (2011) Value predispositions, mass media, and attitudes toward nanotechnology: The interplay of public and experts. *Science Communication* 33:167–200.

Hwang M., Lee E.J., Kweon S.Y., Park M.S., Jeong J.Y., Um J.H., Kim S.A., Han B.S., Lee K.H., Yoon H.J. (2012a) Risk assessment principle for engineered nanotechnology in food and drug. *Toxicological Research* 28:73–79.

Hwang M., Park M.S., Kweon S.Y., Lee E.J., Jeong J.Y., Um J.H., Kim K.J., Yun H.J. (2012b) Information profiling for exposure assessment of engineered nanomaterials in nano-food and drug. Available from: http://iufost.org.br/sites/iufost.org.br/files/anais/01991.pdf. (Accessed 2017 June 13).

Institute I.S.R. (2012) *Community Attitude Towards Emerging Technologies Nanotechnology Issues: Nanotechnology.* Department of Industry, Innovation, Science, Research and Tertiary Education, Richmond, Victoria, Australia.

Jones K., Morton J., Smith I., Jurkschat K., Harding A.H., Evans G. (2015) Human in vivo and in vitro studies on gastrointestinal absorption of titanium dioxide nanoparticles. *Toxicology Letters* 233:95–101. doi:10.1016/j.toxlet.2014.12.005.

Jovanović B. (2015) Critical review of public health regulations of titanium dioxide, a human food additive. *Integrated Environmental Assessment and Management* 11:10–20.

Kim J.S., Song K.S., Sung J.H., Ryu H.R., Choi B.G., Cho H.S., Lee J.K., Yu I.J. (2013) Genotoxicity, acute oral and dermal toxicity, eye and dermal irritation and corrosion and skin sensitisation evaluation of silver nanoparticles. *Nanotoxicology* 7:953–960. doi:10.3109/17435390.2012.676099.

Kim Y.R., Lee S.Y., Lee E.J., Park S.H., Seong N.W., Seo H.S., Shin S.S. et al. (2014a) Toxicity of colloidal silica nanoparticles administered orally for 90 days in rats. *International Journal of Nanomedicine* 9:67–78. doi:10.2147/ijn.s57925.

Kim Y.R., Park J.I., Lee E.J., Park S.H., Seong N.W., Kim J.H., Kim G.Y. et al. (2014b) Toxicity of 100 nm zinc oxide nanoparticles: A report of 90-day repeated oral administration in Sprague Dawley rats. *International Journal of Nanomedicine* 9:109–126. doi:10.2147/ijn.s57928.

Kovvuru P., Mancilla P.E., Shirode A.B., Murray T.M., Begley T.J., Reliene R. (2015) Oral ingestion of silver nanoparticles induces genomic instability and DNA damage in multiple tissues. *Nanotoxicology* 9:162–171.

Kumari A., Yadav S.K. (2014) Nanotechnology in agri-food sector. *Critical Reviews in Food Science and Nutrition* 54:975–984.

Lankveld D., Oomen A., Krystek P., Neigh A., Troost–de Jong A., Noorlander C., Van Eijkeren J., Geertsma R., De Jong W. (2010) The kinetics of the tissue distribution of silver nanoparticles of different sizes. *Biomaterials* 31:8350–8361.

Larsen J.C., Nørby K.K., Reffstrup T.K., Beltoft V.M. (2008) Opinion of the Scientific Panel on Food Additives, Flavourings, Processing Aids and Materials in Contact with Food (AFC) on a Request from the Commission, Flavouring Group Evaluation 61 (FGE. 61): Consideration of aliphatic acyclic acetals evaluated by JECFA (57th meeting) structurally related to acetals of branched and straight-chain aliphatic saturated primary alcohols and branched and straight-chain saturated aldehydes, and an orthoester of formic acid evaluated by EFSA in FGE. 03: Question No EFSA-Q-2008-032M. *The EFSA Journal*:1–12.

Li N., Duan Y., Hong M., Zheng L., Fei M., Zhao X., Wang J., Cui Y., Liu H., Cai J. (2010) Spleen injury and apoptotic pathway in mice caused by titanium dioxide nanoparticles. *Toxicology Letters* 195:161–168.

Loeschner K., Hadrup N., Qvortrup K., Larsen A., Gao X., Vogel U., Mortensen A., Lam H.R., Larsen E.H. (2011) Distribution of silver in rats following 28 days of repeated oral exposure to silver nanoparticles or silver acetate. *Particle and Fibre Toxicology* 8:18–31.

MacNicoll A., Kelly M., Aksoy H., Kramer E., Bouwmeester H., Chaudhry Q. (2015) A study of the uptake and biodistribution of nano-titanium dioxide using in vitro and in vivo models of oral intake. *Journal of Nanoparticle Research* 17(66):1–20. doi:10.1007/s11051-015-2862-3.

Martirosyan A., Schneider Y.J. (2014) Engineered nanomaterials in food: Implications for food safety and consumer health. *International Journal of Environmental Research and Public Health* 11:5720–5750.

Nanoportal. (2017) *Proposed Regulatory Framework for Nanomaterials.* Environment Canada, Health Canada, Ottawa, Canada.

Nanoproduct. (2017) *Iran Nanotechnology Product.* Iran Nanotechnology Initiative Council, Tehran, Iran. Available at http://http://nanoproduct.ir/en. (Accessed on June 24, 2017).

Nanosafety. (2017) *Iran Nanotechnology Safety.* Iran Nanotechnology Initiative Council, Tehran, Iran. Available at http://nanosafety.ir//en#. (Accessed on June 24, 2017).

Nano Malaysia. (2017) Nano Malaysia Programmes-NANOVerify. Nano Malaysia, Kolalumbur, Malaysia. Available at http://www.nanomalaysia.com.my/html/inanovation.aspx?ID=5&PID=17. (Accessed on June 20, 2017).

NIOSH. (2009) *Managing the Health and Safety Concerns Associated with Engineered Nanomaterials.* Approaches to Safe Nanotechnology, Centers for Disease Control and Prevention, Cincinnati, OH.

Park E.J., Bae E., Yi J., Kim Y., Choi K., Lee S.H., Yoon J., Lee B.C., Park K. (2010) Repeated-dose toxicity and inflammatory responses in mice by oral administration of silver nanoparticles. *Environmental Toxicology and Pharmacology* 30:162–168.

Pidgeon N., Harthorn B.H., Bryant K., Rogers-Hayden T. (2009) Deliberating the risks of nanotechnologies for energy and health applications in the United States and United Kingdom. *Nature Nanotechnology* 4:95–98.

Priest S., Greenhalgh T. (2012) Attitudinal communities and the interpretation of nanotechnology news: Frames, schemas, and attitudes as predictors of reader reactions. In: Lente H.V., Coenen C., Fleisher T., Konrad K., Krabbenborg L., Milburn C., Thoreau F., Zülsdorf T.B. (Eds) *Little by Little: Expansions of Nanoscience and Emerging Technologies.* IOS Press, Heidelberg, Germany.

Rossi M., Cubadda F., Dini L., Terranova M.L., Aureli F., Sorbo A., Passeri D. (2014) Scientific basis of nanotechnology, implications for the food sector and future trends. *Trends in Food Science & Technology* 40:127–148. doi:10.1016/j.tifs.2014.09.004.

Sang X., Li B., Ze Y., Hong J., Ze X., Gui S., Sun Q., Liu H., Zhao X., Sheng L., Liu D., Yu X., Wang L., Hong F. (2013) Toxicological mechanisms of nanosized titanium dioxide-induced spleen injury in mice after repeated peroral application. *Journal of Agricultural and Food Chemistry* 61:5590–5599. doi:10.1021/jf3035989.

Schnettler B., Crisóstomo G., Mora M., Lobos G., Miranda H., Grunert K.G. (2014) Acceptance of nanotechnology applications and satisfaction with food-related life in southern Chile. *Food Science and Technology (Campinas)* 34:157–163.

Shinohara N., Danno N., Ichinose T., Sasaki T., Fukui H., Honda K., Gamo M. (2014) Tissue distribution and clearance of intravenously administered titanium dioxide (TiO_2) nanoparticles. *Nanotoxicology* 8:132–141. doi:10.3109/17435390.2012.763001.

Shumakova A., Arianova E., Shipelin V., Sidorova I., Selifanov A., Trushina É., Mustafina O., Safenkova I., Gmoshinskiĭ I., Khotimchenko S. (2014) Toxicological assessment of nanostructured silica. I. Integral indices, adducts of DNA, tissue thiols and apoptosis in liver. *Voprosy Pitaniia* 83:52–62.

Siegrist M., Cousin M.E., Kastenholz H., Wiek A. (2007) Public acceptance of nanotechnology foods and food packaging: The influence of affect and trust. *Appetite* 49:459–466.

Siegrist M., Stampfli N., Kastenholz H., Keller C. (2008) Perceived risks and perceived benefits of different nanotechnology foods and nanotechnology food packaging. *Appetite* 51:283–290.

Stampfli N., Siegrist M., Kastenholz H. (2010) Acceptance of nanotechnology in food and food packaging: A path model analysis. *Journal of Risk Research* 13:353–365. doi:10.1080/13669870903233303.

Tager J. (2014) Nanomaterials in food packaging: FSANZ fails consumers again. *Chain Reaction* 1–16.

Takeuchi M.T., Kojima M., Luetzow M. (2014) State of the art on the initiatives and activities relevant to risk assessment and risk management of nanotechnologies in the food and agriculture sectors. *Food Research International* 64:976–981.

Tavares P., Balbinot F., de Oliveira H.M., Fagundes G.E., Venâncio M., Ronconi J.V.V., Merlini A., Streck E.L., da Silva Paula M.M., de Andrade V.M. (2012) Evaluation of genotoxic effect of silver nanoparticles (Ag-NPs) in vitro and in vivo. *Journal of Nanoparticle Research* 14:791.

Umbreit T., Francke-Carroll S., Weaver J., Miller T., Goering P., Sadrieh N., Stratmeyer M. (2012) Tissue distribution and histopathological effects of titanium dioxide nanoparticles after intravenous or subcutaneous injection in mice. *Journal of Applied Toxicology* 32:350–357.

US EPA. (2010) *State of the Science Literature Review: Everything Nanosilver and More.* Scientific, Technical, Research, Engineering and Modeling Support (STREAMS): Final report EPA/600/R-10/084, Washington, DC.

US EPA. (2014a) *Research in Action: Nanosilver and Consumer Products.* United States Environmental Protection Agency, Washington, DC.

US EPA IRIS. (1996) *Silver (CASRN 7440-22-4) Oral RfD Assessment.* United States Environmental Protection Agency Integrated Risk Information System, Washington, DC.

US FDA. (2014a) *Guidance for Industry—Considering whether an FDA-regulated Product Involves the Application of Nanotechnology.* United States Food and Drug Administration, Silver Spring, MD.

US FDA. (2014b) *Guidance for Industry—Assessing the Effects of Significant Manufacturing Process Changes, Including Emerging Technologies, on the Safety and Regulatory Status of Food Ingredients and Food Contact Substances, Including Food Ingredients that Are Color Additives.* United States Food and Drug Administration, Silver Spring, MD.

US FDA. (2014c) *Guidance for Industry—Use of Nanomaterials in Food for Animals.* United States Food and Drug Administration, Silver Spring, MD.

van der Zande M., Vandebriel R.J., Groot M.J., Kramer E., Rivera Z.E.H., Rasmussen K., Ossenkoppele J.S., Tromp P., Gremmer E.R., Peters R.J. (2014) Sub-chronic toxicity study in rats orally exposed to nanostructured silica. *Particle and Fibre Toxicology* 11:1–8.

van der Zande M., Vandebriel R.J., Van Doren E., Kramer E., Herrera Rivera Z., Serrano-Rojero C.S., Gremmer E.R., Mast J., Peters R.J., Hollman P.C. (2012) Distribution, elimination, and toxicity of silver nanoparticles and silver ions in rats after 28-day oral exposure. *ACS Nano* 6:7427–7442.

Vandermoere F., Blanchemanche S., Bieberstein A., Marette S., Roosen J. (2011) The public understanding of nanotechnology in the food domain: The hidden role of views on science, technology, and nature. *Public Understanding of Science* 20:195–206.

Weir A., Westerhoff P., Fabricius L., Hristovski K., Von Goetz N. (2012) Titanium dioxide nanoparticles in food and personal care products. *Environmental Science & Technology* 46:2242–2250.

Wesley S.J., Raja P., Raj A., Tiroutchelvamae D. (2014) Review on-nanotechnology applications in food packaging and safety. *International Journal of Engineering Research* 3(11):645–651.

Wolterbeek A., Oosterwijk T., Schneider S., Landsiedel R., de Groot D., van Ee R., Wouters M., van de Sandt H. (2015) Oral two-generation reproduction toxicity study with NM-200 synthetic amorphous silica in Wistar rats. *Reproductive Toxicology* 56:147–154. doi:10.1016/j.reprotox.2015.03.006.

Xie G., Wang C., Sun J., Zhong G. (2011) Tissue distribution and excretion of intravenously administered titanium dioxide nanoparticles. *Toxicology Letters* 205:55–61.

Yamashita K., Yoshioka Y., Higashisaka K., Mimura K., Morishita Y., Nozaki M., Yoshida T., Ogura T., Nabeshi H., Nagano K. (2011) Silica and titanium dioxide nanoparticles cause pregnancy complications in mice. *Nature Nanotechnology* 6:321–328.

Yoshida T., Yoshioka Y., Takahashi H., Misato K., Mori T., Hirai T., Nagano K., Abe Y., Mukai Y., Kamada H. (2014) Intestinal absorption and biological effects of orally administered amorphous silica particles. *Nanoscale Research Letters* 9:532.

Yun J.W., Kim S.H., You J.R., Kim W.H., Jang J.J., Min S.K., Kim H.C., Chung D.H., Jeong J., Kang B.C. (2015) Comparative toxicity of silicon dioxide, silver and iron oxide nanoparticles after repeated oral administration to rats. *Journal of Applied Toxicology* 35:681–693.

Index

Note: Page numbers in italic and bold refer to figures and tables respectively.